Healthcare Safety for Nursing Personnel

An Organizational Guide to Achieving Results

Healthcare Safety for Nursing Personnel

An Organizational Guide to Achieving Results

James T. Tweedy

MS, CHSP, CPSO, CHEP, CHCM

CRC Press
Taylor & Francis Group
Boca Raton London New York

CRC Press is an imprint of the
Taylor & Francis Group, an **informa** business

A PRODUCTIVITY PRESS BOOK

CRC Press
Taylor & Francis Group
6000 Broken Sound Parkway NW, Suite 300
Boca Raton, FL 33487-2742

Printed on acid-free paper
Version Date: 20141028

International Standard Book Number-13: 978-1-4822-3027-7 (Hardback)

Visit the Taylor & Francis Web site at
http://www.taylorandfrancis.com

and the CRC Press Web site at
http://www.crcpress.com

Contents

Preface

This text provides a survey of key safety issues, risks, and hazards confronting nurses on a daily basis. Much of the text uses the active voice to present information more succinctly and in fewer words. The text focuses on preventing accidents and controlling hazards in hospitals and other healthcare settings and provides an overview of safety management concepts as they relate to real-world challenges facing nursing personnel. The author presents healthcare safety as an organizational function and not just another program. The text focuses on achieving results because safety is more than just meeting accreditation requirements or regulatory compliance. Achieving safety is the right thing to do. This text provides a solid foundation for working nurses. It should serve as a valuable on-the-job resource. It addresses the need for good leadership and management. The author also briefly addresses the importance that practicing human relation and communication skills can have on healthcare hazard control efforts. The text also contains some very helpful appendices and could serve as a text for nursing schools desiring to offer a credit course in healthcare safety.

Acknowledgments

I would like to thank Jan Mosier for proofreading and editing each of the chapters. Her tireless efforts identified many of my errors, duplications, and omissions. Without her help and personal motivation this revision project would have taken much longer to complete. I would also like to express my thanks to my son, Aaron, daughter-in-law, Mary, and grandsons, Alex and Patrick, for their support and understanding during the project. I also want to acknowledge Harry Stern for his continued support and encouragement during the completion of this project.

About the Author

James (Jim) Tweedy has served as the Executive Director for the International Board for Certification of Safety Managers (IBFCSM), which is also known as the Board of Certified Hazard Control Management (BCHCM), since 2007. IBFCSM was founded in 1976 and offers the following personal credentials: Certified Healthcare Safety Professional (CHSP), Certified Patient Safety Officer (CPSO), Certified Hazard Control Manager (CHCM), Certified Healthcare Emergency Professional (CHEP), Certified Product Safety Manager (CPSM), and Certified Hazard Control Manager-Security (CHCM-SEC).

Jim founded TLC Services, a healthcare, hazard control, organizational performance, and educational consulting organization, in 1996. He has more than 25 years of experience with expertise in the areas of credentialing, hazard control, healthcare safety, leadership and team development, education, and compliance. He holds an MS in Safety Management from Central Missouri University and a BS in Liberal Studies from Excelsior College. He holds Master Level designations as a CHCM, CPSO, CPSM, CHSP, and CHEP. He also holds a Professional Membership in the American Society of Safety Engineers (ASSE) and is a member of the American Society of Healthcare Engineering (ASHE).

Jim is a polished speaker who presents safety-related topics at seminars, conferences, and in the workplace. He taught more than six years at the college level and is an experienced curriculum developer. He has developed and presented original training programs at locations in more than 40 states. He is a recognized leader in the area of healthcare safety and hazard control management. You may contact Jim at IBFCSM, 173 Tucker Road, Suite 202, P.O. Box 515, Helena, AL 35080. The Board phone number is 205-664-8412 and the website can be accessed at www.ibfcsm.org. You may email the Board at info@ibfcsm.org.

1 Nursing Safety

I. INTRODUCTION

Nursing personnel serve as an integral part of any healthcare or medical delivery organization. Nurses not only work to keep patients safe, but also encounter a number of safety and health risks. The Occupational Safety and Health Administration (OSHA) classifies healthcare safety risks in five hazard categories: biological, chemical, physical, ergonomic-environmental, and psychosocial. Registered nurses, licensed vocation or practical nurses, certified medical technicians, and certified nursing assistants provide patient care and support a variety of healthcare and medical settings. Nurses can experience safety and health risks while performing duties such as monitoring patients, accomplishing direct care tasks, assessing and recording symptoms, assisting physicians with treatments, administering medications, and moving or positioning patients.

In December 2007, the Environmental Working Group published an online survey that looked at the health issues of nurses, their respective areas of work, and the amount of exposure to common chemicals and hazardous materials. The results were alarming. The survey looked at 1,500 nurses and their exposure to 11 different common healthcare hazards, including chemo drugs, radiation, sterilizing agents, housekeeping chemicals, anesthetic agents, and certain other therapeutic drugs. While some of these hazards are known to have immediate adverse reactions allergic reactions, others have a more compounded effect over longer periods of time. The survey defined long-term exposure as being at least weekly for 10 years or more. Reported health conditions among the nurses ranged from asthma and cancer to birth defects in their children after exposure during pregnancy. Chemotherapeutic agents have been widely known to be toxic and require special precautions during preparation and administration. Nurses who worked with chemo agents for long periods of time reported 42 higher rates of cancer when compared to other nurses. Of those who reported working frequently with radiation, there was a 16 percent higher incidence of cancer.

Senior healthcare leaders must learn to promote hazard control and safety as an organizational value. Hazard control effectiveness impacts both the overt and covert cultures of any healthcare organization. The safety culture of healthcare organizations must be recognizable by those served. Healthcare organizations seeking to maintain revenues, minimize losses, serve their communities, and meet regulatory or accreditation requirements need effective safety functions. Healthcare is one of the fastest growing sectors of the U.S. economy, employing over 12 million workers with women representing about 80 percent of the healthcare work force. Nursing professions make up a large portion of the medical and healthcare organizational workforce. Rates of occupational injury to healthcare workers continued to rise over the past decade. Hazards facing nursing professionals include needle sticks, back injuries, slips and falls, laser hazards, chemical exposure, biological hazards, and workplace violence. Home health nursing personnel must also face community safety issues. An increased emphasis on topics such as emergency management, indoor air quality, and patient safety indicates that safety will remain a key function in healthcare organizations. Advances in medical technology and clinical treatment techniques expose workers and patients to a variety of potential hazards. See Tables 1.1 and 1.2 for information about OSHA incidence rates and top citations for healthcare.

TABLE 1.1
OSHA Comparative (Nonfatal) Incidence Rates for 2011

Industry	Rate
Private Industry	3.5
Construction	3.9
Manufacturing	4.4
Health Services	8.2
Hospitals	8.5
Nursing Facilities	12.7

Source: OSHA
Note: Incidence rates are per 100 employees.

BOX 1.1   TOP 10 OSHA CITATIONS FOR 2011–2012

Health Services Industries (All Categories)

- Bloodborne Pathogens (1910.1030)
- Hazard Communication (1910.1200)
- Formaldehyde (1910.1200)
- Recordkeeping Forms (1904.0029)
- Medical Services and First Aid (1910.0151)
- Maintenance, Safeguards, and Operational Features for Exit Routes (1910.0037)
- Electrical Systems Design, General Requirements (1910.0303)
- Electrical Wiring Methods and Components (1910.0305)
- Personal Protective Equipment, General Requirements (1910.0132)
- Annual Illness/Injury Summary (1904.0032)

A.   ANA NURSE HEALTH & SAFETY SURVEY RESULTS

The latest health and safety survey from the American Nurses Association (ANA) makes it clear that the efforts to protect nurses from occupational injuries remain a crusade-in-progress. The survey indicated that hospitals appear to be safer workplaces today than 10 years ago, when the last ANA survey was conducted. Safe needle devices and patient-lifting equipment are more available today than a decade ago. However, responses of more than 4500 RNs that participated in the ANA *2011 Health and Safety Survey* indicated the same top three concerns were identified by the 2001 participants and in slightly higher percentages:

- 74 percent cited the effects of stress and overwork (versus 70 percent in 2001).
- 62 percent cited disabling musculoskeletal injury (versus 59 percent in 2001).
- 43 percent cited contracting an infectious disease (versus 37 percent in 2001).

The 2011 survey revealed that 64 percent of RNs work for hospitals that provide patient assist devices. Fifty-six percent of 2011 respondents indicated they experienced musculoskeletal pain related to or made worse by their jobs. Eighty percent worked despite musculoskeletal pain. Slightly less than 15 percent reported suffering three or more work related injuries within a year. Just 21 percent of respondents in 2011 listed fear of contracting HIV or hepatitis from a needle-stick event, down from more than 40 percent from the 2001 survey. Only six percent of 2011 respondents indicated concern about latex allergies, as glove alternatives are more available than in 2001. Ten percent of 2011 respondents voiced concern about their exposure to hazardous drugs and other toxic substances. Awareness of hazards appears to be the key reason for increased concerns. Eleven percent of respondents reported a physical assault within the past year, which was down from 17 percent in 2001. However, more respondents ranked assault risks as a top three nursing concern in 2011 than in 2001.

B. SAFE PATIENT HANDLING AND MOBILITY: INTERPROFESSIONAL NATIONAL STANDARDS

The ANA, in collaboration with a national working group and other professional organizations, recently released Safe Patient Handling and Mobility: Interprofessional National Standards. The 40-page outline contains eight evidence-based standards to prevent injury. The standard should provide a foundation for establishing a culture of safety for all caregivers and patients. An interprofessional work group established the standards that apply in any healthcare setting. The standards call for establishing a culture of safety, which includes ensuring safe levels of staffing, creating a nonpunitive reporting environment, and developing a system for communication and collaboration. Other standards address (1) implementing a safe patient handling and mobility program, (2) using ergonomic design principles to provide a safe environment of care, (3) obtaining safe patient handling technology, (4) creating processes for educating, training, and maintaining personal competence, (5) integrating patient-centered assessments, care planning, and technology, (6) requiring safe patient handling while considering reasonable accommodations and post-injury return to work policies, and (7) establishing a comprehensive evaluation system. The publication also includes a glossary of terms and appendices containing tools and resources.

Nurses play a critical role in ensuring patient safety by monitoring patients for clinical deterioration, detecting errors and near misses, understanding care processes and weaknesses inherent in some systems, and performing countless other tasks to ensure patients receive high-quality care. Nurse vigilance at the bedside must remain the key element in ensuring safety. Assigning nurses an increasing numbers of patients eventually compromises nurses' ability to provide safe care. Several seminal studies have demonstrated the link between nurse staffing ratios and patient safety. The nurse-to-patient ratio is only one aspect of the relationship between nursing workload and patient safety. Overall nursing workload is likely linked to patient outcomes as well. Determining adequate nurse staffing is a very complex process that changes on a shift-by-shift basis, and requires close coordination between management and nursing based on patient acuity and turnover, availability of support staff and skill mix, and many other factors.

C. NEW PATIENT SAFETY SURVEY: FEW NURSES CALL THEIR HOSPITALS SAFE

A recently conducted survey indicated a large number of American, British, and Chinese nurses feel that hospitals are falling short in keeping patients safe, according to a recent survey of 900 nurses from the three countries. Although nearly all nurses said that their hospitals had programs in place that promote patient safety, they questioned their impact. About 40 percent of nurses described their hospital as safe and less than 60 percent believed that patient safety efforts in their hospital were effective. They said access to technology, heavy workload, communication with patients and doctors, and punitive systems for reporting errors were at the core of the problem. The survey was conducted by GE Healthcare and the American Nurses Association. Some 90 percent said they felt most responsible for patient safety. A large majority of nurses see data, technology, and innovation as key to identifying early warning signs and alerting staff. Many nurses said that there was a lack of feedback between patient safety data and the nursing staff. The results also suggest that moving away from a culture of punishment for poor practice could help to improve matters. About 40 percent of nurses rated their hospital as excellent at communication with the patient. Only about 30 percent indicated their hospital was excellent at communication between staff.

D. INTERNATIONAL BOARD FOR CERTIFICATION OF SAFETY MANAGERS

The International Board for Certification of Safety Managers (IBFCSM) was founded in 1976 as a not-for-profit credentialing organization, and operated for some time as the Board of Certified Hazard Control Management (BCHCM). The Board offers qualified working healthcare professionals including nurses the opportunity to earn their Certified Healthcare Safety Professional (CHSP), Certified Healthcare Emergency Professional (CHEP), or Certified Patient Safety Officer (CPSO) credential. Many healthcare professionals hold more than one credential. The Board offers CHSP and CHEP credential holders the opportunity to add the healthcare Fire Safety Management (FSM) designation to their primary certification. The IBFCSM motto, *Individual Credentials—The Key to Upgrading the Profession,* reflects the impact that individual certifications have on improving organizational safety and hazard control functions.

II. FUNCTION OF SAFETY

Safety must focus on developing processes or systems that can help prevent harm and loss. An uncorrected hazard or hazardous situation could contribute to an event resulting in property damage, job interruption, personal harm, or adverse health effects. The process of controlling hazards may require development of written policies, plans, or procedures. Never consider safety as a program but as a function of the organization. The safety function must connect with organizational structures and operational philosophies.

BOX 1.2 BASIC SAFETY PRINCIPLES

- Correcting causal factors results in better use of human and material resources.
- Placing individual blame leads to organizational problems being ignored.
- Data collection provides the foundation for effective analysis of hazards.
- Safety efforts must address poor and hazardous behaviors.
- Communication and human relation skills remain key to achieving safety results.
- Hazard control focuses on accomplishing the job with safety.
- Hazard control functions as a quality tool when integrated into all job functions.
- Good hazard control and efficiency function as partners within an organization.

BOX 1.3 SEVEN VALUES OF EFFECTIVE SAFETY

- Never-Ending Process
- People Focused
- Leadership Driven
- Operational Priority
- Benefits Everyone
- Reduces Organizational Losses
- Prevents Human Harm

The term "program" is derived from the French word "programme," which means agenda or public notice. We can also refer to the Greek word "graphein," which means to write. When used with the prefix "pro" it became "prographein," which means to write before. Many organizations develop written safety programs to satisfy organizational mandates or to demonstrate visual compliance with regulatory requirements. Written plans, policies, and procedures should direct the hazard control function. The word "function," first used in the early 16th century, denotes the concept of performance or execution. A function can relate to people, things, and institutions. A function can refer to serving a designated or defined role in some manner. A function can also relate to participation in an ongoing cultural or social system. Considering hazard control as a function of the organization elevates its priority in the minds of everyone.

BOX 1.4   REASONS FOR INEFFECTIVE SAFETY EFFORTS

- Safety efforts focus on activities instead of behavioral elements.
- Safety problems and issues are not addressed using a systems approach.
- Senior leadership fails to define the organizational safety philosophy.
- The organization focuses primarily on compliance and accreditation issues.
- Physicians in many situations do not participate in safety efforts and become an obstacle.
- Safety education and training programs focus too much on simply documenting attendance.
- Performance- and objective-based training and education are rarely provided.
- Competition is allowed to exist among safety program elements (e.g., patient vs. worker safety).
- Leaders many times fail to address or deal with turf kings and queens with their own agendas.
- Lack of good coordination results in poor "buy-in" by organizational leaders.
- Senior leadership does not communicate goals and objectives to all levels.
- Effective accident investigation techniques are not implemented.
- Root cause analysis methods are used only for patient safety, not all safety events.
- The facility believes a "one-size-fits-all" safety program approach will work.

A.   SAFETY RESPONSIBILITIES

Many healthcare and medical organizations fail to outline specific safety and hazard control responsibilities in their plans, procedures, directives, and job descriptions. The concept of responsibility relates to a person's obligation to carry out assigned duties in an efficient, effective, and safe manner. Senior leaders must ensure that managers, supervisors, and frontline nursing personnel understand the importance of their assigned safety responsibilities. Senior leaders must ensure that job descriptions address safety responsibilities inherent with each job position. Safety efforts will yield results when leaders encourage participation and hold key managers accountable. Senior leaders and hazard control managers must learn to focus on the hazards, behaviors, and risks that pose the most potential harm.

BOX 1.5   SENIOR MANAGEMENT RESPONSIBILITIES

- Develop, sign, and publish an organizational safety policy statement
- Describe key expectations related to accomplishing safety-related objectives
- Ensure that all organizational members can explain the major objectives
- Develop methods to track progress and provide feedback to all organizational members
- Require managers and supervisors to visibly support established objectives

---

B.   NURSING SUPERVISOR INVOLVEMENT

Nursing supervisors must possess the knowledge and experience to provide hazard control guidance to those they lead. First-line supervisors occupy a key hazard control position in many organizations. This position of trust can require supervisors to conduct area inspections, provide job training, ensure timely incident reporting, and accomplish initial accident investigations. Supervisors in many healthcare settings possess little control over factors such as hiring practices, working conditions, and equipment provided to them. Supervisors must understand the role that human factors can play in accident prevention and causation. They must ensure that each person they supervise understands the behavior expectations of the job. Some organizations require employees to sign a safe work agreement. Such an agreement requires the individual to commit to working safely and to adhere to organizational policies or procedures. Supervisors must ensure that frontline personnel can access all safety-related directives, plans, policies, and procedures.

BOX 1.6   NURSING SUPERVISOR RESPONSIBILITIES

- Enforce work rules and correct unsafe or at risk behaviors
- Implement mandated safety policies and procedures for their areas of responsibility
- Provide job- or task-related training and education
- Immediately report and investigate all accidents in their work areas
- Conduct periodic area hazard control and safety inspections
- Ensure proper maintenance and servicing of all equipment and tools
- Lead by example and personally adhere to hazard control requirements
- Conduct safety and hazard control meetings on a regular basis
- Work with organizational hazard control personnel to correct and control hazards
- Ensure all personnel correctly use required personal protective equipment (PPE)

C.   ADDRESSING BEHAVIORS

Nursing supervisors must explain work rules and behavioral expectations to all new or transferred employees. Supervisors must never tolerate individuals that encourage others to disregard work rules or established procedures. When disciplining an individual, do so in private, but always document the facts. Senior leaders, managers, and supervisors must set an example for others. They must discourage poor behaviors by reinforcing the importance of acceptable behaviors. Never confuse correcting a behavior with undertaking needed disciplinary action. When correcting an unsafe behavior, always state the facts about the situation but limit personal opinions. Use statements that begin with "I" but never use "they" statements. Take time to recognize good behaviors by using positive reinforcement. Keep in mind that some individuals may not recognize a hazard or hazardous situation. Some may recognize a hazard but not possess the ability to deal with it. Too many injuries occur when a person recognizes a hazard but fails to respect its potential for causing harm.

BOX 1.7   BEHAVIOR CORRECTION PROCESS

- Step 1 – Identify the unsafe action
- Step 2 – State concern for worker's safety
- Step 3 – Demonstrate the correct and safe way
- Step 4 – Ensure the worker understands
- Step 5 – Restate concern for personal safety
- Step 6 – Follow up

D.   EMPLOYEE ENGAGEMENT

Employee engagement occurs when an individual personally feels their connection to their position or job. This engagement also refers to their personal commitment to the success of the organization.

Employee engagement can contribute to individual satisfaction and personal mental wellness. Engaged employees also help improve the productivity, morale, and motivation of others. Today, many organizations realize the need for balancing work demands with a person's family and other life issues. When off the job, organizational members serve in a variety of roles including as a volunteer, caregiver, and parent. Understanding employee engagement helps leaders and hazard control managers deal with the complexity of human behaviors. Conflicting responsibilities can lead to role misunderstandings and work-related overloads, which can impact organizational objectives, including hazard control efforts.

BOX 1.8 WAYS TO PROMOTE SAFE BEHAVIORS

- Requiring everyone to walk the talk, also known as "modeling"
- Rewarding people when appropriate
- Recognizing people for making good efforts
- Correcting unsafe behaviors in a positive manner
- Learning to deal with behaviors and not attitudes
- Motivating through a focus on promoting trust
- Educating others to increase their understanding
- Presenting the "why" of something
- Encouraging people to become engaged and to participate
- Empowering subordinates to make decisions
- Coaching by promoting teamwork and individual improvement
- Consulting to provide guidance from a short distance away
- Coordinating to allow people to buy in and take ownership
- Leading and motivating others to achieve goals by focusing on the process
- Promoting better listening to learn from others

III. UNDERSTANDING HAZARDS

Classifying and defining hazards can vary greatly depending on a number of factors, including type of industry, process, or operation. For example, mechanical energy hazards can involve components that cut, crush, bend, shear, pinch, wrap, pull, and puncture. Biological hazards can include pandemic, bioterrorism agents, bloodborne pathogens, and infectious waste. Chemical hazards include substances such as solvents, flammable liquids, compressed gases, cleaning agents, and even disinfectants. Physical hazards can include risks posed by fire, radiation, machine operation, and noise. Environmental and ergonomic hazards include slip, trip, and fall hazards, walking and working surfaces, lighting, and tasks with repetitive motions. Psychosocial hazards address issues such as workplace violence, work-related stress, sleep deprivation, mental problems, chemical dependency, alcohol abuse, and horseplay on the job.

A. HAZARD IDENTIFICATION

Hazard identification requires the identification of hazards, unsafe conditions, and risky behaviors. Hazard anticipation relies on human intuition, training, common sense, observation, and continuous awareness. To identify hazards, rely on the use of inspections, surveys, analysis, and human recognition reporting. Hazard identification efforts should focus on unsafe conditions, hazards, broken equipment, and human deviations from accepted practices. Require supervisors or unit safety coordinators to conduct periodic area inspections. These individuals should understand hazardous areas and the workers better than anyone. However, supervisors can fall prey to inspection bias, which results in poor survey results. Many supervisors conduct limited ongoing inspections as part of their daily job duties. Periodic inspections and surveys can focus on critical components of equipment, processes, or systems with a known potential for causing serious injury or illness. Some equipment inspections help meet preventive maintenance requirements or hazard control plan objectives. Safety standards can mandate that qualified persons periodically inspect some types of equipment, such as elevators, boilers, pressure vessels, and fire extinguishers, at regular intervals. Establish the frequency of inspections by considering the scope and type of the hazardous operations. Many hazard control plans fail to provide sufficient guidance about how to conduct hazard surveys, inspections, and audits. Inspections, audits, and hazard surveys can only help identify hazards when conducted properly. Providing a checklist to an untrained person can result in his or her failure to properly identify hazards or unsafe conditions. General checklists serve as tools that guide an inspection process. These documents do not contain information about all potential hazards. The effective use of demand response checklists will also require some type of education or training. Demand response checklists address specific operations and complex job processes such as the operation of robotic systems or the control of hazardous energy.

BOX 1.9   HEALTHCARE OCCUPATIONAL HAZARD CATEGORIES

- Biological hazards include bacteria, viruses, infectious waste, and bloodborne pathogens.
- Chemical hazards can pose a variety of risks due to their physical, chemical, and toxic properties.
- Ergonomic and environmental hazards include repetitive motion, standing, lifting, trips, and falls.
- Physical hazards include things such as radiation, noise, and machine-generated hazards.
- Psychosocial hazards include substance abuse, work-related stress, and workplace violence.

Note: Some hazards may fit in more than one category.

BOX 1.10   SOME COMMON FACTORS INHERENT IN GOOD WORK ENVIRONMENTS

- Good workplace design and proper equipment placement, including guards and controls.
- Equipment inspections and preventive maintenance are conducted as scheduled.
- The organization conducts inspections, audits, and hazard surveys on a regular basis.
- Corrective actions and hazard controls are implemented immediately to eliminate risks.
- Employees formally commit to work safely and maintain hazard-free work areas.
- Work areas are equipped with proper lighting, ventilation, and environmental controls.
- Employees must use PPE when mandated.
- Supervisors conduct job instruction, inspections, and initial accident investigations.

B.   PREPARING FOR INSPECTIONS

Conduct education and training sessions about how to conduct inspections. Periodic inspections provide opportunities for hazard control personnel, line supervisors, and top managers to listen to the concerns of those doing the work. Inspections should accurately assess all environments, equipment, and processes. Learn to identify potential hazards by observing individuals accomplish specific job tasks or processes. Learning to identify hazards and recognize unsafe behaviors requires inspectors to use their observation skills. Inspectors must focus on using all five human senses. Look for deviations from accepted work practices and rely on intuition or gut feelings to assist with the identification of hazards. Curiosity can help uncover hidden hazards. Learning to use visualization techniques to connect the dots can create a mind picture of a hazardous situation. Never allow human emotions or personal issues to drive the inspection process. Inspectors should maintain a professional demeanor and rely on logic when assessing tough situations. Inspectors must always point out potential or immediate dangers. They must never operate any equipment unless trained and authorized to do so, and should ask questions about tasks or processes, but refrain from disrupting operations or creating distractions. Well-designed checklists can assist with the documentation of any key findings.

BOX 1.11   SPECIAL INSPECTION SITUATIONS

- When new equipment is installed
- When new operations or procedures are added
- When work or tasks are relocated or revised
- When new construction or remodeling is in progress
- When any special or unusual program arises

Note: Review all contracts to ensure inclusion of safety-related considerations.

BOX 1.12   FUNDAMENTAL ELEMENTS OF HAZARD ANALYSIS

- Understand that hazard analysis deals with the science of and standards relating to hazards
- Evaluate hazard information using a practical approach
- Investigate accidents and near-miss events to discover causes
- Conduct root cause analysis to uncover contributing causes
- Determine worker perceptions about safety in the workplace
- Deal with perceptions (it cannot be avoided)
- Collect sufficient hazard information for analysis
- Remember that effective hazard analysis leads to effective hazard control or correction
- Identify employees at risk of exposure and evaluate control measures
- Establish a baseline to be used throughout a continuous or ongoing process
- Use inspections to identify and assess hazards in each work environment
- Determine potential hazard severity and possible effects on workers
- Evaluate PPE effectiveness
- Develop orderly processes for evaluating frequent and serious hazards

---

C.   JOB HAZARD ANALYSIS

Job hazard analysis (JHA) permits an examination of job-related tasks and processes for the purpose of discovering and correcting inherent risks and hazards. Supervisors and other experienced personnel can perform the process by working sequentially through the steps of a job process or task (see Table 1.2). Job hazard analysis requires an understanding of potential job hazards, and can

TABLE 1.2

Job Hazard Analysis Steps

Step A	Break the Job Down: Examine each step in the process for hazards or unsafe conditions that could develop during the job process.
Step B	Identify Hazards: Document process hazards, environmental concerns, and any anticipated human issues.
Step C	Evaluate Hazards: Assess identified hazards and behaviors to determine their potential roles in an accident event.
Step D	Develop and Design Hazard Controls: Develop or design the best hazard controls based on evaluating each hazard. Coordinate implementation of all feasible controls.
Step E	Implement Required Controls: Coordinate and obtain management approval for implementation.
Step F	Revise and Publish the Job Hazard Analysis Information: Update the JHA and then communicate implementation actions with the organizational members.

help in assessing the tools, machines, and materials used to perform a job. Personnel conducting the analysis must possess knowledge of hazard control, including use of PPE. A well-developed job hazard analysis can also serve as an effective teaching tool. Organizations should develop a job hazard analysis for all tasks, processes, or phase-related jobs. Conduct and update a job hazard analysis when a process changes or an accident occurs. Recommend that each organization develop standardized procedures and formats for conducting the analysis. An effective analysis provides the basis for developing and implementing appropriate control measures. Post analysis results at appropriate workstations and other job or process locations.

D. Job Design

Creating well-designed jobs, tasks, and processes can help reduce worker fatigue, reduce repetitive motion stress, isolate hazardous tasks, and control human factor hazards. The concept of job design refers primarily to administrative changes that help improve working conditions. The design of safe work areas must address workstation layout, tools and equipment, and the body position needed to accomplish the job. Safe work area design reduces static positions, and minimizes repetitive motions and awkward body positions. Consider the importance of human factor issues when designing work processes.

E. Hazard Control and Risk Management

Risk management in any setting can be described as the probability that a hazard will cause injury or damage. In some organizations, risk management operates separately from the hazard control function. For example, hospitals consider risk management to be a separate function from environmental safety efforts. Some other types of organizations may consider risk management an integral element of hazard control function. Risk management from an insurance and loss control perspective can quickly become a reactive managerial element. Risk management views all losses to the organization and not just human injury. Risk assessment relates to the process by which risk analysis results drive decision-making. Risk control efforts address hazardous events by implementing interventions to reduce severity. Risk management includes not only control efforts, but finance as well. Risk control considers all aspects of system safety, hazard control management, and safety engineering. Risk finance considers insurance, risk pooling, and self-insurance.

F. Hazard Control and Correction

Organizations must use the concept known as *hierarchy of controls* to reduce, eliminate, and control hazards or hazardous processes. Hazard controls can also include actions such as using *enclosure, substitution,* and *attenuation* to reduce human exposure risks. An enclosure keeps a hazard physically away from humans. For example, completely enclosing high voltage electrical equipment prevents access by unauthorized persons. Substitution can involve replacing a highly dangerous substance with a less hazardous one. Attenuation refers to taking actions to weaken or lessen a potential hazard. Attenuation could involve weakening radioactive beams or attenuating noise to safer levels. The use of system safety methods, traditional hazard control techniques, and human factors must begin at the initial stages of any design process.

BOX 1.13   HAZARD CORRECTION MONITORING SYSTEM

- Implement processes or systems to report and track hazard correction actions
- Establish a timetable for implementing corrective hazard controls
- Prioritize hazards as identified by inspections, reporting, and accident investigations
- Require employees to report hazards using an established processes
- Provide quick feedback about the status of hazard correction activities
- Delegate responsibility for correcting hazards and documenting completion
- Permit supervisors and experienced employees to initiate hazard control actions

---

1.   Active and Passive Controls

Passive hazard controls do not require continuous or even occasional actions from system users. Active controls require operators or users to accomplish a control-related task at some point during the operation to reduce risks.

BOX 1.14   COMMON "NEVER-EVER" HAZARDS

- Obstacles preventing the safe movement of people, vehicles, or machines
- Blocked or inadequate egress routes and emergency exits
- Unsafe working and walking surfaces
- Using worn or damaged tools/equipment or misusing tools
- Failing to identify hazards and provide proper equipment including PPE
- Operating equipment with guards removed or bypassed
- Permitting the presence of worn, damaged, or unguarded electrical wiring, fixtures, and cords
- Lack of adequate warning, danger, or caution signs in hazardous or dangerous areas

---

2.   Engineering Controls

Seek to eliminate hazards by using appropriate engineering controls. Make the modifications necessary to eliminate hazards and unsafe conditions. The design of machine guards, automobile brakes, traffic signals, pressure relief valves, and ventilation demonstrate engineering controls at work. For example, proper ventilation can remove or dilute air contaminants in work areas. Air cleaning devices can also remove contaminants such as particulates, gases, and vapors from the air. Using engineering, design, and technical innovation remains the top priority for controlling or eliminating hazards. Establishing preventative and periodic maintenance processes can help ensure tools and equipment operate properly and safely. Preventative maintenance must also address engineered hazard controls and emergency equipment. If needed, schedule shutdowns to address preventative and predictive maintenance issues. Ensure the preventive maintenance addresses safety and hazard control issues as well as operational or production requirements.

3. Administrative Controls

Use administrative controls such as scheduling to limit worker exposure to many workplace hazards such as working in hot areas. However, OSHA prohibits employee scheduling to meet the requirement of air contaminant exposure limits. The scheduling of maintenance and other high exposure operations during evenings or weekends can reduce exposures. Use job rotation to limit repetitive motion tasks or reduce the exposure time to occupational noise hazards. Use a work-rest schedule for very hazardous or strenuous tasks.

4. Work Practice Controls

These controls can reduce hazard exposure through development of standard operating procedures. Conducting training and education about the safe use of tools and equipment is an important work practice. Practices can also include knowing emergency response procedures for spills, fire prevention principles, and dealing with employee injuries. Job-related education and training helps individuals work safely and minimize hazard exposure risks. Work practice controls must address task accomplishment and ensure workers understand all job-related hazards.

BOX 1.15   HIERARCHIES FOR CONTROLLING HAZARDS

- Engineering and technological innovation remain the preferred types of hazard control.
- Substitution results in using a less hazardous substance or less dangerous equipment.
- Isolation results in moving either workers or hazardous operations to reduce risks.
- Work practices include policies or rules that can reduce human exposure to a hazard.
- Administrative controls limit human exposure through use of rotation and scheduling.
- Only consider PPE when other controls prove inadequate.

G. Personal Protective Equipment

Consider the use of appropriate PPE and clothing when engineering, administrative, and work practice controls fail to provide adequate or mandated protection for individuals exposed to hazards and unsafe conditions. OSHA can require PPE to protect the eyes, face, head, and extremities. Examples can include protective clothing, respiratory devices, protective shields, and barriers. When employees provide their own PPE, the employer must ensure its adequacy, including proper maintenance and sanitation. Employers must assess the workplace to determine hazards that would require the use of PPE. Employers must select and require the use of PPE that will protect from the hazards identified in the PPE Hazard Assessment. OSHA requires the employer to verify completion of the assessment through a written certification that identifies the workplace, certifying person, and assessment date. Never permit the use of defective or damaged PPE. Train employees on the proper selection and use of PPE. Employees must demonstrate the ability to use PPE properly before using it on the job. Provide retraining whenever employees fail to demonstrate an understanding of proper PPE use. Never use PPE as a substitute for engineering, work practice, or administrative controls. Consider PPE as all clothing and other work accessories designed to create a hazard protection barrier. PPE should comply with applicable ANSI standards. Using PPE can create hazards such as heat disorders, physical stress, impaired vision, and reduced mobility. Review PPE policies at least annually. The review should include evaluation of accident and injury data, current hazard exposures, training effectiveness, and documentation procedures. The employer must verify that affected employees receive and understand required training through a written certification that contains the name of each employee, dates of training, and topics covered. Employers in most situations must provide PPE mandated by OSHA at no cost to employees. OSHA does not require employers to pay for nonspecialty safety-toe footwear including steel-toe shoes or boots and nonspecialty prescription safety eyewear if employees wear them away from the jobsite. The employer must pay for replacement PPE, except when the employee loses or intentionally damages PPE. When employees provide their own PPE, employers may permit use. OSHA does not require reimbursement to the employee for that equipment. Employers cannot require employees to provide their own or pay for PPE.

H. AREAS TO EVALUATE

Evaluate senior leader commitment by determining time and resources allocated to hazard control efforts. Determine which policies and procedures enhanced hazard control effectiveness. Conduct evaluations to determine how well interfacing staff functions support hazard control efforts and accident prevention initiatives. Review submitted cost-benefit analysis reports to determine accuracy and documentation reliability. If conducted, review results of perception surveys completed during the evaluation period. Refer to the sample perception survey located in Appendix 1. Review and assess the information included in organizational property damage reports. Attempt to compare or "benchmark" accident and injury data with similar types of organizations or industries. Refer to compliance and insurance inspections to ensure completion of corrective actions of all recommendations and findings. Organizations can also conduct risk measurements by reviewing historic accident and injury frequency and severity data. This retrospective measurement can provide valuable information about exposure sources that contributed to the injury or illness. Historic accident and inspection reports can also shed light on causal factors and the circumstances contributing to the injury or damage. Knowledge of accident causal factors when connected to specific jobs and tasks can provide insight for recognizing potential risk.

IV. SYSTEM SAFETY

Two early pioneers in system thinking, Daniel Katz and Robert Kahn, viewed most organizations as open social systems. These open systems consisted of specialized and interdependent subsystems. These subsystems possessed processes of communication, feedback mechanisms, and management intervention that linked them together. We can describe a system using three brief definitions: (1) a set of interrelated parts that function as a whole to achieve a common purpose; (2) a piece of software that operates to manage a related collection of tasks; or (3) a design for an organization that perceives sets of processes as a related collection of tasks. Systems operate as either open or closed entities or processes. They can take various forms or shapes and express themselves as mechanical, biological, or social entities. Open systems can interact with other inside subsystems or the outside environment. Closed systems exert little interaction with other systems or the outside environment. Open systems theory originated in the natural science fields. It subsequently spread to fields as diverse as information technology, engineering, and organizational management. Open systems view an organization as an entity that takes input from the environment, transforms those inputs, and releases outputs. This results in reciprocal effects on the organization itself along with the environments in which the organization operates. An organization can become part of the environment in which it operates. The majority of systems operate as open entities. These systems require interaction with the environment for the source of inputs and the destination of outputs.

BOX 1.16   DESIGN-RELATED WEAKNESSES

- Failure to design adequately
- Missing or inadequate policies and rules
- Training and education objectives not developed
- Poorly written plans
- Inadequate processes
- Lack of appropriate procedures

BOX 1.17   OPERATIONAL-RELATED FAILURES

- Not implementing or carrying out required functions or processes
- Not adhering to established policies, procedures, and directives
- Not developing and/or presenting appropriate education and training sessions
- Not ensuring adequate supervision or failing to provide required oversight
- Not conducting comprehensive accident investigations or root cause analysis after an event
- Not evaluating plans, procedures, and processes to determine weaknesses

V.   UNDERSTANDING ACCIDENTS

We can simply define an accident as "an unplanned event that interferes with job or task completion." When an accident occurs, someone will lose valuable time dealing with the event. An accident can result in some kind of measureable loss such as personal injury or property damage. We can also classify an accident event as a near miss with no measurable loss. Accidents normally result from unsafe acts, hazardous conditions, or both. Accident prevention efforts must emphasize development of necessary policies, procedures, and rules. The hazard control plan should outline organizational objectives, goals, and responsibilities. The organization needs to evaluate the priority and effectiveness of accident prevention efforts. The costs of accidents should provide motivation for senior leaders to support hazard control efforts. Accidents resulting in injuries or property damage can cause interruption of production or other operations. Hazard control managers must endeavor to obtain management's attention and support by communicating to them losses in terms of dollars and manpower utilization. We can calculate or closely determine the direct costs associated with an accident. However, determining indirect costs can pose a challenge to the best managers and hazard control managers. Traditionally, many hazard control and safety personnel hold the view that the indirect costs of an accident far exceed the calculated or known direct costs. Fred Manuele wrote a thought provoking article entitled "Accident Costs, Rethinking Ratios of Indirect to Direct Costs" that appeared in *Professional Safety* in January of 2001. His article encouraged safety personnel to refrain from using any ratios that data could not accurately support. He wisely pointed out that the direct costs of accidents had increased significantly in recent years due to indemnity and soaring medical costs.

BOX 1.18   COMMON MYTHS ABOUT ACCIDENTS

- Accidents result from a single or primary cause.
- Accidents must generate injury or property damage.
- Accidents occur when random variables interact.
- Accidents can result from an act of God or nature.
- Accident investigations must determine fault.

> ### BOX 1.19   KEY ACCIDENT PREVENTION PRINCIPLES
>
> - Accident prevention plans and directives must place a strong emphasis on identifying, evaluating, and controlling hazards and unsafe or at risk behaviors.
> - Organizations must place a strong emphasis on working safely since human behaviors and factors remain the most unpredictable aspect of accident prevention efforts.
> - Prevention efforts should rely on the use of systematic processes such as self-inspections and the timely reporting of hazards or accident events.
> - Accident prevention requires the comprehensive analysis of accident and injury data to determine types of events, severity of injuries, and trends within specific units or the entire organization.
> - The prevention of accidents must include process innovations such as machine guarding, use of warning signs, requiring use of PPE, providing real-world safety education and training sessions, and developing adequate administrative policies or procedures.
> - Organizations must develop and implement processes that permit the effective monitoring and evaluation of all accident prevention plans and other initiatives

A.   ACCIDENT REPORTING

The timely and accurate reporting of accidents and injuries permits an organization to collect and analyze loss-related information. This information can help determine patterns and trends of injuries and illnesses. Organizations should encourage reporting by all members. The reporting process should focus on the importance of tracking hazards, accidents, and injuries, including any organizational trends. Educate all personnel to understand the need for maintaining a systematic process that accurately and consistently provides updated information. The system must not only permit data collection, but also provide for a means to display any measure of success or failure in resolving identified hazards. Maintain records that enable managers at all levels to access data. Information made available to managers can assist them with changing policies, modifying operational procedures, and providing job-related training. Senior leaders must ensure the use of a system that meets the needs of the organization.

B.   ACCIDENT INVESTIGATIONS

A successful accident investigation must first determine what happened. This leads to discovering how and why an accident occurred. Most accident investigations involve discovering and analyzing causal factors. Conduct accident investigations with organizational improvement as the key objective. Many times an investigation focuses on determining some level of fault. Some organizations seek to understand the event to prevent similar occurrences. Most large organizations would benefit if the responsible supervisor conducted the initial investigation. Accident causal factors can vary in terms of importance because applying a value to them remains a very difficult challenge. Sometimes investigators can overvalue causal factors immediately documented after the accident. These immediate causal factors reveal details about the situation at the time of occurrence. Investigators can also devalue less obvious causes that remained removed by time and location from the accident scene. Identifying and understanding causal factors must include a strong focus on how human behaviors contributed to the accident. Organizations must conduct comprehensive investigations when preliminary information reveals inconsistencies with written policies and procedures. Using sound investigational techniques can help determine all major causal factors. Investigations should seek to find out what, when, where, who, how, and most of all, why. Organizations must work to reduce the time between the reporting of a serious accident and the start of the investigation.

1. Classifying Accident Causal Factors

Understand the importance of initially documenting and classifying causal factors in one of the following three categories. The first category relates to *operational factors* such as unsafe job processes, inadequate task supervision, lack of job training, and work area hazards. The second category relates to *human motivational factors* including risky behaviors, job-related stress, poor attitudes, drug use, and horseplay. The third category relates to *organizational factors* including inadequate hazard control policies and procedures, management deficiencies, poor organizational structure, and lack of senior leadership.

2. Interviewing Witnesses

Interview witnesses individually and never in a group setting. If possible, interview a witness at the scene of the accident. It also may be preferable to carry out interviews in a quiet location. Seek to establish a rapport with the witness and document information using their words to describe the event. Put the witness at ease and emphasize the reason for the investigation. Let the witness talk, listen carefully, and validate all statements. Take notes or get approval to record the interview. Never intimidate, interrupt, or prompt the witness. Use probing questions that require witnesses to provide detailed answers. Never use leading questions. Ensure that logic and not emotion directs the interview process. Always close the interview on a positive note.

3. Accident Analysis

Organizations can use a variety of processes to analyze accident causal factors. Hazard evaluations and accident trend analysis can help improve the effectiveness of established hazard controls. Routine analysis efforts can also enable organizations to develop and implement appropriate controls in work procedures, hazardous processes, and unsafe operations. Analysis processes rely on information collected from hazard surveys, inspections, hazard reports, and accident investigations. This analysis process can provide a snapshot of hazard information. Effective analysis can then take the snapshots and create viable pictures of hazards and accident causal factors.

When attempting to understand accident causes, hazard control personnel must identify, catalog, and then analyze the many factors contributing to an adverse event. Analyze to determine how and why an accident occurred. Use findings to develop and implement the appropriate controls. Don't overlook information sources such as technical data sheets, hazard control committee minutes, inspection reports, company policies, maintenance reports, past accident reports, formalized safe-work procedures, and training reports. When using accident investigation evidence, remember the information can exist in a physical or documentary form. It can come from eyewitness accounts or from documentary evidence. The analysis must evaluate the sequence of events, extent of damage, human injuries, surface causal factors, hazardous chemical agents, sources of energy, and unsafe behaviors. Consider factors such as horseplay, inadequate training, supervisory ineffectiveness, weak self-inspection processes, poor environmental conditions, and management deficiencies. A good accident analysis should create a word picture of the entire event. Refer to photos, charts, graphs, and any other information to better present the complete accident picture. The final analysis report should include detailed recommendations for controlling hazards discovered during the investigation and analysis.

4. Root Cause Analysis

Root cause analysis processes can help connect the dots of accident causation by painting a picture that includes beneath the surface causes. Organizations many times fail to use effective and systematic techniques to identify and correct system root causes. Best-guess corrective actions do not address the real causes of accidents. Ineffective quick-fix schemes don't change processes to prevent future incidents. Root cause analysis focuses on identifying causal factors and not placing blame. A root cause process must involve teams using systematic and systemic methods. The focus must remain on the identification of problems and causal factors that fed or triggered the unwanted event. When analyzing a problem we must understand what happened before discovering why it happened. Don't overlook causal factors related to procedures, training, quality processes, communications, safety, supervision, and management systems. An effective RCA process lays a foundation for designing and implementing appropriate hazard controls. Root causes always preexist the later-discovered surface causes. When root causes go unchecked, surface causes will manifest in the form of an unwanted event. For simplicity reasons, system-related root causes fall into two major classifications. The first class concerns design flaws such as inadequate or missing policies, plans, processes, or procedures that impact conditions and behaviors. The other category, known as operational weaknesses, refers to failures related to implementing or carrying out established policies, plans, processes, or procedures. When discovered and validated, specific root causes can provide insight to an entire process or system. The process can also help identify what fed the problem that impacted the system. Finally, RCA provides insight for developing solutions or changes that will improve the organization.

BOX 1.20   CONDUCTING TEAM ROOT CAUSE ANALYSIS SESSIONS

- Use individual creativity and expertise of all team members to solve problems effectively
- Design processes that focus on discovering causal factors beneath the surface
- Determine root causes and solutions through the use of good communication skills
- Achieve innovative results, suggestions, and solutions through team synergy
- Ensure all team members understand what happened before attacking the why
- Remember that accidents and other adverse events involve multiple causes

---

BOX 1.21   COMMON CAUSAL FACTORS

Poor Supervision
- Lack of proper instructions
- Job, task, or safety rules not enforced
- Inadequate PPE, incorrect tools, and improper equipment
- Poor planning, improper job procedures, and rushing the worker

Work Practices
- Use of shortcuts and/or working too fast
- Incorrect use of or failure to use protective equipment
- Horseplay or disregard of established safety rules
- Physical or mental impairment on the job
- Using improper body motion or technique

Unsafe Materials, Tools, and Equipment
- Ineffective machine guarding
- Defective materials and tools
- Improper or poor equipment design
- Using wrong tool or using tool improperly
- Poor preventive maintenance procedures

Unsafe Conditions
- Poor lighting or ventilation
- Crowded or poorly planned work areas
- Poor storage, piling, and housekeeping practices
- Lack of exit and egress routes
- Poor environmental conditions such as slippery floors

C.   PREPARING ACCIDENT REPORTS

When preparing an accident investigation report, use the analysis results to make specific and constructive recommendations. Never make general recommendations just to save time and effort. Use a previously drafted sequence of events to describe what happened. Photographs and diagrams may save many words of description. Identify clearly if evidence is based on facts, eyewitness accounts, or assumptions. State the reasons for any conclusions and follow up with the recommendations. An accident analysis process must consider all known and available information about an event. Clarify any previously reported information and verify any data or facts uncovered during the investigation. Review and consider witness information and employee statements or suggestions.

VI.   HUMAN FACTORS

Hazard control personnel and top management must recognize that human behavior can help prevent and cause accidents. Understanding human behaviors can pose a challenge to most individuals. Behavior-based accident prevention efforts must focus on how to get people to work safely. Hazard control managers, organizational leaders, and frontline supervisors deal with human behaviors on a daily basis. The definition of behavior-based hazard control could read as follows: "the application of human behavior principles in the workplace to improve hazard control and accident prevention." Organizational issues such as structure and culture, along with personal considerations, impact individual behavior.

BOX 1.22   UNDERSTANDING BASIC HUMAN ISSUES

- Consider character as moral and/or ethical structures of individuals or groups.
- Belief refers to a mental act and habit of placing trust in someone or something.
- Human values focus on what people believe and things of relative worth or importance.
- Culture refers to socially accepted behaviors, beliefs, and traditions of a group.
- A person's attitude or their state of mind relates to their feelings about something.
- Behaviors relate to the open manifestation of a person's actions in any given situation.

Human factors can address a wide range of elements related to the interaction between individuals and their working environment. Management styles, the nature of work processes, hazards encountered, organizational structure, cultures, and education or training can influence individual behaviors. Hazard control efforts must address factors that impact human conduct or behavior. Organizations must ensure all members possess the knowledge, skill, and opportunity to act in preventing accidents. Motivation to act responsibly relates to a person's desire to do the right thing. Human factors can refer to personal goals, values, and beliefs that can impact an organization's expectations, goals, and objectives. Flawed decision-making practices and poor work practices can create atmospheres conducive to human error. Causal links between the accidents and unseen organizational factors may not appear obvious to most investigators. However, these factors could easily interact with trigger mechanisms to contribute to accident events.

A.   ERROR

An essential component of accident prevention relates directly to understanding the nature, timing, and causes of errors. Error, a normal part of human behavior, many times is overlooked during accident investigations and analysis processes. Errors can result from attention failure, a memory lapse, poor judgment, and faulty reasoning. These types of errors signify a breakdown in an individual's information-processing functions. When analyzing accident causes, hazard control personnel may never determine individual intent. However, focusing on the nature of behaviors in play at the time of error occurrence can provide some insight. When categorizing errors, attempt to differentiate between those occurring during accomplishing skilled behavior tasks and those related to unskilled tasks such as problem solving. Rule-based errors can occur when human behavior requires the use of certain situational actions. Process errors occur when individuals lack an understanding of procedures or complex systems. Knowledge-based errors occur when an individual lacks the skill or education to accomplish a certain task correctly. Planning errors occur when individuals fail to use a proper plan to accomplish a task. Finally, execution errors occur when a correct plan is used, but an individual fails to execute the plan as required.

BOX 1.23   COMMON UNSAFE ACTS

- Failure to follow established job or safety procedures
- Cleaning or repairing equipment with energy hazards not locked out
- Failing to use prescribed PPE
- Failing to wear appropriate personal clothing
- Improperly using equipment
- Removing or bypassing safety devices
- Operating equipment incorrectly
- Purposely working at unsafe rates or speeds
- Accomplishing tasks using incorrect body positions or postures
- Incorrectly mixing or combining chemicals or other hazardous materials
- Knowingly using unsafe tools or equipment
- Working impaired under the influence of drugs or alcohol

B.   MOTIVATING PEOPLE

Motivating individuals to work safely requires the use of various approaches depending on the situation. Little evidence exists that supports using punitive measures to motivate safe behaviors. The use of good human relations and effective communication skills can help improve individual motivation. The development of policies, procedures, and rules can never completely address all unsafe behaviors. Taking risks depends on individual perceptions when weighing potential benefits against possible losses. Some professionals believe that providing incentives to work safely coupled with appropriate feedback can enhance hazard control efforts in at least the short term. However, employee incentives with tangible rewards can also cause some individuals not to report hazards, accidents, and injuries.

BOX 1.24   PROMOTING EMPLOYEE INVOLVEMENT

- Develop an open-door policy to provide employee access to supervisors
- Implement an easy-to-use accident, injury, and hazard reporting system
- Require supervisors to conduct periodic safety meetings
- Develop off-the-job safety and health education objectives
- Encourage employee participation in job hazard analysis sessions
- Disseminate hazard control and accident information in a timely fashion
- Place more emphasis on people and less emphasis on compliance or accreditation
- Solicit employee suggestions on how to best use hazard control resources and funds
- Mandate hourly employee representation on safety committees and improvement teams
- Provide special education and training sessions for those working odd shifts
- Use periodic safety perception surveys to obtain information about what workers think

REVIEW EXERCISES

1. What were the top two health service organization OSHA citations for FY 2011–2012?
2. What was the OSHA (nonfatal) injury incident rate in 2011 for hospitals?
3. According to the 2011 ANA Health and Safety Survey, what were the top two items of concern for nurses?
4. List seven standards addressed in the ANA's Safe Patient Handling and Mobility publication.
5. In your own words, describe healthcare safety as an organizational function.
6. List the seven fundamental values of an effective safety function.
7. List the six steps of an effective behavior correction process.
8. Define the following terms:
    a. Employee engagement
    b. Safety modeling
    c. Coaching
9. List the five categories of healthcare occupational hazards and provide an example of each.
10. Provide two basic definitions of a "system."
11. List at least four basic design-related weaknesses.
12. Define the following terms:
    a. Accident
    b. Near miss
    c. Measurable loss

2 Leadership and Management Overview

I. INTRODUCTION

Understanding some of the basic concepts related to effective management and leadership can prove valuable to those working or supporting organizational accident prevention efforts. This chapter provides a brief review of management theory and functions. It also provides a quick overview of key leadership principles that should complement management efforts. Some people think of leadership as just another function of management. However, I prefer to view management as the key support function of sound leadership. Organizational leaders and those that manage hazard control functions must consider the roles that people play in preventing or contributing to accidents and injuries. Many people view managing as an art that anyone can master with practice. Others view managing as a learned discipline or social science. Good managers seem to demonstrate that managing consists of learning and practice. The word "manager" can also refer to an individual's job or position title. How often do we meet someone with the title of manager who demonstrates little understanding of even basic management principles? Individuals do not become effective managers because they hold the title or position. Too often society misuses the term "manager" much like it also overuses the word "safety." We tend to use both words out of context or without much thought to their true meanings.

Leading simply refers to taking actions to influence others toward attainment of organizational goals and objectives. Effective leaders understand the importance that human engagement plays in goal accomplishment. Human engagement refers to the concept where individuals personally feel their connection to their position and to organizational success. Human engagement contributes to personal satisfaction, which helps increase productivity, morale, and motivation. Some organizations now realize the need for balancing organizational demands with a person's family and other life issues. Individuals, when away from the organization, serve in a variety of roles including volunteer, caregiver, and parent. Understanding the concept of human engagement helps leaders and hazard control managers better understand the behaviors and reactions of individuals to organizational issues and decisions. Conflicting responsibilities can lead to role misunderstandings and overloads that can impact their support for organizational objectives including hazard control efforts.

II. LEADERSHIP

Effective leading requires a manager to motivate subordinates, communicate effectively, and effectively use power. To become effective at leading, managers must first understand their subordinates' personalities, values, attitudes, and emotions. Leaders need to identify opportunities to reward personal and team success. Place the emphasis on improving organizational systems and processes instead of blaming individuals. True leaders focus on processes and must learn to educate followers instead of dictating to them. Leaders who use conditional statements to encourage others must also listen closely before taking actions. Leaders must promote ownership or *buy in* of hazard control as an organizational value. Encourage creativity to increase responsible actions of those being led. However, leaders must establish and communicate expectations in clear and concise terms. Taking these actions will reduce the need for any future mandates. True leaders learn to trust in their people skills while remaining uncertain about *how* to best meet objectives. Effective leaders learn to look beyond numbers if possible and resist trying to quantify everything. Leaders must learn to make both tactical and strategic decisions. They must possess a vision of the organization structure and the path that it's traveling. System thinking helps leaders to see the big picture or the true organizational cultures that impact failure and success. Formal leaders must become effective ambassadors of the hazard control message.

BOX 2.1   CONCEPTS LEADERS MUST UNDERSTAND

- Character refers to the moral or ethical structure of an individual or group.
- Belief refers to the mental act or habit of placing trust in someone or something.
- Value refers to an individual's perception of worth or importance assigned to something.
- Culture reflects the socially accepted behaviors, beliefs, and traditions of a group.
- Attitude refers to an individual's personal state of mind or feeling about something.
- People can perform a job or task and hide their attitude from others.
- Behavior relates to an open manifestation of a person's actions in a given situation.

BOX 2.2   SAFETY LEADERSHIP BASICS

- Keep work areas safe from risks and hazards
- Emphasize the importance of safe behaviors
- Educate employees regarding safety performance
- Communicate clearly the organization's safety goals and objectives
- Make safety part of every job or task
- Promote a total safety culture
- Keep all employees informed regarding successes and failures
- Reward safe behaviors
- Correct unsafe worker actions
- Never use safety as a disciplinary tool
- Never promote safety as simply a compliance issue
- Promote safety as the right thing to do
- Remember that attitudes are caught never taught
- Emphasize good judgment and common sense
- Understand the relationships of culture, communication, and coordination

---

A.   PRACTICAL LEADERSHIP

Effectiveness refers to taking the right actions to achieve a desired or expected outcome. Emotional issues can impact and even sidetrack the best hazard control efforts. True leaders must learn to use logic when seeking to reduce accidents and injuries. The appropriateness and quality of decisions at every organizational level can impact hazard control effectiveness. Leaders should understand that repetition acts as the mother of learning. Leaders should also consider human learning abilities, including retention of information, when developing orientation, education, and training sessions. Hazard control promotes the importance of recognizing and identifying unsafe work conditions and behaviors. Stress the importance of improving worker awareness of hazards and exposures encountered on the job. Supervisors should continuously stress the need for improving awareness on the job. High reliability organizations that use system safety methods place a strong emphasis on both task and situational awareness on the job.

B. Leadership Ethics

When addressing leadership, consider ethics as vital since it provides the foundation of any hazard control management function. Without ethics, hazard control loses its organizational and personal value. Webster's Collegiate Dictionary defines ethics as the (1) "discipline of dealing with good and bad with moral duty and obligation," and (2) "a set of moral principles or values or a theory or system of moral values." Ethics finds its roots in natural law, religious tenets, parental or family influence, educational experiences, life experiences, cultural norms, and societal expectations. Business ethics refers to the application of the discipline, principles, and ethical theories to the organizational context. We can define business ethics as the "principles and standards that guide behaviors in the business world." Ethical behaviors remain an integral part of conducting business affairs. The three considerations that impact and influence ethical decision-making in business include individual difference factors, operational or situational factors, and issue-related factors.

III. MANAGEMENT

It seems so easy to create and use the phrase "safety management" as if communicating a definitive concept or process. However, safety management can mean different things to different people. The National Safety Management Society defines the term as "... that function which exists to assist all managers in better performing their responsibilities for operational system design and implementation through either the prediction of management systems deficiencies before errors occur or the identification and correction of management system deficiencies by professional analysis of accidental incidents (performance errors)." This definition stresses the importance of identifying and correcting management-related deficiencies. Many people never think of addressing management deficiencies when referring to the term "safety management." Poor and inefficient management provides opportunities for accidents to occur. Some individuals also think of management and leadership as synonymous terms. You can manage materials, projects, and processes, but you must lead people. Organizations need all managers and supervisors to provide hazard control leadership to their subordinates. I often made this statement while teaching safety educational sessions: "Good leaders who can't or don't want to be burdened down with the managerial details must quickly find someone who can help them." I then follow up with this statement: "Good managers who can't lead or don't want to lead need to quickly learn the art of delegating to someone who can." Dealing with too many details can cause operational managers to become overburdened with objectives, goals, and time constraints. Many organizations now use "project management" personnel to oversee the activities of long-term or complex projects. Leaders must move toward meeting established objectives but do so by considering both people and processes during the journey. Hazard control management efforts without leadership can easily fail. However, hazard control professionals not using good management techniques can also fail in meeting expectations or objectives. Managers must set an example by following work rules and behavioral expectations established for their subordinates. Organizational members must see management personnel consistently setting a positive example in the area of hazard control. Managers must make practicing safety a priority and lead by example. They must learn to help others achieve personal success while meeting established organizational objectives. People perform better when provided with the proper information, the necessary tools, and the delegated authority to get a job done. People must view themselves as participants in a project and never as a pawn of a manager.

A. Knowledge Management

The concept of *knowledge management* goes beyond the concept of information technology. Many organizations manage information but neglect to manage knowledge. Managing knowledge requires the identification, analysis, and understanding of all operational processes. Knowledge provides no organizational value unless the information is relevant, available, and disseminated to end users. The failure to communicate accident and hazard information can result in poor analysis of hazards or accident experience. Each organizational process, department, or function must contribute information about accidents, hazards, and unsafe behaviors. This aggregate knowledge provides the basis for determining appropriate hazard controls, the need for training or education, or required innovations to improve performance.

BOX 2.3   TIPS FOR IMPROVING ORGANIZATIONAL KNOWLEDGE MANAGEMENT

- Identify the knowledge that the organization already possesses
- Determine the kinds of knowledge that the organization needs
- Evaluate how knowledge can add value to organizational effectiveness
- Create processes to help the organization achieve objectives
- Maintain and effectively use knowledge assets to improve the organization

---

B. DECISION-MAKING

People employ management concepts to help them make good decisions. Managers must constantly evaluate alternatives and make decisions regarding a wide range of matters. Decision-making involves uncertainty and risk, and decision-makers possess varying degrees of risk aversion when making decisions. Decision-making may require evaluating information and data generated by qualitative and quantitative analyses. It must rely not only on rational judgments, but factors such as decision-maker personality, peer pressure, organizational situations, and a host of other issues. Management "icon" Peter Drucker identified several key decision-making practices that successful executives used. Leaders should consider the following questions before making decisions. The first question simply asks, "What needs accomplishment?" The second question seeks to find out, "What's best for the organization?" When a decision-maker gets the answers, he or she can then proceed with developing a plan of action. Decision-makers must take responsibility for their actions. True decision-makers use team pronouns such as "we" and never the self-gratifying pronoun "I."

C. PSYCHOLOGICAL SAFETY

Workplace-related psychological safety demonstrates itself when employees feel unable to put themselves on the line, ask questions, seek feedback, report problems, or propose a new idea without fearing negative consequences to themselves, their jobs, or their career. A psychologically safe and healthy workplace actively promotes emotional well-being among employees while taking all reasonable steps to minimize threats to employee mental health.

D. CRISIS MANAGEMENT

The study of crisis management originated with the large-scale industrial and environmental disasters in the 1980s. Three elements common to most definitions of crisis include determining threats to the organization, planning for the elements of surprise, and making decisions in short time frames. The fourth element relates to addressing the need for change. When change does not take place, some could view the event as a failure or incident. Crisis management uses response methods that address both the reality and perception of crises. Organizations should develop guidance to help define what constitutes a crisis and what triggers would require immediate response.

E. TRADITIONAL ORGANIZATIONAL STRUCTURE

Traditional organizational structure follows two basic patterns. The first structure, referred to as a *line organization*, permits top management to maintain complete control with a clearly defined chain of command. This basic line structure works well in small companies with the owner or top manager functioning at the top of the organizational structure. Everyone understands the clear lines of distinction between the owner or manager and subordinates. A *line-and-staff organization* combines the line organization with appropriate staff departments that provide support and advice to the line functions of the organization. Many medium and large organizations use the line-and-staff structure with multiple layers of staff managers supporting overall operations. An advantage of the line-and-staff organizational structure relates to the availability of technical and managerial functions. The organization incorporates these needed staff and support positions into the formal chain of command. However, conflict can arise between line-and-staff personnel, creating disruptions within the organization. This conflict can at times impact the effectiveness of the hazard control management function. Hazard control managers must remain focused on identifying and correcting the causes of accidents regardless of the organizational structure. However, they must understand that organizational structure can hinder accident prevention efforts.

Organizational leaders must integrate the function of hazard control into the organizational management structure with clearly defined responsibility and authority. Top management must focus on identifying and correcting operational and staff management-related deficiencies that could hinder hazard control efforts. The organizational structure must consider developing processes that help identify and analyze system deficiencies that contribute to accidents. Leaders must ensure that support functions such as human resources, facility management, and purchasing receive information about their management deficiencies that could impact hazard control. Consider the following scenario: a human resource department mistakenly assigns a new employee to a hazardous job position without properly screening or evaluating the person's qualifications. This could contribute to an accident or mishap. Senior leaders of staff and departments must understand their roles and responsibilities related to hazard control. Many organizational structures permit and even unknowingly encourage support or staff department managers to create their own "little dynasty." This can result in the self-coronation of "turf kings and queens." Once crowned, these rulers may not see the need to coordinate or communicate important issues with other functions.

IV. ORGANIZATIONAL CULTURE

Organizational culture consists and exists based on assumptions held by a particular group. These assumptions can include a mix of values, beliefs, meanings, and expectations held in common by its members. Cultures can determine acceptable behaviors and problem solving processes. Organizational trust refers to the positive and productive social processes existing within the workplace. Trust can encourage group members to engage in cooperative and expected organizational behaviors. Trust also provides the foundation for demonstrating commitment and loyalty. For example, an organization with a safety and health focused culture enhances the well-being, job satisfaction, and organizational commitment of all members. A culture with social support systems can enhance a member's well-being by providing a positive work environment for those dealing with depression or anxiety. The established culture sets the tone for an organization. Negative cultures can even hinder the effectiveness of the best plans and policies. Unhealthy cultures create stressful environments, which can lower employee well-being and impact organizational productivity.

BOX 2.4   ELEMENTS FOR CREATING SAFETY CULTURES

- Positive perception of teamwork
- Safe behaviors exist as the norm
- Job satisfaction
- Perception of senior management effectiveness
- Recognizing the reality of job-related stress
- Adequacy of supervision, education, and training
- Opportunities for effective organizational learning
- Nonpunitive response to error by leaders

A. COVERT AND OVERT CULTURES

Many times senior managers fail to acknowledge the existence of the covert, informal, and hidden cultures. They incorrectly hold the belief that the established overt, formal, or open culture drives organizational success and productivity. They also fail to acknowledge the tremendous influence of hidden cultures on organizational behaviors. Why do these hidden cultures exist? They exist to meet the needs of its members. Failing to acknowledge these hidden cultures can hinder an organization's ability to change or improve the formal culture. The actual climate, not the established structure, exerts the most influence on organizational performance. Hidden cultures can also support a very effective organizational communication process known as the grapevine. The grapevine serves as the informal and confidential communication network that quickly develops within any organization to supplement the formal channels. The actual function of the grapevine will vary depending on the organization. For example, it could communicate information inappropriate for formal channels. The grapevine can carry both good and bad organizational news. The grapevine in some instances serves as a medium for translating top management information into more understandable terms. The grapevine also serves as a source of communication redundancy to supplement formal channels. When formal communication channels become unreliable, the grapevine can quickly operate as the more trusted communication system.

BOX 2.5   IMPROVING THE SAFETY CULTURE

- Never use orientation sessions to conduct performance-based safety training or education
- Conduct safety orientation topics during the morning
- Require supervisors to conduct and document job-related training
- Centralize all respirator fit testing activities to ensure proper training
- Do not insist on a sign-in for training (documentation becomes more important than learning)
- Have all workers document their training by completing, at a minimum, a 10-question quiz
- Train second- and third-shift workers on their shift
- Orient second- and third-shift workers on their shift
- Provide mandatory education for third-shift workers regarding stress and sleep deprivation
- Establish an off-the-job safety education program
- Address an off-the-job safety topic at every safety function

Some organizations use socialization processes to educate new members about the organization's cultures. Socialization can occur in both formal and informal culture arenas. The socialization process will determine which culture, formal or informal, will exert the most influence on an individual. Organizational members then must decide to remain in or leave the organization. Many that stay may experience isolation. Leading culture change requires a sound understanding of organizational behaviors, attitudes, expectations, and perceptions. Changing the organizational culture must also impact the behaviors, attitudes, and perceptions of all organizational members. Leaders must use all available sources to communicate the "change" message. Change can have both a negative and positive impact on leaders as well. It can dethrone the turf kings and queens. However, it can encourage development of teamwork and continuous improvement processes.

When leading change within any organization, never forget the importance of communication and feedback throughout the entire journey. Most people naturally resist change, which causes organizations to change very slowly. First-level change deals with people, structure, policies, and procedures. Second-level change deals with complex systems, cultures, and processes. Many people now view change as an inherent and integral part of organizational life. Some new trends in organizational dynamics emerged during recent years. These emerging trends can create conflict and concern for organizational leaders and members. They can create both opportunities and threats in the minds of people. Any change creates tensions that leaders must address to prevent unwanted or dysfunctional change. Many organizations now operate on a global scale with increased competition. These organizations also must embrace economic interdependence and increased collaboration. This globalization results in a wide range of consumer needs and preferences.

Change does not occur just because a top manager writes a memo and declares it done. Organizational change must consider how to best transform the organization from within. When leaders communicate need and intent to change, they must provide information to address issues such as behaviors and expectations. Communicate the reasons for change so that everyone understands. Provide guidance and leadership to ensure change happens. Before change can occur in large organizations, leaders must acknowledge the existence and influence of hidden or covert cultures operating in the organization. Always articulate how change will impact all organizational members and operational functions. When educating others about change, provide sessions that focus on real situations. Work to change or shift the culture before implementing other innovations or interventions. Not to do so would provide no supporting foundation for the innovations or interventions. Leaders must promote trust and ensure the involvement of organizational members in decision-making processes. Provide team members with the opportunity to voice concerns or make suggestions that would help build trust. Encourage team members to express some kind of choice and allow them flexibility to make decisions related to their job tasks. Migrating decision-making permits individuals with the appropriate expertise, education, or experience, regardless of rank or position, to make an informed decision.

When changes in worksites, processes, materials, and equipment occur, hazards can emerge. During any change process, hazard surveillance and self-inspections must become frequent. Organizations should develop an enhanced hazard review process when undertaking major changes in processes, systems, or operations. Maintain sound coordination and communication systems among all parties involved in the change process.

Change may require revising existing job hazard analyses, reviewing standard operating practices, evaluating lockout methods, and assessing personal protective equipment requirements. Change-related hazard analyses could prove cost-effective in terms of preventing accidents, injuries, and other organizational losses. Individuals respond differently to change. Some organizational members may require additional time to adapt and accept the change.

V. INTERFACING SUPPORT FUNCTIONS

The noninvolvement of staff functions and service components in hazard control can hinder success. Don't ignore a natural interface of hazard control management with other support functions such as facility management, purchasing, and human resource management. Virtually every department and function of a modern organization contributes in some way to the effectiveness of the hazard control management function. There should be a greater interrelationship among staff and support functions that interface with accident prevention such as personnel, procurement, maintenance, etc. Too often, these functions operate in parallel tracks with little or no interaction. They must work in harmony to make an impact on preventing accidents and controlling hazards.

A. OPERATIONAL AND SUPPORT FUNCTIONS

Operational and line elements must remain conscious of their roles in accident prevention in terms of organizational policy, regulations, procedures, safety inspections, and other activities to support the hazard control function. Frequently, technical advances result in acceleration of organizational activities without the provision for accompanying related safeguards by management. Planning, research, budget, and legal functions must interface with accident prevention and hazard control efforts.

B. HUMAN RESOURCES

As far as practical, the human resource (HR) function should recruit, evaluate, and place the right person in the right job in terms of physical ability and psychological adaptability. Incorporate into job descriptions specific physical requirements, known hazards, and special abilities required for optimum performance. Human resource professionals must identify all hazardous occupations and determine the knowledge, skills, abilities, physical requirements, and medical standards required to perform the job in a safe manner. All employees must receive appropriate orientation, training, and education necessary to support safe job accomplishments.

C. FACILITY MANAGEMENT

Facility management functions should ensure proper design layout, lighting, heat, and ventilation in work areas. Review specifications for new facilities, major renovations to existing facilities, and any plans for renting or leasing new work or storage areas. Maintenance activities should provide preventive maintenance service to avoid breakdown of equipment and facilities. Coordinate efforts with engineering, purchasing, and safety in reporting obsolete and/or hazardous equipment. Ensure that maximum safety is built into the work environment. It is much more efficient to correct a hazardous situation than to guard it or instruct employees to avoid it.

D. EMPLOYEE HEALTH

Occupational health professionals and hazard control managers must coordinate and communicate issues on a continuous basis. Provide prompt emergency treatment of all injuries and illnesses. The coordination of safety and health functions helps workers learn how to protect themselves from hazards. Recommend a multidisciplinary approach to manage health, risks, and costs. Report all work-related incidents of injury and exposure allegations immediately to employee health. Employee health should monitor, manage, or coordinate all workers' compensation injuries, reports on progress, imposition of necessary work restrictions, and return to work evaluations. Pre-employment placement evaluations should focus on job-related issues with a thorough job analysis as part of the evaluation. If the evaluation indicates no medical causes for performance problems, refer the employee back to management for appropriate administrative action. A pre-placement assessment develops a baseline for medical surveillance and helps determine capability of performing essential job functions. The Americans with Disabilities Act (ADA) requires job descriptions for all job offers needing pre-placement (post-job offer) physical capacity determinations. Functional capacity evaluations may help in determining job placement and modifications. Essential job functions can determine capability of a prospective employee to perform those functions with or without reasonable accommodations. Assessments may include an update of the occupational and medical histories, biological monitoring, and medical surveillance. Conduct a post-exposure assessment following any exposure incident. Determine the extent of exposure and develop measures for prevention. Rehabilitation involves facilitating the employee's recovery to a pre-injury or illness state. Inform occupational health about the rehabilitation of workers with any illnesses or injuries including those not considered work related. The goal of case management is to work with the employee to facilitate a complete and timely recovery.

BOX 2.6   COMPONENTS OF THE EMPLOYEE HEALTH FUNCTION

- Bloodborne Exposure Control Plan (29 CFR 1910.1030)
- Fitness-For-Duty (Local Policies)
- Personal Protective Equipment (29 CFR 1910.132)
- Eye Protection (29 CFR 1910.133)
- Fire Safety (NFPA 101, 29 CFR 1910.38, Local and State Codes)
- TB Policy (CDC Guidelines and Health Department Requirements)
- Immunizations (CDC and Health Department Recommendations)
- Radiation Safety (29 CFR 1910.1096)
- Reproductive Hazards (OSHA, NRC, & NIOSH Recommendations)
- Confidentiality of Medical Records (HIPAA and OSHA Standards)
- Hazard Communication (29 CFR 1910.1200)
- Substance Abuse (Local Policies)
- Work-Related Injuries (29 CFR 1904) & Workers' Compensation Statutes
- OSHA Record Keeping (29 CFR 1904)
- Hearing Protection (29 CFR 1910.95)
- Work-Related Stress & Shift Work (NIOSH Publications)

VI. WORKERS' COMPENSATION

Workers' compensation laws ensure that employees injured or disabled on the job receive appropriate monetary benefits, eliminating the need for litigation. These laws also provide benefits for dependents of those workers killed because of work-related accidents or illnesses. Some laws also protect employers and fellow workers by limiting the amount an injured employee can recover from an employer and by eliminating the liability of coworkers in most accidents. State workers' compensation statutes establish this framework for most employment. The injury or illness must result from employment. Workers' compensation provides benefits to the injured worker including medical coverage and wages during periods of disability. Employers can obtain coverage through commercial insurance carriers, establishing their own self-insurance program, or by being placed in a state-controlled risk fund. The state or the National Council of Compensation Insurance (NCCI), an independent rating organization, normally determines basic rates paid by employers. Factors affecting rates can include (1) company or fund quoting the coverage, (2) classification code(s) of the employer, (3) payroll amount for the work force covered, and (4) experience rating.

The Federal Employment Compensation Act provides workers' compensation for nonmilitary, federal employees. Many of its provisions remain typical of most workers' compensation laws. Many times awards remain limited to disability or death sustained while in the performance of the employee's duties. The act covers medical expenses due to the disability and may require the employee to undergo job retraining. A disabled employee receives two-thirds of his or her normal monthly salary during the disability. The Long Shore and Harbor Workers' Compensation Act provides workers' compensation to specified employees of private maritime employers. The Black Lung Benefits Act provides compensation for miners suffering from black lung or pneumoconiosis. The Act requires liable mine operators to pay disability payments and establishes a fund administered by the Secretary of Labor providing disability payments to miners when mine operators can't pay. The World Health Organization (WHO) defines impairment as any loss or abnormality of psychological, physiologic, or anatomic structures or functions. The American Medical Association (AMA) defines impairment as loss, loss of use, or derangement of any body part, system, or function. WHO defines a disability as any restriction or lack of ability, resulting from an impairment, to perform an activity in the manner of within the range considered normal. Most states require examiners to use the AMA "Guides to the Evaluation of Permanent Impairment" to determine accurate impairment ratings. The AMA guidelines limit the range of impairment values reported by different examiners.

A.   RETURN TO WORK/MODIFIED DUTY POSITIONS

Establishing a realistic return to work function can save organizations financial losses due to fraudulent claims. Any "Return to Work Initiative" should accommodate injured workers by modifying jobs to meet their work capabilities. This action permits employees to become productive assets during their recovery. Early return to work options can accelerate an employee's return by addressing the physical, emotional, attitudinal, and environmental factors that otherwise inhibit a prompt return. Senior management must commit to returning injured workers to productive roles. Develop profiles of jobs considered suitable for early return participants. A profile should define the job in terms of overall physical demands, motions required, environmental conditions, the number of times performed each week, and duration. Conduct a systematic analysis of specific jobs for the purpose of modifying them to accommodate the unique needs of the injured worker. Individuals skilled in ergonomic task analysis, engineering, safety, and biomechanics can help perform the job analysis. Managed care providers can also assist in job modifications. Communicate the availability of early return jobs with care providers, claims adjusters, and the injured worker. Work with your managed care provider and worker to move them to full production status in their assigned jobs as quickly as possible.

B.   SUBSTANCE ABUSE

Substance use, misuse, abuse, and coping strategies can significantly impact mental health at work. Generally, substance use becomes a problem when an individual loses control over their use and/or continues to use despite experiencing negative consequences. Employers should look for warning signs for employees struggling with substance abuse. Signs of substance abuse can appear similar to those caused by stress, lack of sleep, and physical or mental illness. Identify abuse by establishing pre-employment, random, and "for cause" testing. Ensure the development and implementation of a testing policy. Refer and evaluate employees addicted to performance-impairing drugs such as alcohol, narcotics, sedatives, or stimulants to qualified assistance or treatment facilities. Establish an agreement between the organization and the individual to address rehabilitation and random testing upon return to work. Measurable losses attributed to substance abuse can include absenteeism, overtime pay, tardiness, sick leave abuse, health insurance claims, and disability payments. Some of the hidden costs of substance abuse can include low morale, poor performance, equipment damage, diverted supervisory time, and low production quality. Losses can include legal claims, workers' compensation payments, disciplinary actions, security issues, and even dealing drugs in the workplace. Supervisors play the key role in maintaining an effective substance abuse policy.

BOX 2.7   SIGNS OF SUBSTANCE ABUSE

- Increased absenteeism
- Poor decisions and ineffectiveness on the job
- Poor quantity and/or quality in production
- High accident rates
- Resentment by coworkers who pick up the slack
- Poor morale in the department
- Late three times more often than other workers
- Uses three times more sick leave than others
- Five times more likely to file a workers' compensation claim
- Involved in accidents four times more often than other employees

---

VII.   ORIENTATION, EDUCATION, AND TRAINING

Orientation relates to the indoctrination of new employees into the organization. Orientation can be defined as the process that informs participants how to find their way within the organization. Usually safety and hazard control topics only make up a portion of any new employee orientation session. Many well-meaning organizations attempt to present detailed safety and hazard control information during new employee orientation sessions. However, attempting to provide too much performance-based safety education during orientation can prove ineffective due to time constraints. Meeting the learning objectives must take precedence over simply documenting an educational session. New employee orientation sessions must address the importance of safety, management's commitment, and worker responsibilities to practice good hazard control principles. Some performance-based Occupational Safety and Health Administration (OSHA) standards require that employees receive more detailed instruction that orientation sessions can provide. Present this information in other education or training sessions. Education refers to the incorporation of knowledge, skills, and attitudes into a person's behavior and includes the connotation of thinking. Education can provide information on topics previously trained. System safety methods use Instructional System Design (ISD) educational and training methods to ensure the competency of individuals working in or supporting operations. ISD requires development of competencies before designing educational and training sessions. The sessions presented focus on the competencies both in the classroom and in realistic operational or job settings. Another example of ISD usage occurs in the construction industry, which developed "tool box" safety presentations many years ago to ensure workers practice safety on the job. Sometimes we use the phrases "job-related training" or "job safety training" to refer to instructional system design. Training relates to the acquisition of specific skills while education refers to the incorporation of knowledge, skills, and attitudes into a person's behavior. Consider training as the process of presenting information and techniques that leads to competency of those participating. Conduct hands-on training outside of a classroom if possible unless using realistic simulation processes. Effective training must strive to promote understanding, positively impact worker attitudes, and improve individual performance. Training must facilitate the transfer of knowledge and skills that relate to real-world activities. Many organizations do not dedicate sufficient time, allocate sufficient resources, or require attendance at training and education sessions.

A. PROVIDING ADEQUATE SESSIONS

An around-the-clock operation makes the education and training of shift employees even more challenging. Educate and train shift workers prior to or during their shift but never after the shift. Many organizations do not honestly evaluate training and education effectiveness. They maintain attendance or participation documentation. However, this documentation may not document and validate retention or competency. Some professional educators recommend documenting training and education attendance at the conclusion of the session. Use a short quiz or performance assessment to document learning. Consider the use of employee safety meetings to educate workers about on- and off-the-job safety topics. Publish an education and training policy statement to outline goals and objectives. Use various methods such as posters, flyers, bulletins, newsletters, classroom presentations, on-the-job training sessions, professional seminars, safety education fairs, and computer-assisted training to communicate hazard control and safety topics. Some organizations delegate a number of training responsibilities to the individual departments. Other organizations employ a full-time educational coordinator. Large or specialized departments in some organizations, such as laboratories, conduct most of their own training.

B. SAFETY TRAINING

Safety and training personnel must coordinate education and training objectives to ensure they meet organizational needs. Conduct training for employees transferring to new jobs or work areas. Train those returning from an extended period away from the job and those new to the workforce. Schedule training sessions to match the needs of the organization and needs of learners. Always view education and training as organizational functions and never as programs. When implementing an effective hazard control education and training function, consider the following elements: (1) identify needs, (2) develop objectives, (3) determine learning methods, (4) conduct the sessions, (5) evaluate effectiveness, and (6) take steps to improve the process. Training must compliment and supplement other hazard controls and address rules and work practices. Some ways to evaluate training can include the following: (1) student opinions expressed on questionnaires, (2) conducting informal discussions to determine relevance and appropriateness of training, (3) supervisors' observations of individual performance both before and after the training, and (4) documenting reduced injury or accident rates. Revise the content of the session when an evaluation reveals that those attending did not demonstrate the knowledge or competency expected. Recurring sessions should cover on-the-job training and refresher sessions to ensure employees remain current in worker-related issues, including safety topics. Changes covered might include updated technology procedures, new government regulations, and improved practice standards. Engineering controls remain the preferred way of preventing accidents involving hazards related to unsafe mechanical and physical hazards. However, education and training serves as the most effective tool in preventing accidents by human causes. Through adequate instruction, people can learn to develop safe attitudes and work practices. Design education and training sessions by using clearly stated goals and objectives that reflect the knowledge and skills needs of people.

VIII. EFFECTIVE SPEAKING AND WRITING

Healthcare professionals including nurses must learn to understand the communicative process and demonstrate their ability to speak and write effectively. Communication consists of the sender, the message, and the audience. For successful communication, the audience must not only get the message, but must interpret the message in the sender's intended way. Communication refers to the purposeful act or instance of transmitting information using verbal expressions or written messages. To function effectively, all leaders and managers need to know how to effectively communicate with all organizational members. Managers and leaders must understand the different communication channels available. Downward communication involves more than passing information to subordinates. It can involve managing the tone of the message and effectively demonstrating skill in delegation.

When communicating upward, tone becomes more crucial along with timing, strategy, and audience adaptation. A sender wants to transmit an idea to a receiver through using signs capable of perception by another person.

A. COMMUNICATION

Communication refers to the sharing or exchanging of thought by oral, written, or nonverbal means. Communication signs can include the printed or spoken word, a gesture, a handshake, or a facial expression. The receiver takes those signs, interprets them, and then reacts with feedback or simply ignores the message. When communicating, a sender encodes a message using some tangible sign. A sign may consist of anything seen, heard, felt, tasted, or smelled. The receiver decodes the message to comprehend its meaning. The meaning of the message can differ since both the senders and receivers can assign their own meanings. Each individual's unique set of experiences can function as a perceptual filter. The filter blends the education, upbringing, and life experiences of the perceiver. Leaders must learn to use effective communication to provide vision and direction to others. Motivating, inspiring, and persuading others to work together requires effective communication skills. Miscommunication can result in expensive mistakes, organizational embarrassment, and in some cases accidents or even death. Today, communication effectiveness can suffer from too much information. Around-the-clock media coverage, e-mails, and web-based information sources make it difficult to filter the valuable information needed to accomplish our goals and objectives. We must learn to communicate with clarity and focus. Failure to communicate relates to answering the wrong question, answering only part of the question, and adding irrelevant information. Many communicators answer the question but provide unnecessary feedback information.

B. COMMUNICATION BARRIERS

Communication barriers, often also called noise or static, can complicate the communication process. While unavoidable, both the sender and receiver must work to minimize them. Interpersonal communication barriers can arise within the realm of either the sender or receiver. If an individual holds a bias against the topic under discussion, anything said in the conversation can affect perception. Organizational barriers can occur because of interactions taking place within another larger work unit. The serial transmission effect takes place when a message travels along the chain of command path. As it moves from one level to the next, it changes to reflect the person who passed it on. By the time a message travels from the bottom to the top of the chain, it changes and is not likely recognized by the person who initiated it. Nonverbal communication occurs when information exchanges through nonlinguistic signs. Many consider body language synonymous with nonverbal communication. Body language provides a rich source of information during interpersonal communication. The gestures that people make during an interview can emphasize or contradict what he or she says. Posture and eye contact can indicate respect and careful attention.

C. EFFECTIVE WRITING

When preparing to communicate using written correspondence, organize information using a logical and systematic process. This helps the recipients to understand the message without reading it over and over. When communicating in a clear manner, place emphasis on the rules of language, including proper spelling and correctly pronouncing words. Good communicators also learn how to assemble and punctuate sentences. Effective communicators never hide their ideas or information in a jungle of unnecessary verbiage. Use of incorrect language can cripple credibility and limit acceptance of ideas. Developing strong language skills requires commitment. Many writers and speakers cripple their attempt at communication by using bureaucratic jargon, big words, and too much passive voice. Good writers and speakers want to inform or persuade their audience. Building credibility with the targeted audience requires the use of support and logic. Nothing can cripple a clearly written and correctly punctuated correspondence quicker than a fractured fact or a distorted argument. However, properly using logic remains a challenge for many individuals to master since it challenges the mind's ability to think in the abstract. Never attempt to hide intellectual shortcomings with verbal overdose. Communicators need to consider a seven-step approach that will support communication success. Good communication requires preparation, and the first four steps lay the groundwork for the drafting process.

BOX 2.8   SEVEN STEPS FOR EFFECTIVE COMMUNICATION

1. Analyze Purpose and Audience
2. Research Your Topic
3. Support Your Ideas
4. Organize and Outline
5. Draft
6. Edit
7. Get Feedback and Approval

REVIEW EXERCISES

1. What are at least five concepts that all leaders must learn?
2. In your own words, define the concept of effectiveness.
3. How does ethical conduct impact leadership?
4. Explain the primary difference between leadership and management.
5. Why would knowledge management be a vital part of effective safety efforts?
6. What three elements are common to most definitions of a crisis?
7. What would be the foundational element of organizational culture?
8. List the four key factors that influence workers' compensation rates.
9. List seven signs of substance abuse by an employee.
10. Define the following terms:
    a. Orientation
    b. Education
    c. Training

3 Nursing Hazards

I. INTRODUCTION

Hazards of all types can exist in healthcare facilities and organizations must take steps to identify and control these hazards. Conducting periodic tours, inspections, and surveys can help identify and control hazards. Organizations with established safety cultures can rely on staff vigilance to help identify hazards and prevent accidents. Healthcare supervisors must also focus on correcting unsafe acts and behaviors. Facility personnel at all levels should learn to observe hazards and behaviors that could contribute to accidents. Senior leaders should stress the importance of job safety education and training. Supervisors should communicate the need for personal involvement in safety and hazard control efforts.

BOX 3.1   NURSING SAFETY RESPONSIBILITIES

- Advise leaders, supervisors, and the safety committee about hazards and safety issues
- Use good communication and human relations skills to help achieve safety objectives
- Seek ways to help in the development, implementation, and maintenance of safety plans
- Promote safety awareness throughout the nursing unit and the organization
- Get involved with organizational hazard surveys and environmental tour efforts
- Provide input and suggestions that address the development and evaluation of safety training
- Understand the importance of timely accident reporting, investigations, and hazard analyses
- Ensure compliance with safety and fire regulatory standards and codes
- Participate in organizational root cause analysis sessions as necessary to identify causal factors
- Attend professional education sessions to stay current with healthcare safety and health issues
- Consider earning healthcare safety-related certifications
- Understand the requirements of accreditation standards and regulatory compliance
- Promote patient safety as an integral part of the total organizational safety system

---

BOX 3.2   BASIC SAFETY TIPS FOR NURSES

- Follow good hygiene and handwashing practices when exposed to hazardous substances
- Use proper devices, equipment, and techniques to prevent needle stick and sharps injuries
- Properly select and use required personal protective equipment (PPE) and other appropriate barriers
- Adhere to safe patient moving and lifting techniques, including the use of assist devices

(Continued)

BOX 3.2   BASIC SAFETY TIPS FOR NURSES   *(Continued)*

- Avoid awkward positions when moving patients and take frequent breaks for repetitive tasks
- When working second and third shifts, develop proper sleep and diet patterns
- Learn to identify hazards associated with various shifts or job assignments
- Seek assistance from organizational sources to deal with job-related stress
- Help keep all corridors and other passages clear of clutter and equipment
- Practice good electrical safety in the performance of all duties
- If exposed to radiation, follow established procedures and wear a monitoring device
- Learn to identify safety hazards including fire prevention

BOX 3.3   NURSE SAFETY EDUCATION TOPICS

- Bloodborne pathogens and healthcare infection risks
- Standard precautions and handwashing requirements
- How to select, use, maintain, and store personal protection equipment (PPE)
- How to select the proper footwear and other protective clothing
- Methods and engineering innovations for preventing needle stick injuries
- Patient and material handling or lifting techniques
- Detailed information about shift work and sleep deprivation
- Protections to take against workplace violence
- How to work safely with compressed gases
- Importance of following established safety procedures
- Emergency response, egress, and evacuation procedures
- Accident, incident, hazard, and injury reporting procedures
- Hazard communication about chemicals and hazardous wastes
- Location, availability, and use of chemical Safety Data Sheets
- Medical equipment procedures and adverse event reporting
- Patient safety issues including organizational and clinical topics

---

A.   ADMINISTRATIVE AREA SAFETY

Many healthcare organizations overlook the administrative areas during safety surveys. These areas contain a number of hazards including lifting, climbing, repetitive motions, tripping, electrical hazards, and others. Office areas can also experience workplace violence in locations such as admissions, emergency departments, gift shops, patient affairs, and business offices. This section briefly addresses some common hazards found in administrative areas. Healthcare personnel must learn to close desks and filing cabinet drawers after every use. Take actions to ensure electrical cords and wires don't create walkway hazards. Ensure aisles and passageways remain clear at all times. Do not permit individuals to use furniture or chairs to reach overhead items. Establish cleanup procedures for spills and communicate that everyone must take responsibility for reporting loose carpeting or damaged flooring. Prohibit the placement of storage boxes and other materials on top of storage or file cabinets. When storing materials, place heavy items at the lowest level possible to prevent overreaching. Keep fire extinguishers accessible and emergency egress routes unobstructed. Never permit the storage of materials closer than 18 inches from fire sprinkler heads. Provide a good lighting system that supports the tasks being performed. Take actions to minimize glare from ceilings, walls, and floors. Clean and replace old bulbs and faulty lamp circuits. Never position workstations facing windows, unshielded lamps, or other sources of glare. Take actions to protect personnel from noise generated by video display terminals, copiers, telephones, fax machines, hallway pedestrian traffic, and break areas. Identify electrical hazards posed by faulty equipment, unsafe installation, or misuse of equipment. Ensure the proper grounding of equipment and machines that could pose a shock hazard. Provide sufficient outlets to eliminate the need for extension cords. Position all floor outlets carefully to prevent the creation of tripping hazards. Inspect all electrical equipment and cords regularly. Repair or replace defective, frayed, and improperly installed cords. Never store anything in front of electrical panels.

II. SHIFT WORK AND JOB STRESS

A. SHIFT WORK

Healthcare organizational leaders, supervisors, and safety personnel should realize that shift work and sleep deprivation affect not only task accomplishment, but also personal safety. People function best during daylight hours. Performance is decreased during periods of rapid eye movement (REM). This REM sleep is known as the dream period and normally occurs in the early morning hours. During this period, the body temperature is at its lowest. Humans have what is called "circadian rhythms" or a 24-hour body clock. This body clock can vary with individuals but can also receive undue influence from environmental factors. Shift work and lack of sleep can contribute to health problems or the increased risk of health problems. Shift workers tend to experience high stress levels and family problems. Healthcare organizations should strive to make shift work safer and educate workers on adjustment strategies. Supervisors should know how to evaluate workers. They must look for the signs of sleep deprivation, stress, and fatigue. Evaluate the relationship between work shifts and personal health. Some factors to consider include the time and length of the work shifts, scheduled days off, demands of the job, personal characteristics, and work environments. Working extended hours increases exposure to environmental, chemical, biological, ergonomic, and psychosocial hazards. Research indicates bright lights in work settings can improve worker alertness. Some studies indicate that power naps of no more than 35 to 40 minutes in length can increase alertness during the sleepiest times of a shift. Taking a short nap before a late night shift can help a person stay alert and awake. Sleepiness occurs most during night shifts. Poor daytime sleeping habits of many shift workers contributes to their on-the-job sleepiness. However, sleeping during the day can interfere with social and family activities. Personnel working long or extended shifts must consider ways to better cope with fatigue, lack of sleep, and family situations. The Institute of Medicine recommends that nurses work no more than 60 hours in a seven-day period and no more than 12 hours on single day. Most nurses working 12-hour shifts tend to lose alertness during the last two hours of their schedule.

BOX 3.4   SLEEP DEPRIVATION EDUCATIONAL TOPICS

- Stress the importance of getting six to seven hours of uninterrupted sleep
- Explain the importance of sleeping in a dark room free from distractions
- Provide tips on how to deal with noise during sleep periods
- Encourage the eating of a nutritious meal during the shift
- Avoid caffeine late in a shift since it disrupts sleep patterns
- Address the importance of exercising on a regular basis

B. STRESS

Nurses experience stress complicated by understaffing, paperwork, tight schedules, equipment malfunctions, demanding families, dependent patients, and even death. Some workers feel that the depersonalized nature of healthcare leaves them feeling alone, isolated, angry, and frustrated. Stress contributes to worker apathy, lack of confidence, and absenteeism. Studies have shown that healthcare workers have a high rate of hospital admission for mental disorders. Stress can cause loss of appetite, mental disorders, migraines, sleeping disorders, and emotional instability. It can also increase the use of tobacco, alcohol, or drugs. Stress can affect a person's attitude, motivation, and behavior, which can impact the quality of patient care. Healthcare organizations should educate employees and management about job stress by establishing employee assistance initiatives and organizational change education sessions. Employee assistance initiatives can improve the ability of workers to cope with difficult work situations. Stress management plans should focus on teaching workers about the nature and sources of stress, the effects of stress on their health, and learning personal skills that can help reduce stress.

BOX 3.5   COMMON HEALTHCARE NURSE STRESS FACTORS

- Understaffing and unbearable workload
- Inadequate resources to accomplish the task
- Working in an unfamiliar area, job, or role
- Rotating work shifts or shifts longer than eight hours
- Little or no input or participation in schedule planning
- No recognition by senior leadership for doing a good job
- Personal talents and expertise not utilized
- Exposure to biological, physical, and chemical hazards
- Increased potential for workplace violence
- Poor departmental or unit organization

BOX 3.6   WAYS TO COPE WITH STRESS

- Conduct staff meetings and allow open communication
- Implement a formal stress management plan
- Provide accessible counseling from a nonjudgmental source
- Promote flexibility and creativity within the department
- Ensure adequate staffing and sufficient resources
- Organize work areas and departments to function efficiently
- Work to provide reasonable and flexible work schedules
- Emphasize the importance of worker safety and health
- Provide regular in-service education and training sessions
- Implement a complaint and suggestion system

III. SLIP, TRIP, AND FALL PREVENTION

Occupational Safety and Health Administration (OSHA) Standard 29 CFR 1910.22, Walking-Working Surfaces, and ANSI A1264.2-2006, Standard for the Provision of the Slip Resistance on Walking/Working Surfaces, provide guidance on preventing slips, trips, and falls. Safety personnel must identify, evaluate, and correct any hazards that could contribute to these types of events.

Educate staff members about the causal and behavioral aspects of fall prevention efforts. Establish procedures to analyze trends related to slip and fall incidents. Slip, trip, and fall incidents can frequently result in serious disabling injuries. Slips and falls can result in lost workdays, reduced productivity, expensive workers' compensation claims, and diminished ability to care for patients. In 2009, the Bureau of Labor Statistics (BLS) reported a hospital incidence rate of 38.2 per 10,000 employees for slips, trips, and falls occurring on the same level. The 2009 rate was 90% greater than the average rate for all other private industries combined. These events resulted most often in sprains, strains, dislocations, and tears for healthcare personnel, including nurses. Contaminants on the floor contribute to most healthcare facility slip, trip, and fall incidents. Implementing effective housekeeping procedures, conducting proper floor cleaning, using walk-off mats, posting safety signs, and requiring the wearing of slip-resistant shoes minimize the risk of slipping.

IV. SAFETY SIGNS, COLORS, AND MARKING REQUIREMENTS

OSHA Standard 29 CFR 1910.145 addresses accident prevention signs and tags. These specifications apply to the design, application, and use of signs/symbols used to prevent accidental injuries or property damage. These specifications do not cover plant bulletin boards, safety posters, or any signs designed for streets, highways, railroads, or marine applications. OSHA standards do not address sign design for Danger, Caution, and Safety Instruction signs except for purpose and colors. OSHA requires that signs be designed with rounded or blunt corners; they must be free from sharp edges, burrs, splinters, or other sharp projections. The ends or heads of fastening devices cannot create a hazard. The size of the sign, height and width of the letters, and viewing distances must meet ANSI Z535.2 requirements. Ensure signs contain concise and easy-to-read wording. Use letters large enough to meet determined intended viewing distances. Place signs in locations so that individuals can take action to avoid the hazard. Use legible signs that do not cause distraction or create a hazard. Never place signs on moveable objects or adjacent to moveable objects such as doors and windows. If necessary, equip signs with emergency or battery-operated illumination. OSHA Standard 29 CFR 1910.144 requires the use of red to mark fire protection equipment and apparatus. Use red danger markings for safety cans or other portable containers of flammable liquids, excluding shipping containers. Red safety cans must contain some additional clearly visible identification either in the form of a yellow band around the can or the name of the contents conspicuously stenciled or painted on the can in accordance with 1910.1200. OSHA mandates the use of yellow as the basic color for designating caution. Use yellow for the marking of physical hazards such as striking against, stumbling, falling, and getting caught in between.

V. ELECTRICAL SAFETY (29 CFR SUBPART S)

OSHA now considers the National Electrical Code (NEC) or National Fire Protection Association (NFPA)/ANSI 70 as a national consensus standard. Article 517 of NFPA 70 contains special electrical requirements for healthcare facilities. Refer to 29 CFR 1926.401–449 for OSHA construction-related electrical requirements. In addition, state and local regulations may apply. Electricians and maintenance personnel must understand OSHA electrical safety standards published in 29 CFR 1910.301–399. NFPA/ANSI 70 applies to every replacement, installation, or utilization of electrical equipment. Supervisors must inspect work areas for possible electrical hazards. Electrical current travels through electrical conductors; its pressure is measured in volts. Resistance to the flow of electricity is measured using ohms, which can vary widely. Resistance determination considers the nature of the substance itself, the length and area of the substance, and the temperature of the substance. Some materials, like metal, offer very little resistance and become conductors very easily. Other substances, such as porcelain and dry wood, offer high resistance. Insulators prevent the flow of electricity. Water that contains impurities such as salts and acids make a ready conductor. Electricity travels in closed circuits and follows its normal route through a conductor. Electrical equipment can cause shock, electrocution, and catastrophic property damage due to fire or explosion risks. Electrical fires in healthcare facilities many times result from short circuits, overheating equipment, and failure of current safety devices. Explosions may occur when flammable liquids, gases, and dusts interact with ignition sources generated by electrical equipment.

VI. THREE COMMON HAZARDS

A. BLOODBORNE PATHOGENS

The OSHA Bloodborne Pathogens Standard requires workers to take precautions when dealing with blood and other potentially infectious materials (OPIM). Implement engineering and work practice controls to eliminate or minimize exposure to bloodborne pathogens. Engineering controls can reduce employee exposure either by removing, eliminating, or isolating a hazard. Ensure employees wear appropriate PPE or clothing such as gloves, gowns, and facemasks.

Ensure employees properly discard contaminated needles and other sharp instruments immediately or as soon as feasible. Consider all blood and other potentially infectious body fluids as infected. The Bloodborne Pathogens Standard does allow healthcare organizations to use acceptable alternatives to Universal Precautions. Consider using Centers for Disease Control and Prevention (CDC) Standard Precautions or Body Substance Isolation techniques as appropriate for the setting. Safer needle devices must contain integrated safety features designed to prevent needle stick injuries. OSHA considers safer needle devices as passive or active. Passive needle devices offer the greatest protection because they contain safety features that automatically trigger during use. For example, consider a spring-loaded retractable syringe or self-blunting blood collection device. Intervention studies show that the use of safer needles systems can reduce injuries.

B. HAZARDOUS MATERIAL SAFETY

Hazardous chemical exposures can occur from aerosols, gases, and skin contaminants. Exposures can occur on an acute basis or result from chronic long-term exposures. Some substances commonly used in the healthcare setting can cause asthma or trigger attacks. Studies indicate scientific evidence linking cleaners and disinfectants, sterilants, latex, pesticides, volatile organic compounds, and pharmaceuticals to asthma. Many medications and compounds used in personal care products have known toxic effects. Although many medications pose hazards to workers, those most commonly identified as hazardous to healthcare workers include antineoplastic drugs and anesthesia substances. Anesthetic gases can pose problems when escaping and can create occupational inhalation hazards.

C. NURSING PERSONNEL AND WORKPLACE VIOLENCE

A recent survey of nurses and physicians revealed that about three-fourths of the respondents reported witnessing physicians engage in disruptive behaviors. Most of the incidents involved verbal abuse of another staff member. OSHA considers these types of issues as workplace violence. The same survey revealed that about two-thirds of respondents witnessed disruptive nurse behaviors. Most respondents believed that such unprofessional actions increase potential for medical error occurrence. Disrespectful behavior and in some cases, physical violence by physicians, can contribute to nursing dissatisfaction. Physicians practicing in high-stress specialties such as surgery, obstetrics, and cardiology appear more at risk to disruptive behaviors. The Joint Commission requires that organizations develop a code of conduct policy for all staff. The Joint Commission also recommends a "zero tolerance" approach to all disruptive behaviors. It issued a Sentinel Alert to address the seriousness of these risks. Patient and family assaults on healthcare workers can occur, especially during times of increased stress. Poor workplace security and unrestricted movement by the public in the facility can increase these types of risks. Emergency department personnel also face significant risks from assaults by patients or their families. Those carrying weapons in emergency departments create the opportunity for severe or fatal injuries. However, no location within a healthcare setting is immune from workplace violence. The OSHA publication *Guidelines for Preventing Workplace Violence for HealthCare and Social Service Worker,* provides an outline for developing a violence prevention plan. Develop performance-based violence prevention plans. Facilities must meet the challenge of developing a specific process that will yield results and protect healthcare personnel. The elements of any prevention plan must include management commitment, employee involvement, worksite analysis, hazard prevention and controls, and training and education.

VII. HELICOPTER SAFETY

Many acute care and trauma facilities receive patients brought in by helicopters. Staff responsible for meeting these airborne ambulances must know proper safety procedures. Restrict the heliport to trained and authorized personnel. Personnel should never approach the landing zone until the craft lands. Wait for crew permission before approaching the craft. Require the wearing of hearing protection when aircraft engines are running. Prohibit smoking in the heliport area. Personnel should never shine lights directly in front of the aircraft during landing. Approach the craft from the front or side as required, but never approach from the rear. Stay in a crouched stance when approaching the aircraft. The wind created by the rotor blades can create hazardous situations due to flying dust, litter, and loose clothing. Place stethoscopes in pockets and secure hats or scarves. Tape sheets securely to the stretcher. Place the portable oxygen bottles so that they do not extend beyond the cart. The flight crew retains responsibility for opening and closing the aircraft doors. In most situations, the flight crew takes responsibility for offloading the patient. The flight crew should direct hospital personnel on proper patient handling. They also provide directions about departure from the heliport. Not following instructions could result in an intravenous line being pulled or a change in the patient's skeletal alignment. Offloading with the rotor blades turning requires caution at all times. Ensure the availability of sufficient fire extinguishers and that all personnel can operate the extinguishers.

VIII. NONCOMPENSATED OR VOLUNTARY WORKER SAFETY

Volunteers work in a variety of capacities in most healthcare organizations. Each volunteer needs training and education on all potential exposure risks. Many organizations limit volunteers from patient contact tasks that could expose them to hazards, including bloodborne pathogens. Healthcare organizations using volunteers should provide a comprehensive safety orientation session. Some key education topics should include fire safety, emergency evacuation procedures, infection control precautions, patient safety topics, radiation exposure controls, general safety orientation, hazard identification, and accident reporting procedures.

IX. HOME HEALTH SAFETY

Nurses who provide in-home care to patients face many of the same on-the-job safety concerns as their colleagues who work in controlled facility environments, like a hospital. Universal safety concerns for nurses include needle sticks, musculoskeletal injuries, verbally or physically aggressive patients or family members, and driving to and from a job located in a neighborhood with higher than average crime-related incidents. Safety concerns that are unique to working in a home environment may include traveling from home to home, encountering pets, observing drug use, or walking on poorly maintained sidewalks to a patient's home. Providing care in a small space and lifting without the convenience of an adjustable-height bed or mechanical lift can create challenges.

Proper safety for the home healthcare nurse and patient should begin with a thorough assessment of the home to identify potential risks. Prior to the patient's discharge from the facility, someone should complete the assessment and develop a plan to mitigate or eliminate any risks. Patient and family education about maintaining equipment and a hazard-free environment will help facilitate a safe environment for all. Expert training can equip home healthcare nurses with the ability to assess and manage safety risks. Annual safety training and ongoing in-service education should include personal protection techniques, fire safety, body mechanics, infection prevention measures, proper use of equipment, and instruction on how to report any suspicious or threatening behaviors within the home setting. Agencies must provide ongoing clinical supervision and support for all personnel night and day.

BOX 3.7   HOME HEALTH SAFETY TIPS

- Know the surroundings and stay alert when traveling to and from a patient home.
- Obtain accurate directions and addresses.
- Park your car in a well-lit area, away from trees or bushes.
- Notify your employer if you feel threatened or have concerns about personal safety.
- Keep all medical supplies out of view and locked in the trunk of your car between visits.
- Use mechanical lifting devices, if available, when transferring your patients.
- Use proper body mechanics when lifting household items or manually transferring your patient. Keep knees bent, feet apart, and the object or patient close to your body.
- Notify the employer about safety issues related to lifting and transferring a patient.
- Always keep extra supplies of PPE, including gloves, gowns, eye protection, and masks, in your car or travel bag.

Notify your agency if PPE supplies are low in the patient's home. Home healthcare industry personnel encounter a number of hazards not found in traditional healthcare settings. They spend a great deal of time traveling and entering environments where they have little or no control. Many patients live in unsafe neighborhoods that could expose providers to violence. However, home healthcare providers are exposed to bloodborne pathogens and patient moving hazards much like their hospital counterparts. Nursing aides, nurses, and therapists often work alone with no one to support them in a time of crisis. Home-based healthcare remains unpredictable and the agency must take responsibility for worker safety. Home health personnel must ensure care plans identify hazards and care requirements. Use well-lit and common walkways when visiting patients. Instruct patients and family members on the importance of infection control. Always knock or ring the doorbell before entering a patient's home. Know injury and emergency reporting procedures and learn to document unsafe behaviors, threats, and menacing pets. If possible, schedule joint visits in unsafe neighborhoods or homes. Always request security escorts for night visits in potentially unsafe areas. If threatened, home health personnel should scream, kick, and use chemical spray or a whistle. Lock automobiles and keep them in top mechanical condition. Park vehicles as near to the patient's home as possible.

BOX 3.8 HOME HEALTH WORKER TRAINING TOPICS

- Understanding the patient's age and cultural, economic, and social factors
- Disease manifestations
- Mental, emotional, and spiritual needs
- Personal security precautions, including travel safety
- How to conduct a hazard assessment during a consultation visit
- Protection guidelines for bloodborne pathogen exposure
- Sharps and needle safety precautions
- Medical waste disposal procedures
- Back injury prevention techniques
- Safe lifting and transfer techniques
- Care plan development to identify risks

X. SURGICAL DEPARTMENT

All personnel should know the location of all emergency equipment. This includes drugs, cardiac arrest equipment, and resuscitators. Ensure the use of explosion-proof electrical equipment and plugs.

Develop written schedules of inspections and maintenance of all electrical equipment. Operating room personnel should receive annual training on bloodborne pathogens and other safety issues. Implement policies and procedures for sharps injury reporting. OSHA requires an appropriate sharps safety plan that includes the evaluation and selection of safer sharps and needles. Operating rooms must document the evaluation of commercially available safety products such as safe suturing devices, safety needles, safety scalpels, sharps, and passing containers. Offer a Hepatitis B vaccination series at no cost to employees with exposure to blood or OPIM. The surgical staff must maintain a sharps injury log that includes the type and brand of device involved in an exposure, the work area where the exposure occurred, and an explanation of how it happened. Minimize the hazards of exposures in surgery suites by promoting the use of

- Safer needles and other sharps devices
- Blunt suture needles
- Needleless IV connectors
- Proper containerization of sharps
- A no pass zone for surgical instruments

The Centers for Medicare & Medicaid Services (CMS) recently decided to change the minimum hospital operating room relative humidity from 35 percent to 20 percent. CMS currently requires hospitals to comply with the 2012 edition of NFPA 101: *Life Safety Code®*. That edition references the 1999 edition of NFPA 99: *HealthCare Facilities Code*, which requires operating room humidity to be at least 35 percent. This *Life Safety Code* waiver CMS permits hospitals to keep operating room relative humidity level at a minimum of 20 percent. A most recent edition of NFPA 99 shifted to a minimum requirement of 20 percent humidity. The waiver does not apply if more state or local laws/regulations have more stringent requirements. Facilities must monitor relative humidity levels in anesthetizing locations and must take action to ensure levels remain at or above 20 percent. Facilities that elect to use this categorical waiver do not have to apply in advance. However, they must document their decision to use the waiver.

Chemicals employees could be exposed to in surgical facilities include peracetic acid, used in cold sterilizing machines, and methyl methacrylate, used to secure prostheses to bone during orthopedic surgery. Recommend mixing methyl methacrylate only in a closed system. Employees should carefully read and follow instructions and warnings on labels. Employees should follow all Safety Data Sheet (SDS) instructions regarding safe handling, storage, and disposal of hazardous chemicals. According to the Hazard Communication Standard, employers must inform employees of chemical hazards and have on hand SDSs for all hazardous chemicals used in their facilities.

Staff is at risk of trip and fall hazards such as falling over portable equipment that easily blends into the floor or slipping on debris. Electrical cords crossing floors can create trip hazards. OSHA requires that work areas are kept clean, orderly, and in a sanitary condition. Keep aisles and passageways clear and in good repair, with no obstruction across or in aisles that could create a hazard. Provide ceiling or floor plugs for equipment, so power cords need not run across pathways. Static postures from continuously standing in one position during lengthy surgical procedures may cause muscle fatigue and pooling of blood in the lower extremities. Standing on hard work surfaces such as concrete can create trauma and pain to feet. Avoid awkward postures such as tilting the head forward for long periods of time. Provide stools where their use is possible. Use shoes with well-cushioned insteps and soles. Provide a footrest bar or a low stool. Use height-adjustable work surfaces. Ensure that all electrical service near sources of water is properly grounded. Use appropriate PPE and safe work practices for assessed hazards. Develop procedures to routinely monitor the condition of equipment and address work practices of employees.

Recommend a hands-free technique for passing instruments. The National Institute for Occupational Safety and Health (NIOSH) published *The Effectiveness of the Hands-Free Technique in Reducing Operating Room Injuries* in November 2001. The use of blunt needles when appropriate and suturing devices with needle stick protection offer the highest protection against suture needle sticks. Avoid placing hands unnecessarily near sharps. Surgical personnel should avoid unnecessary handling of sharps. Never hold any sharp simultaneously with another instrument. Contain sharps in designated zones at all times. Never cross the room with a sharps instrument in hand. Some operating rooms use trays or basins where the instrument is placed before being picked up by the second person. Sometimes they use a designated area on a cart or table. Verbally announce the transfer of sharps into the neutral zone. Keep eyes on sharps until placed in designated zones. Use safety transfer trays and magnetic drapes to transfer sharps between nurse and surgeon during surgical procedures. Recent studies indicate that more than 60 percent of scalpel blade injuries were inflicted by the user on assistants, typically during equipment transfer. The thumb and index finger of the nondominant hand commonly experience injury with scalpels and suture needles because they reposition or hold tissue. Develop alternatives including the use of retractors instead of hands, rounded scissors instead of pointed tips, and staples for skin closure, and also use electrocautery instead of standard scissors. Dispose of sharps immediately after use. Make puncture-resistant containers available nearby to hold contaminated sharps. The OSHA Bloodborne Pathogens Standard also requires the discarding of contaminated needles and other sharp instruments immediately or as soon as feasible after use into appropriate containers.

Never bend, recap, or remove contaminated needles and other contaminated sharps except as noted in 29 CFR 1910.1030. Employers must provide readily accessible handwashing facilities and ensure that employees wash their hands immediately or as soon as feasible after removal of gloves. Surgical staff should wear appropriate protective clothing and equipment, which can protect from unwanted fluid splash or sharps injuries. PPE includes gloves, facemasks, soak-proof gowns, impervious boots or shoe covers, face shields, and other eye protection devices. Make safety scalpels with movable shields or retracting blades available to surgeons and other operating room personnel.

XI. PERINATAL NURSING

Perinatal nurses educate pregnant women about delivery options, nutrition, and other important issues. They also assist during the delivery of babies and help new mothers learn how to bond with their children. Because perinatal nurses work in healthcare facilities, they are exposed to a number of risks on a daily basis. Assisting with the delivery of babies puts perinatal nurses at risk of coming into contact with blood and body fluids that contain infectious organisms. Using proper handwashing techniques and wearing gloves, masks, and gowns protects perinatal nurses against these organisms. Perinatal nurses use needles and other sharp instruments to assist in the care of pregnant women and the delivery of babies. This puts them at risk for sharps injuries, which occur when a used needle or other sharp object pierces the skin. These injuries increase the risk for exposure to disease-causing organisms found in the blood. Following universal precautions for sharps disposal protects perinatal nurses from this risk. Nurses should never break, bend, or recap needles. Needles, disposable razors, and other sharp objects should also be disposed of in puncture-resistant containers. Perinatal nurses regularly have contact with equipment and tools powered by electricity. Labor and delivery beds have electrical circuits that allow nurses to adjust the height of the bed and the position of the patient to make it easier to deliver a baby. IV pumps and defibrillators also have electrical components. Perinatal nurses must use basic safety precautions to protect themselves and their patients from shocks, burns, and electrocution. Any machine with a damaged cord should be taken out of service until it has been repaired and inspected. A nurse should report any loose wires, electrical shorts, or related problems to a biomedical equipment technician. Nurses should never use electrical equipment while their hands are wet, as this increases the risk for injuries. Surgeons perform Caesarean sections in operating rooms that contain flammable materials. Perinatal nurses are at risk for burns and smoke inhalation if a fire breaks out during one of these procedures. Following safety precautions related to the use of oxygen and anesthetic gases helps prevent these fires and protect workers from fire-related injuries. Perinatal nurses should keep flammable liquids away from oxygen tanks, as oxygen promotes rapid combustion. Operating room personnel should use caution when working with cauterizing tools, endoscopes, lasers, and other electrical devices. Keeping heat sources away from dressings, sponges, gauze, and surgical drapes also reduces the risk of fires. Perinatal nurses come into contact with hazardous substances on a regular basis. Some products used to disinfect medical equipment irritate the respiratory system and burn the eyes. Epinephrine, which is used to treat anaphylactic shock and cardiac arrest, may be fatal if swallowed, absorbed through the skin, or inhaled. Wearing masks and gloves protects perinatal nurses from inhaling or absorbing hazardous chemicals. Nurses should always wash their hands after handling chemicals and medications. Proper disposal of contaminated materials also reduces the risk of contact with dangerous chemicals.

XII.   INTENSIVE CARE UNITS

ICU workers experience the risk of exposure to blood, OPIM, and bloodborne pathogens because of the immediate, life-threatening nature of treatment. The Bloodborne Pathogens Standard requires precautions when dealing with blood and OPIM. Engineering and work practice controls must serve as the primary means to eliminate or minimize exposure to bloodborne pathogens. Employees should wear appropriate PPE when anticipating blood or OPIM exposure. Ensure employees discard contaminated needles and other sharp instruments into appropriate containers immediately or as soon as feasible after use. Maintain exposure control plan documentation of consideration and implementation of appropriate commercially available and effective engineering controls designed to eliminate or minimize exposure to blood and OPIM. Treat all blood and other potentially infectious body fluids as infected and take appropriate precautions to avoid contact with these materials. The Bloodborne Pathogens Standard does allow hospitals to practice acceptable alternatives to Universal Precautions such as Standard Precautions or Body Substance Isolation. Intensive care units, particularly neonatal units, may be designed without walls between patient spaces. This may allow employees to unknowingly experience exposure to aerosolized chemicals and x-ray radiation that escapes from neighboring areas. Ensure that all rooms can remove contaminants through normal ventilation means. When using recirculated air, ensure installation of adequate filtering mechanisms. Because of the ICU atmosphere, potential slip and fall hazards exist if water or other fluids remain on floors. Never permit electrical cords to run across pathways or place emergency equipment or supplies in passageways. Provide safe cleanup of spills and keep walkways free of obstruction.

Injury may occur to employees from improper training or use of medical equipment. Implement procedures that routinely monitor the status of equipment and proper training of employees to use equipment safely. Workplace violence is an issue in ICUs because of the crowded, emotional situations that can occur with critical patients. Good work practice recommends a security management plan that addresses workplace violence issues. Train staff to recognize and diffuse violent situations and patients. Require reporting of suspicious behaviors. Provide intervention measures including verbal, social, physical, and pharmacological interventions. Warning signs of anger and violence include (1) pacing and/or restlessness, (2) clenched fist, (3) increasingly loud speech, (4) excessive insistence, (5) threats, and (6) cursing.

Studies suggest work stress may increase a person's risk for cardiovascular disease, psychological disorders, workplace injury, and other health problems. Early warning signs may include headaches, sleep disturbances, difficulty concentrating, job dissatisfaction, and low morale. Ensure workloads remain in line with workers' capabilities and resources. Design jobs to provide meaning, stimulation, and opportunities for workers to use their skills.

XIII.   EMERGENCY DEPARTMENT

Emergency department personnel experience continuing risks for exposure to blood, OPIM, and bloodborne pathogens because of the immediate, life-threatening nature of treatment. Employees can be exposed to hazardous chemicals or hazardous drugs. Emergency departments should implement approved decontamination procedures. Because of the emergency department's busy atmosphere, slips, trips, and falls pose a major risk. Provide safe cleanup of spills, and keep walkways free of obstruction. Keep access to exits clear and unobstructed at all times. Injury may occur to employees from improper training or use of equipment such as defibrillators. Electric shock may also occur as a result of lack of maintenance or misuse of equipment. Workplace violence is an issue in emergency departments because of the crowded and emotional situations that can occur with emergencies. Good work practice recommends a security management plan to address workplace violence. Train staff members to recognize and diffuse violent situations and patients. Stay alert for potential violence and suspicious behavior and report it. Provide intervention measures including verbal, social, physical, and pharmacological interventions. Install concealed panic buttons in the department and at triage areas. Install proper lighting and video surveillance equipment. Limit access to the area by implementing a waiting room area with controlled access points. Patients must enter through a secure door. Consider the use of metal detectors to determine if individuals possess weapons of some type. Provide a secure room for patients identified as violent. This room could include controls such as video camera surveillance. Exposure to tuberculosis (TB) and other infectious agents can occur from patients in waiting rooms and treatment areas. Provide and practice early patient screening for TB during admission to identify potentially infectious patients. Provide isolation to prevent employee and other patient exposure. Provide engineering, work practice, and administrative procedures to reduce the risk of exposure. Ask patients with a productive cough to wear a mask to prevent the spread of infection. Post waiting rooms signs that state, "If you are coughing you may be asked to wear a mask." Isolate patients until verification testing is negative. Some departments provide an isolation room to safely isolate potentially infectious patients. Others can designate an isolation area for infectious patients. Maintain Isolation rooms under negative pressure. Acid-fast bacilli (AFB) isolation refers to a negative-pressure room or an area that exhausts room air directly outside or through high-efficiency particulate air filters (HEPA) if recirculation is unavoidable. Protect employees from exposure to the exhaled air of an individual with suspected or confirmed TB. Isolate patients with suspected or confirmed TB.

   Exposure of department personnel can occur from patients exposed to biological agents, chemical agents, and mass causalities as a result of terrorist attacks or events. Provide and plan for emergency response for healthcare employers and emergency responders. The Department of Health and Human Services (DHHS), the CDC, the American Hospital Association (AHA), the Department of Defense (DoD), and OSHA publish resources for hospitals to use when planning for terrorist events.

XIV. DIALYSIS UNIT SAFETY

Hazards present in this unit include exposure to sterilizing solutions and bloodborne pathogens including Hepatitis B. During recent years a number of reports indicate blood contamination incidents related to the internal components of dialysis equipment. The possibility of cross-contamination could permit the transfer of bloodborne pathogens from patient to patient. Under certain conditions, cross-contamination is possible despite the use of new blood tubing sets and external transducer protectors. Please also note that routine maintenance is not adequate to detect internal machine contamination. Qualified personnel should inspect all machines, including the internal pressure tubing set and pressure sensing port, for possible blood contamination. Always use an external transducer protector and utilize pressure alarm capabilities as indicated in the manufacturer's instructions. If contamination occurs, take the machine out of service. Ensure employees working in dialysis units know the adverse health effects of Glutaraldehyde. Staff should wear proper protective equipment whenever handling sterilizing solutions. Protective equipment should include rubber gloves, protective aprons, and eye and face protection. Appoint a dialysis staff member as the safety coordinator with authority to enforce biological safety policies within the dialysis unit. Require all personnel to follow the requirements of the OSHA Bloodborne Pathogens Standard 29 CFR 1910.1030. Hospital-based dialysis units should coordinate infection control exposure plans with the hospital infection control function. Refer to the appropriate CDC guidelines for additional information. Appoint a dialysis staff member as the safety coordinator with authority to enforce biological safety policies within the dialysis unit. Isolate patients who are HBsAg-positive in a separate room or unit designated for HBsAg-positive patients if possible. Otherwise, segregate these patients from hepatitis B seronegative patients in a separate area. Assign staff members with the most dialysis experience or best technique to care for the HBsAg-positive patients.

A. EQUIPMENT

Never use dialysis equipment for both HBsAg-positive and seronegative patients. If this is impossible, make the staff aware that the chance for cross-contamination increases significantly. All patients should receive specific assignments for dialysis chairs or beds and machines. Change linen used on chairs and beds for each patient. Clean chairs and beds after each use. All patients should receive an assigned supply tray including a tourniquet, marking pencils, and antiseptics. Never use clamps, scissors, and other nondisposable items for more than one patient unless autoclaved or appropriately disinfected.

B. PERSONAL PROTECTIVE EQUIPMENT

Staff must wear disposable gloves while handling patients or dialysis equipment and accessories. Staff should wear gloves at all times including taking blood pressure, injecting saline or heparin, or touching dialysis machine knobs to adjust flow rates. The staff should never touch any surfaces with gloved hands that will subsequently be touched with bare hands. The staff should wear gloves whenever handling blood specimens and whenever working in the laboratory area. Wearing protective eyeglasses and surgical-type masks during any procedure with potential for spurting or splattering of blood is recommended. Staff should wear gowns or scrub suits at all times and properly dispose of clothing at the end of each day. Follow the housekeeping practices required by the OSHA Bloodborne Pathogens Standard.

XV. MEDICAL EQUIPMENT MANAGEMENT

Medical equipment management is also known as biomedical equipment management or clinical engineering. It includes the business processes used in interaction and oversight of the medical equipment involved in the diagnosis, treatment, and monitoring of patients. The related policies and procedures govern activities from the selection and acquisition through the incoming inspection, acceptance, maintenance, and eventual retirement and disposal of medical equipment. Medical equipment management is a recognized profession within the medical logistics domain. The medical equipment management professional's purpose is to ensure that equipment used in patient care is operational, safe, and properly configured to meet the mission of the medical treatment facility. Some medical equipment professional functions include

- Equipment Control & Asset Management
- Equipment Inventories
- Work Order Management
- Data Quality Management
- Quality Assurance
- Patient Safety and Risk Management
- Hospital Safety Planning
- Radiation Safety
- Medical Gas Systems
- In-Service Education & Training
- Accident Investigation
- Safe Medical Devices Act (SMDA) of 1990
- Health Insurance Portability and Accountability Act (HIPAA)

A. MEDICAL EQUIPMENT MANAGEMENT PLAN

Healthcare facilities should implement a management plan that promotes the safe and effective use of medical equipment in support of patient care. This medical equipment management plan maintains complete and continuous compliance with the accreditation organization medical equipment management requirements. These requirements help assess and control the clinical and physical risks of fixed and portable equipment used for the diagnosis, treatment, monitoring, and care of patients. The medical equipment management policy includes all of the current medical equipment management requirements, as well as requirements from other care functions that pertain to medical equipment. The management plan consists of policies and procedures designed to address the following components and issues:

- Selecting and acquiring medical equipment
- Evaluating and identifying equipment to be included in the plan
- Determining scheduled maintenance intervals and procedures
- Repairing and maintaining patient care equipment
- Performance standards for equipment inspection, scheduled maintenance, and testing
- Scheduling inspections, maintenance, and tests on equipment included in the plan
- Development of a policy to provide after-hours service
- Coordinating and documenting services of manufacturers and third-party providers
- Acting on hazard alerts and recalls of medical equipment
- Complying with the provisions of the SMDA of 1990
- Reporting and investigating problems associated with failures and user errors
- Educating and training operators and maintainers about medical equipment issues
- Providing emergency procedures for medical equipment failures

B. Joint Commission Requirements

Establish criteria for identifying, evaluating, and inventorying all equipment before placing in use. Management should provide guidance on monitoring and acting during equipment recall situations. Describe processes for managing effective, safe, and reliable equipment. Identify and implement all processes for selecting and acquiring medical equipment. Evaluate the condition and function of the equipment when received. The organization may choose to include all medical equipment in the management plan. Organizations may use any appropriate strategy including predictive maintenance processes, interval-based inspections, corrective maintenance, or metered maintenance to ensure reliable performance. Define intervals for inspecting, testing, and maintaining appropriate equipment. Minimize clinical and physical risks by using criteria such as the manufacturer's recommendations, risk levels, and current organization experience. Implement processes for monitoring equipment hazard notices and recalls. Develop procedures for monitoring and reporting equipment incidents that must be reported by the SMDA of 1990.

BOX 3.9 MEDICAL EQUIPMENT RISK ASSESSMENT CRITERIA

- Equipment function (diagnosis, care, treatment, or monitoring)
- Clinical use or application
- Maintenance requirements
- Equipment incident history

C. Maintenance, Testing, and Inspecting

- Maintain an up-to-date inventory of all equipment identified in the medical equipment management plan, regardless of ownership
- Document performance and safety testing of all equipment before initial use
- Document maintenance of equipment used for life support consistent with maintenance strategies to minimize any clinical and physical risks identified in the equipment plan
- Document maintenance of non-life support equipment on the inventory consistent with maintenance strategies to minimize clinical and physical risks
- Document performance testing of all sterilizers used
- Document chemical and biological testing of water used in renal dialysis based upon regulations, manufacturer's recommendations, and organization experience

D. Medical Equipment Reporting

The FDA Modernization Act of 1997 changed medical device adverse event reporting as of February 19, 1998. On January 26, 2000, the Food and Drug Administration (FDA) published in the Federal Register changes to the implementing regulations, 21 CFR 803 and 804, to reflect these amendments and the removal of Part 804. The user facility semiannual reporting requirement has been changed to annual reporting. The annual report is now due on January 1 of each year. Reports can protect the identity of user facilities except in connection with certain actions brought to enforce device requirements under the act. Report to the FDA any of the following: death, serious illness or injury, or other significant adverse experience.

E. Safe Medical Device Act of 1990

This act requires healthcare facilities to report serious or potentially serious device-related injuries or illness of patients and/or employees to the manufacturer of the device. The FDA wants to obtain important information on device problems. The act applies to all inpatient facilities, ambulatory surgery care centers, peri-operative facilities, diagnostic units, and outpatient treatment centers. It does not apply to physician offices. Failure to comply can result in civil penalties. Healthcare workers that provide care, review patient care, repair devices, or provide preventive maintenance must report device-related incidents. The incidents include device failure, malfunction, design problems, user errors, and inadequate labeling. Reporting responsibilities extend to physicians, nurses, allied health professionals, students, and other organizational personnel. Examples of a medical device include an anesthesia machine, pacemaker, heart valve, suture, surgical sponge, wheelchair, hospital bed, catheter, infusion pump, dialysis machine, artificial joint, and implant devices.

F. SMDA Reportable Events

Facilities must report medical device events involving patient deaths to the FDA. Report serious injuries caused by devices or in which devices played a role to the manufacturer. The FDA requires that hospitals maintain documentation of all reportable events. Identify the person that completed the investigation and the information used to form an opinion about the causes of the event. When reporting, use the following forms:

*   FDA Form 3500, MedWatch Voluntary Reporting
*   FDA Form 3500A, MedWatch Mandatory Reporting Medication and Device Experience
*   FDA Form 3419, Medical Device Reporting Annual User Facility Report

BOX 3.10   RESPONDING TO DEVICE-RELATED INCIDENTS

*   Protect the device, including packaging material and related parts.
*   Document the equipment or device engineering and/or serial numbers.
*   Remove the equipment from use and tag as defective.
*   Notify the patient's physician as appropriate.
*   Notify Safety, Risk Management, and the appropriate response department.
*   Complete an incident report as required by the policies but within 24 hours.
*   The facility must file the report with the manufacturer or the FDA within 10 days.

G. Other Reporting Requirements

The FDA mandates the reporting and tracking of designated devices. The designated devices include vascular grafts, ventricular bypass devices, pacemakers, and implant-type infusion pumps. The FDA requires that the receipt of tracked devices be reported to the manufacturer and that patient demographic and medical information be reported to the manufacturer upon implanting or use of the device within five working days. This enables the manufacturer to trace specified medical devices to patients and to facilitate patient notification and/or device recall.

XVI. SECURITY

Healthcare facilities must establish a zero tolerance philosophy regarding personnel and property security, including physical violence. Any effective hospital-wide security function must place a strong emphasis on the employment of adequate human and physical resources to protect the facility and people from suspicious, dangerous, or illegal activities. The Security Coordinator should develop, implement, and monitor a comprehensive security management function. The security function must provide a variety of services focused on safeguarding and improving the physical security of patients, staff, and visitors while protecting the rights of individuals in accordance with all applicable laws. Security personnel must conduct regular patrolling of hospital campus and buildings. They should maintain a close liaison with local law enforcement officials and report/ investigate any suspicious or criminal activity. The security function must also conduct traffic control during emergency operations plan implementation and during normal operations of the hospital. An important objective involves controlling patient and visitor access to the emergency department and other hospital areas. The security function must also serve as goodwill ambassadors by providing assistance to employees and visitors with problems such as dead batteries or lost children. Security must report and assess all potential security concerns and identify incidents and issues for investigation and follow-up.

A. NFPA 99-2012, Security Management (Chapter 13)

Chapter 13 of NFPA 99-2012 addresses security issues in emergency departments, pediatric locations, infant care units, medication storage locations, clinical labs, forensic patient treatment areas, and behavioral units. The chapter also addresses communications, data infrastructure, security of medical/health records, media relations, crowd control, employee practices, and security operations. Facilities must conduct a Security Vulnerability Analysis and plan for protection of people and resources beyond a disaster event. Security education should address customer relations, emergency procedures, use of force issues, importance of effective de-escalation of tense tactics, and restraint usage. The new code requires development of policies, plans, and procedures to address hostage situations, bomb threats, workplace violence, disorderly conduct, and restraining order policies.

B. Resources

The hospital must allocate and dedicate adequate physical and human resources to provide a reasonable level of protection from illegal acts. These physical resources could include two-way radio monitoring stations and security personnel. Silent "panic" alarms located in the emergency department can provide for faster response. The issuance of identification cards for employees, volunteers, physicians, ministers, and others requiring access can greatly improve security. Human resources can improve security by ensuring sufficient staffing of security personnel during overnight shifts and at times when there are other security-related concerns.

C. Incident Reporting

Leaders must require that all hospital employees report all suspicious individuals and activities to the appropriate security function or office for dispatch of security response personnel. Maintain accurate logs for all calls and dispatches. Ensure a management review of all security incident reports is completed by response personnel. Submit these reports to the Chief or Coordinator of Security Management. Report any incident resulting in injury to an individual, any assault on a hospital patient (whether an injury results or not), and any incident involving the brandishing or use of a weapon immediately to security and hospital leadership or a nurse supervisor. Analyzing all reports and providing appropriate information to the safety and/or risk management committees is recommended. The appropriate committee would then present a summary report of security issues to the governing board at least on an annual basis. Provide an evaluation of the objectives, scope, performance, and effectiveness of the Security Plan to the care environment director and/or committee on an annual basis.

D. Property Protection

Part of the security function involves safeguarding hospital and individuals' property. Encourage patients to send valuables and other personal property home. Patients receive written information about safeguarding their belongings upon admission. Require hospital employees to report any theft of which they become aware, and any individual carrying objects or packages that appear to belong to the hospital. Hospital security personnel may not conduct searches of patients' and visitors' persons or possessions. The hospital must investigate all thefts and other criminal acts reported. The organization should support the prosecution to the full extent of the law when a suspect is charged with a crime committed on hospital property.

E. SENSITIVE AREA ACCESS

The hospital must restrict and protect access to the facility including sensitive areas. All exterior doors, with the exception of the emergency department lobby and ambulance entrance doors, must be locked 24 hours or locked during certain hours in accordance with specific policy. Security personnel must make rounds to verify locked doors. Report unlocked doors and investigate as required. Security personnel must conduct periodic walk-through tours of construction areas unless the contractor employs construction site security officers. Sensitive areas should install additional controls such as fire stairwell alarms. Control patient flow in the emergency department through access (door) controls and intercom communication. Keep pharmacy doors locked at all times and limit the number of personnel with access. Locate medication rooms throughout the hospital in very visible areas near nursing stations and keep doors locked at all times.

F. IDENTIFICATION OF PATIENTS, VISITORS, AND STAFF

The hospital must provide identification badges to all employees and volunteers. Outside contractors or vendors whose employees visit the hospital must use hospital-issued ID cards. Other vendors should report to materials management/purchasing prior to visiting other departments. Visitors who seek to enter the hospital after regular visiting hours must be screened by security. Provide for only one public entrance after 9 p.m. Suggest issuing color-coded visitor passes with the intended destination noted on it. Encourage all personnel to report any suspicious individuals seen on a hospital campus. Dispatch security personnel to question such individuals and escort them off hospital grounds.

G. TRAFFIC CONTROL AND VEHICLE ACCESS

The hospital must maintain designated traffic and parking controls. These controls include establishing fire lanes and crosswalks, posting speed limits, providing handicapped parking areas, and placing "no parking" signage in appropriate areas. The emergency department should provide a separate ambulance entrance and loading area. Security personnel must regularly patrol this area and other loading zones and fire lanes. When necessary, locate drivers leaving vehicles unattended and require them to move the vehicles.

H. WEAPONS ON CAMPUS

Establish a firearms policy for the entire campus, including a policy for all security personnel. Security personnel authorized to carry a firearm must know the deadly force policy. The only occasion in which an officer should use a firearm is if he or she believes it is the only way that an individual can be stopped before using deadly force against another person. The hospital should define weapons as firearms (including air guns), knives (other than ordinary penknives and pocketknives), explosives, and any other deadly weapon as determined by the hospital. Publish policies and procedures for reporting and informing visitors, patients, and employees of this policy (noncompliant individuals will be asked by security to leave the hospital grounds). Train security personnel to restrain violent individuals and do so upon request of nursing personnel/physicians or when, in their professional judgment, restraint is necessary to protect others.

I. Handling Civil Disturbances

Determine if security personnel can make arrests or must call local law enforcement authorities for assistance. Ban unwanted visitors and arrest violators for trespassing. Use additional security to seal off areas of the hospital such as the emergency department and screen anyone wishing to enter.

J. Handling Situations Involving "VIPs" or the Media

Plan to handle situations that might result in an influx of media representatives, a large patient entourage (e.g., Secret Service personnel accompanying a federal officer), and/or large numbers of curious onlookers. Additional security personnel should be called in as needed by the administrator on duty. Contact local law enforcement authorities and/or use private security firms as needed.

K. Access and Crowd Control

Access control will depend in part on what area of the hospital that is affected. Officers can be stationed at each unlocked hospital entrance to screen visitors and issue incident-specific visitor passes. Plant operations personnel can be utilized to help set up physical barriers/controls as needed. Facilities not operating an isolated patient suite can limit access to certain areas/floors by way of guards. Hospital personnel can help detect and deter unauthorized attempts to gain access.

L. Communications

Consider using the hospital's boardroom as a media work center. Facilities should provide a dedicated phone line cable in that area for such situations. An alternative site for media workspace is the cafeteria or a dining area. The hospital telephone system allows for blocking calls to specific patient rooms.

M. Orientation and Education

Security personnel and other identified staff members should undergo appropriate education and training to ensure they possess and maintain the skills and knowledge necessary to safeguard the security of patients, visitors, and staff. All hospital employees must receive instruction on security issues as part of their general orientation. This includes instruction on how to report security incidents involving patients, visitors, and employees, and how to summon security assistance. In addition, employees in security-sensitive areas of the hospital receive additional education in their departments to identify specific mechanisms or procedures designed to minimize security risk.

N. Evaluation of Security

Security personnel must conduct ongoing assessments of security needs and issues. The safety and/or care environment committees should establish performance standards and review effectiveness annually. The review should address the following areas:

- Staff security management knowledge and skill
- Level of staff participation in security management activities
- Security monitoring and inspection activities
- Detailed security and incident reporting procedures communication
- Inspection, preventive maintenance, and testing of security equipment

O.   USING PROVEN PRACTICES TO IMPROVE SECURITY

Hospital security professionals today better understand how to manage security risks than at any time in history. Increased and changing risks can compromise the effectiveness of security operations. OSHA workplace safety regulations and Joint Commission or other accreditation standards provide guidance on minimum requirements. Other organizations such as the American Society for Information Science (ASIS), the International Association for Healthcare Security and Safety (IAHSS), and NFPA also provide information on practices and guidelines that can help hospitals provide excellent security services. Learn from the successes—and mistakes—of other hospitals. Look for ways to improve security department capacity, reduce compensation expenses through better scheduling and management, and improve security officer recognition/performance.

P.   FORENSIC PATIENTS

Classify forensic patients into four categories: medical clearance, police hold, police custody, and emergency detention. The following four examples comprise the majority of forensic patients who interact with the general public every day in hospitals. Hospitals should run a risk assessment on police hold patients prior to intake. Hospitals must maintain responsibility for all patients and retain the right to ask how much of a danger a given patient presents to their facility. Healthcare security officers should continuously evaluate the status of forensic patients throughout their shift. All information on these patients should be passed on to relieving shifts. If possible, methods of tracking and flagging forensic prisoners should be integrated into the registration process. Nursing staff should report any concerns or suspicious activities involving their forensic patients.

Q.   WORKPLACE VIOLENCE

NIOSH defines workplace violence as any physical assault, threatening behavior, or verbal abuse occurring in the workplace. Violence includes overt and covert behaviors ranging in aggressiveness from verbal harassment to murder. Nurses and nursing assistants suffer the most nonfatal assaults resulting in injury. BLS rates measure the number of events per 10,000 full-time workers—in this case, assaults resulting in injury. In 2000, health service workers overall had an incidence rate of 9.3 for injuries resulting from assaults and violent acts. Healthcare workers face an increased risk of work-related assaults stemming from several factors including handguns and weapons among patients, their families, or friends. Other risks include the following:

- The increasing use of hospitals by police and the criminal justice system for criminal holds and the care of acutely disturbed, violent individuals
- The increasing number of acute and chronic mentally ill patients released from hospitals without follow-up care
- The availability of drugs or money at hospitals, clinics, and pharmacies, making them likely robbery targets
- Factors such as the unrestricted movement of the public in clinics and hospitals and long waits in emergency or clinic areas that lead to client frustration over an inability to obtain needed services promptly
- The increasing presence of gang members, drug or alcohol abusers, trauma patients, or distraught family members
- Low staffing levels during times of increased activity such as mealtimes, visiting times, and when staff must transport patients
- Isolated work with clients during examinations or treatment

- Solo work, often in remote locations with no backup or way to get assistance, such as communication devices or alarm systems
- Lack of staff training in recognizing and managing escalating hostile and assaultive behavior
- Poorly lit parking areas

The healthcare sector leads all other industries with 45 percent of all nonfatal assaults against workers resulting in lost work days in the United States. From 1993 to 1999, approximately 765,000 assaults occurred against healthcare workers resulting in days away from work. From 2003 to 2009, eight registered nurses (RNs) were FATALLY injured at work. In 2009, there were 2050 assaults and violent acts reported by RNs, requiring an average of four days away from work. In 2009, the Emergency Nurses Association (ENA) reported that more than 50 percent of emergency center (EC) nurses had experienced violence by patients on the job and 25 percent of EC nurses had experienced 20 or more violent incidents in the past three years.

Lateral violence, also called "horizontal violence," refers to acts that occur between workers and has been a long-term issue for nurses for decades, where nurses inflict psychological injury on each other. Horizontal violence, also called bullying, can be covert or overt acts of verbal and nonverbal aggression causing enough psychological distress to nurses to cause them to leave the profession. Rather than wait for healthcare employers to volunteer to establish such programs, some states have sought legislative solutions including mandatory establishment of a comprehensive prevention program for healthcare employers, as well as increased penalties for those convicted of an act of violence against a nurse.

1. Workplace Violence Protection for Nurses by Accrediting Bodies

Although there is no federal standard that requires workplace violence protections, effective January 1, 2009, The Joint Commission on Accreditation of Healthcare Organization (JCAHO) created a new standard in the "Leadership" chapter that addresses disruptive and inappropriate behaviors in two of its elements of performance. First of all, there is an organization code of conduct that defines acceptable and disruptive and inappropriate behaviors. Secondly, leaders must create and implement a process for managing disruptive and inappropriate behaviors.

2. Workplace Violence Prevention (NIOSH Publication No. 2002-101)

All hospitals should develop a comprehensive violence prevention plan. No universal strategy exists to prevent violence. The risk factors vary from hospital to hospital and from unit to unit. Hospitals should form multidisciplinary committees that include direct care staff as well as union representatives (if available) to identify risk factors in specific work scenarios and to develop strategies for reducing them. All hospital workers should be alert and cautious when interacting with patients and visitors. They should actively participate in safety training and be familiar with their employers' policies, procedures, and materials on violence prevention. NIOSH defines workplace violence as "violent acts (including physical assaults and threats of assaults) directed toward persons at work or on duty." This includes terrorism, as illustrated by the terrorist acts of September 11, 2001, which resulted in the deaths of 2886 workers in New York, Virginia, and Pennsylvania. Although these guidelines do not address terrorism specifically, this type of violence remains a threat to US workplaces. Healthcare and social service workers continue to face significant risk of job-related violence. Assaults represent a serious safety and health hazard within these industries. OSHA's violence prevention guidelines provide recommendations for reducing workplace violence. OSHA suggests developing public and private violence prevention plans with input from stakeholders following a careful review of workplace violence studies and tracking their progress in reducing work-related assaults. Although not every incident can be prevented, many can, and the severity of injuries sustained by employees can be reduced. Adopting practical measures such as those outlined here can significantly reduce this serious threat to worker safety.

BOX 3.11   WORKPLACE VIOLENCE PREVENTION

- Create and disseminate a clear policy of zero tolerance for workplace violence, verbal and nonverbal threats, and related actions. Ensure that managers, supervisors, coworkers, clients, patients and visitors know about this policy.
- Ensure that no employee who reports or experiences workplace violence faces reprisals.
- Encourage employees to promptly report incidents and suggest ways to reduce or eliminate risks. Require records of incidents to assess risk and measure progress.
- Outline a comprehensive plan for maintaining security in the workplace. This includes establishing a liaison with law enforcement representatives and others who can help identify ways to prevent and mitigate workplace violence.
- Assign responsibility and authority to individuals or teams while ensuring appropriate training and skill development. Make adequate resources so the team or responsible individuals can develop expertise on workplace violence prevention in healthcare and social services.
- Affirm management commitment to a worker-supportive environment that places as much importance on employee safety and health as on serving the patient or client.
- Set up a company briefing as part of the initial effort to address issues such as preserving safety, supporting affected employees, and facilitating recovery.

BOX 3.12   ELEMENTS OF VIOLENCE PREVENTION

- Management commitment and employee involvement
- Worksite analysis
- Hazard prevention and control
- Safety and health training
- Recordkeeping and evaluation

3. Management Commitment and Employee Involvement

Management commitment and employee involvement remain the essential elements of effective safety and health plans. Management and frontline employees must work together using a team or committee approach. If employers opt for this strategy, they must be careful to comply with the applicable provisions of the National Labor Relations Act. Employee involvement and feedback enable workers to develop and express their own commitment to safety and health and provide useful information to design, implement, and evaluate the efforts. A worksite analysis involves a step-by-step, commonsense look at the workplace to find existing or potential hazards for workplace violence. This entails reviewing specific procedures or operations that contribute to hazards and specific areas where hazards may develop. A threat assessment team, patient assault team, or similar task force or coordinator may assess the vulnerability to workplace violence and determine the appropriate preventive actions to be taken. This group may also be responsible for implementing workplace violence prevention plans. The team should include representatives from senior leadership, risk management, security, safety and health, and human resources. The team or coordinator should periodically inspect the workplace and evaluate employee tasks to identify hazards, conditions, operations, and situations that could lead to violence. After identifying hazards through systematic worksite analysis, the next step is to design measures through engineering or administrative and work practices to prevent or control these hazards. If violence does occur, post-incident response can be an important tool in preventing future incidents. Engineering controls remove the hazard from the workplace or create a barrier between the worker and the hazard. Base the selection of any control measure on the hazards identified in the workplace. Administrative and work practice controls affect the way staff performs jobs or tasks. Changes in work practices and administrative procedures can help prevent violent incidents.

Post-incident response and evaluation can help prevent future violence. All workplace violence efforts should provide comprehensive treatment for employees victimized personally or traumatized by witnessing a workplace violence incident. Injured staff should receive prompt treatment and psychological evaluation whenever an assault takes place, regardless of its severity. Provide the injured transportation to medical care if not available onsite. Every employee should understand the concept of "universal precautions for violence"—that is, that violence should be expected but can be avoided or mitigated through preparation. Frequent training also can reduce the likelihood of being assaulted. Employees who may face safety and security hazards should receive formal instructions on the specific hazards associated with the unit or job and facility. This includes information on the types of injuries or problems identified in the facility and the methods to control the specific hazards. It also includes instructions to limit physical interventions in workplace altercations whenever possible. In addition, train all employees to behave compassionately toward coworkers when an incident occurs. Training and education should involve all employees, including supervisors and managers. New and reassigned employees should receive an initial orientation before being assigned their job duties. Visiting staff, such as physicians, should receive the same training as permanent staff. Qualified trainers should instruct at the comprehension level appropriate for the staff. Effective training should involve role-playing, simulations, and drills. Employees should receive required training annually. In large institutions, refresher education may be needed more frequently, perhaps monthly or quarterly, to effectively reach and inform all employees.

Supervisors and managers need to learn to recognize high-risk situations to ensure not placing employees in assignments that compromise their safety. They also need training to ensure that they encourage employees to report incidents. Supervisors and managers should learn how to reduce security hazards and ensure that employees receive appropriate training. Following training, supervisors and managers should be able to recognize a potentially hazardous situation and to make any necessary changes. Security personnel need specific training from the hospital or clinic, including the psychological components of handling aggressive and abusive clients, types of disorders, and ways to handle aggression and defuse hostile situations. Training sessions should also provide an opportunity for an evaluation. At least annually, the team or coordinator responsible for the plan should review its content, methods, and the frequency of training. Plan evaluation may involve supervisor and employee interviews, testing, and observing and reviewing reports of behavior of individuals in threatening situations.

XVII. ERGONOMICS

The word ergonomics comes from the Greek words *ergo*, which means work, and *nomos*, which means law. It can also be referred to as the science or art of fitting the job to a worker. A mismatch between the physical requirements of a task and the physical capacity of the worker can result in musculoskeletal disorders. Ergonomics should focus on designing equipment and integrating work tasks to benefit the ability of the worker. Healthcare facility work environments expose patient and resident caregivers to ergonomic stressors. Successful ergonomic interventions must deal with personal issues instead of attempting to solve problems with universal solutions. Healthcare organizations should address ergonomic issues, risks, and injuries by developing a written ergonomics safety management plan. Ergonomic hazards refer to workplace conditions that pose the risk of injury to the musculoskeletal system of the worker. They include repetitive and forceful movements, vibration, temperature extremes, and awkward postures that arise from improper work methods and improperly designed workstations, tools, and equipment. Ergonomics addresses issues related to the "fit" between people and their technological tools and environments. Ergonomics draws on many disciplines in its study of humans and their environments, including anthropometry, biomechanics, mechanical engineering, industrial engineering, industrial design, kinesiology, physiology and psychology.

Organizations should identify existing and potential ergonomics hazards. Assessment of work tasks must include an examination of duration, frequency, repetition, awkward postures, and magnitude of exposure to force in all lift tasks. Conduct environmental walk-through tours to ask workers about lifting or stressful tasks. OSHA logs and workers' compensation injury reports can provide data related to ergonomic hazards. Use administrative controls to ensure adequate staffing. Emphasize the importance of patient or resident assessment to help determine level of risk for lifting, moving, or transferring tasks. Implement engineering controls to help isolate or remove the hazards by providing proper selection, training, and use of assist devices or equipment. Stress the early identification and treatment of injured employees. Develop a modified or transitional duty plan for workers recovering from an injury. Healthcare personnel experience a great variety of activities involving manual lifting, laterally transferring between two horizontal surfaces, ambulating, repositioning in beds or chairs, or manipulating extremities. Healthcare workers can also experience risks when transporting patients or equipment, performing activities related to daily living, stopping falls or transfers from the floor, and assisting in surgery.

Many organizations develop and implement an ergonomics policy with written goals, objectives, and accountability policies. Leaders should encourage worker involvement in ergonomics improvement efforts. NIOSH recommends reducing or eliminating potentially hazardous conditions by using engineering controls or implementing work practices and improved management policies. To meet ergonomics challenges, equipment should comply with ergonomics principles.

Effective training covers the problems found in each employee's job. Training and education can go a long way toward increasing safety awareness among both managers and employees, and can keep employees informed about workplace hazards. Soliciting suggestions from workers about ergonomic hazards can help improve work practices. Effective training can ensure employees properly use equipment, tools, and machine controls. Reactive ergonomics only takes corrective actions when required to do so by injury or complaint. Proactive ergonomics seeks to identify all areas needing improvement. Attempt to solve problems by changing equipment design, modifying job tasks, and improving environmental designs. Healthcare providers need to be familiar with worker jobs and tasks and participate in matching jobs and work environments to worker needs. Use information obtained from job hazard analyses, job descriptions, photographs, and videotapes to identify ergonomic hazards. According to the International Ergonomics Association, physical ergonomics addresses human anatomical, anthropometric, and physiological issues that relate to physical activity. Cognitive ergonomics addresses the concern with mental processes such as perception, memory, reasoning, and motor response. Macro-ergonomics emphasizes a broad system view of design considering organizational environments, culture, history, and work goals. It deals with the physical design of tools and the environment. It is the study of the society and technology interface and considers human, technological, and environmental variables and their interactions.

BOX 3.13   EXAMPLES OF ERGONOMIC RISK FACTORS

- Jobs requiring identical motions every three to five seconds for more than two hours
- Work postures such as kneeling, twisting, or squatting for more than two hours
- Use of vibration or impact tools or equipment for more than a total of two hours
- Lifting, lowering, or carrying more than 25 pounds more than once during a work shift
- Piece rate or machine-paced work for more than four hours at a time
- Workers' complaints of physical aches and pains related to their work assignments

A. EVALUATION OF ERGONOMICS EFFORTS

Leaders should evaluate the effectiveness of ergonomics efforts and follow up on unresolved problems. Evaluation and follow-up are central to continuous improvement and long-term success. Good medical management can help eliminate or reduce development of ergonomic-related problems. The goal should be early identification, evaluation, and treatment of problems. Elements of medical management should include the following: (1) accurate reporting and recording, (2) responding to complaints and symptoms, (3) providing employee education, (4) conducting periodic surveys, (5) establishing baseline health assessments, and (6) implementing surveillance procedures.

B. WORKSTATION EVALUATIONS

Evaluations should assess prolonged work in any posture that may result in harm or injury. Assess offices, computer areas, and nursing stations. Evaluate force, duration, position, frequency, and metabolic expenditure of workers. Workers should be provided with good chairs that have arm and leg rests if required. Provide workstations that permit posture variations and have sufficient space for knees and feet. Workers such as admission personnel, appointment clerks, transcriptionists, medical coding personnel, and other data entry personnel that work on computers four hours or more each are at risk for developing hand, arm, shoulder, neck, or back disorders.

C. WORKSTATION INTERVENTIONS

Signs of problems can include complaints of pain, tingling, numbness, swelling, and other discomforts. Employers should analyze trends, absenteeism, and turnover rates for those involved in data entry tasks. Workers should take short breaks often to allow the eye muscles to relax. Teach workers to glance at an object about 20 feet away. Some workers get relief by blinking or shutting their eyes for just a few seconds. Other interventions include padded keyboards, adjustable tables, and tilting screens. Allow workers to experiment to find a position that is comfortable to them. Ensure lighting is sufficient to help prevent glare and eyestrain. Provide glare control devices if necessary. Data entry personnel should take two or three short breaks for every hour of continuous work. Consider chair height as correct when the sole of a person's foot can rest on the floor or a footrest with the back of the knee slightly higher than the seat of the chair. Workers should arrange desk accessories to reduce twisting and turning. The body is most relaxed with arms loose, wrists straight, elbows close to the body, and neck and spine straight. Any standing workstation should have an anti-fatigue mat, work surface below the elbows, and a footrest so the worker can elevate one foot. A sitting station should have a surface at least 18 inches wide and rounded in the front. Chairs should allow unrestricted movement, be adjustable, and support the lower back.

D. HUMAN FACTORS

Human factors as a science concerns understanding the properties of human capabilities. The application of this understanding to the design, development, and deployment of systems and services relates to human factors engineering. Human factors can include sets of human-specific physical, cognitive, or social properties. These human factor sets can interact in a critical or dangerous manner with technological systems, the human natural environment, or human organizations. Human factors engineering applies knowledge about human capabilities and limitations to the design of products, processes, systems, and work environments. It also relates to the design of all systems having any type of human interface. Its application to system design improves ease of use and performance while reducing errors, operator stress, training, user fatigue, and product liability. It is the only discipline that relates humans to technology. Human factors engineering focuses on how people interact with tasks, machines or computers, and the environment with the consideration that humans have limitations and capabilities.

> **BOX 3.14   FACTORS IMPAIRING HUMAN PERFORMANCE**
>
> - Limited short-term memory
> - Running late or being in a hurry
> - Inability to multitask
> - Interruption of the job or task
> - Stress or lack of sleep
> - Fatigue or effects of shift work
> - Environmental factors
> - Personal or home distractions
> - Drug and substance abuse

E.   MUSCULOSKELETAL DISORDERS

Changes in healthcare during recent years resulted in increasingly heavy demands on nurses and other healthcare workers. Extended schedules along with increased physical and psychological demands can increase the risk of experiencing musculoskeletal disorders (MSDs) and injuries. Healthcare workers experience more upper extremity workers' compensation claims than workers in other industries. Nursing staff levels can impact physical and postural risk factors related to impaired sleep, pain medication use, and absenteeism. Encourage nurses to participate in ergonomics interventions. Traditional methods used to prevent work-related musculoskeletal injuries associated with patient moving include use of proper body mechanics, training personnel about safe lifting techniques, and the use of lumbar support belts. Evidence suggests that these three interventions, by themselves, prevent worker injuries. Many healthcare organizations now follow evidence-based practices such as providing patient handling equipment, implementing no-lift policies, and creating patient lift teams.

Early indications of MSDs can include persistent pain, restriction of joint movement, or soft tissue swelling. Activities outside of the workplace that involve substantial physical demands may also cause or contribute to MSDs. In addition, development of MSDs may be related to genetic causes, gender, age, and other factors. There is evidence that reports of MSDs may be linked to certain factors such as job dissatisfaction, monotonous work, and limited job control. Encourage workers to participate in the design of work, equipment, procedures, and training. Evaluate equipment regularly and respond to employee surveys. Effective solutions usually involve workplace modifications that eliminate hazards and improve the work environment. Work-related MSDs should be managed in the same manner and under the same process as any other occupational injury or illness. Like many injuries and illnesses, employers and employees can benefit from early reporting of MSDs. Early diagnosis and intervention, including return to duty procedures, can improve the effectiveness of employee treatment. Return to duty procedures can also minimize the likelihood of disability and reduce workers' compensation costs.

F.   ADMINISTRATIVE ISSUES

Conduct a review that evaluates if the equipment is appropriate for the specific lifting or moving activity. The review should involve onsite testing of a variety of equipment by the end users. Provide for the convenient storage of assist and institutional equipment. This can ensure that equipment is easy to find and, in turn, help encourage healthcare workers to use it. Use flexible purchasing procedures that allow for the evaluation and purchase of up-to-date equipment with the most appropriate features. Administrative issues affect the equipment available to employees, the types of work tasks they perform, and the methods of accomplishment.

G. Equipment Maintenance

A regular maintenance plan can help ensure sufficient quantities of equipment in all units or floors and avoid shortages and breakdowns. Some maintenance-related problems include jammed or worn wheels, which make it harder to move and steer or which cause chairs or other equipment to shift during transfers. Hard-to-reach controls or manual cranks on beds, chairs, or equipment can create risks and cause workers to assume awkward postures or make forceful exertions. Handles on beds, carts, or other equipment of the wrong size or placed at an inappropriate height can also contribute to injuries. Missing attachable IV/med poles can lead to workers awkwardly pushing gurneys or wheelchairs with one hand and holding free-standing poles with the other hand. Older mechanical lift devices can become hard to operate, uncomfortable, unstable, or even dangerous. High or heavy medical, food, or linen carts can result in unnecessary bending, reaching, or twisting when loading or unloading. Use systematic preventative maintenance techniques to keep all assist and moving equipment in proper working condition.

H. Facility Design Issues

Healthcare workers may need to assume awkward postures because rooms, bathrooms, hallways, and other spaces are small, crowded, or contain obstructions. These factors may also prevent getting help from other employees or using assist equipment. Poorly maintained floors can cause slipping, tripping, and abrupt movements when lifting or moving patients, residents, or equipment. Well-designed and maintained institutional equipment and facilities remain important factors in reducing or preventing back injuries. Institutional equipment should allow the user to maintain neutral body postures and reduce forceful motions. Beds, wheelchairs, cardiac chairs, and other equipment must be easy to adjust and move. Facilities should provide easy to operate equipment.

I. Mechanical Lift and Assist Devices

Train personnel on lifting equipment and proper procedures before permitting use of mechanical lifting devices. Always explain the lift to the resident or patient before beginning the procedure. Ensure the resident or patient is positioned correctly in the sling before continuing the lift procedure. One person must ensure that the patient remains stable during the entire lifting procedure. Never allow the sling to swing and never leave a patient or resident suspended in the sling. Mechanical assist devices or lifts can help reduce injury by avoiding unnecessary manual transfers, awkward postures, forceful exertions, and repetitive motions.

J. Understanding the Body

The first seven vertebrae, called cervical vertebrae, form the neck. Areas of the spine such as the neck, where flexible, can experience strains and sprains. The shoulder consists of a ball and socket joint where the ball of one bone fits into a hollow crevice of another. The shoulder joint allows movement and rotation of the arms inward, outward, forward, or backward. There are several different tendons attached to bones in the shoulder. Bursar reduces friction and cushions the tendons as they slide back and forth. The spine is a column of approximately 30 bones called vertebrae that run from the neck to the tailbone. These vertebrae stacked on top of one another in a shaped column form spinal joints, which move independently. Health spines contain three natural curves: a forward curve in the neck, a backward curve in the chest area, and another forward curve in the lower back. The back's three natural curves should align correctly when ears, shoulders, and hips form a straight line. At the end of the spine, the vertebrae fuse together to form the sacrum and the tailbone. The lower back or lumbar area provides the workhorse capacity of the back. It carries most of the weight and load of the body. Aligning and supporting the lumbar curve properly helps prevent injury to vertebrae, discs, and other parts of the spine. The spine contains various types of associated soft tissues like the spinal cord, nerves, discs, ligaments, muscles, and blood vessels. Discs, the soft shock-absorbing cushions located between vertebrae, allow these joints to move smoothly and absorb shock as you move. Each disc contains a spongy center and tough outer rings. The vertebrae are connected by a complex system of ligaments that knit them together. Strong flexible muscles maintain the three natural spinal curves and help in movement. The most important muscles that affect the spine include the stomach, hip flexors, hamstrings, buttocks, and back muscles.

K. INJURIES AND DISORDERS

A sprain refers to damage to ligament fibers caused by moving or twisting a joint beyond its normal range. A strain occurs when a muscle or a muscle tendon unit is overused. Bursitis is an irritation of bursa in the shoulder areas caused by rubbing on adjacent tendons. Tendinitis occurs when a tendon is overused and becomes inflamed. When the tendon sheath is involved, the condition is called tenosynovitis. Neck tension syndrome occurs where the last neck vertebra meets the first mudpack vertebra and is a major site of acute back pain, muscle tension, and other injuries. Common symptoms can include muscle tightness, soreness, restricted movement, headaches, and numbness/tingling in the hands, wrists, arms, or the upper back. Over time, discs wear out or degenerate from natural aging. The discs dry out and become stiffer and less elastic. The outer fibrous rings can crack and the discs narrow. They become less able to handle the loads put on them. If the inner jelly-like center bulges into the outer rings, it may compress nearby nerves or blood vessels. If the inner jelly-like center breaks through the outer rings, the condition is called a ruptured or herniated disc.

BOX 3.15   COMMON ERGONOMIC-RELATED DISORDERS

- *Tenosynovitis:* This malady results in the inflammation of the tendons and their sheaths. It often occurs at the wrist and is associated with extreme wrist movement from side to side.
- *Trigger Finger:* A condition caused by any finger being frequently flexed against resistance.
- *Tendinitis:* A condition where the muscle-tendon junction becomes inflamed due to repeated abduction of a body member away from the member to which it is attached.
- *Tennis Elbow:* This form of tendinitis is an inflammatory reaction of tissues in the elbow region caused by palm upward hand motion against resistance, such as the violent upward extension of the wrist with the palm down.
- *Carpal Tunnel Syndrome:* A common affliction caused by the compression of the median nerve in the carpal tunnel. It is often characterized by tingling, pain, or numbness in the thumb and first three fingers. It is often associated with repeated wrist flexion.
- *Reynaud's Syndrome:* A condition where the blood vessels in the hand constrict from cold temperature, vibration, emotion, or unknown causes. It is easily confused with the one-sided numbness of carpal tunnel syndrome.

L. BACK-RELATED PROBLEMS

Common causes of back pain can relate to poor physical condition and being unaccustomed to a task. Other factors that contribute to pain include poor posture and lifting objects beyond a person's ability. Contributing factors for back injuries include understaffing, inadequate training, poor body mechanics, inadequate safety precautions, and not using assist devices. The natural curves of the spine are held in place and supported by muscles in the back and abdomen. These muscles must be strong and healthy. If standing for a prolonged period, one foot should rest on a low stool to support the lower back. Keep the head up and chest lifted. Select a chair that supports the lower back but is not too high. Tuck the buttocks and keep feet flat on the floor. Sleep on the back if possible with a small pillow under knees, or sleep on the side with knees bent. Never sleep on the stomach, or on the back with legs straight out. A fitness program that improves aerobic capacity while strengthening back muscles can help prevent back pain. Each individual should choose an exercise program that fits their needs and abilities. NIOSH has now concluded that the use of lumbar support belts to reduce the risk of injury remains unproven. NIOSH previously concluded that the lumbar supports do not reduce spinal compression during heavy lifting tasks. NIOSH also expressed concerns that the belts might give workers a false sense of security and result in some lifting excess weights. The NIOSH study only reviewed data from other studies and did not do any original research. Several recent scientific studies conducted at leading universities indicate that correctly fitted lumbar support belts could help alleviate pressure on the soft tissue of the back and spine. Some associations and insurance groups claim that the use of support belts has resulted in a significant reduction in workers' compensation costs. Use back support belts only when included as an integral part of total back care management efforts.

The lower discs can experience more damage than other discs because they bear most of the load in lifting, bending, and twisting. Sciatica occurs when bulging or ruptured discs constrict the sciatic nerve of nearby blood vessels causing pain to the hips, buttocks, or legs. Degenerative or osteoarthritis simply means the wearing out of joints, vertebrae, discs, facets, or other structures over time. Osteoarthritis is associated with loads put on the spine over long time periods. As the discs dry out and narrow, they lose their shock-absorbing ability. The vertebrae become closer together, irritated, and may produce bony outgrowths. Facet joint syndrome occurs when the facets interlock with the vertebrae above and below to form joints in the spine. The facets can become misaligned from bending, lifting, and twisting while working. Slipped vertebrae occur when the vertebrae in the lower back pushes forward so they don't line up with other vertebrae. This condition disrupts the proper natural curves of the spine and causes joints, ligaments, and muscles to become overburdened. Spinal canal narrowing can occur in the canal that the spinal cord runs through or in the gap at the sides of vertebrae where nerves exit.

M. Back Injury Prevention

Management and prevention efforts should focus on eliminating lifts wherever possible. Use patient handling, transfer, and lifting equipment. Establish patient lift guidelines to help workers safely assess patient handling situations. Redesign the workplace to increase efficiency and decrease the potential for injuries. Educate workers about back anatomy and personal back care responsibilities. Provide recurring education and training on proper body mechanics and patient transfer techniques. Require employees to participate in exercise and/or stretching routines before lifting. Establish and train two-person lift and transfer teams. Use physical or occupational therapy professionals to instruct workers in patient handling techniques. Assess the patient or resident before lifting or moving them. Eliminate or reduce manual lifting and moving of patients or residents whenever possible. Get patients or residents to help as much as possible by giving them clear, simple instructions with adequate time for response. Know your own limits, do not exceed them, and get help whenever possible. Never transfer patients when off-balance. Never permit workers to lift alone. Require team lifting for fallen patients and when using assistive devices. Limit the number of allowed lifts per worker per day. Investigate all accidents and make changes to prevent recurrence. Assign a case management worker to oversee medical treatment and return to work efforts. Never move or lift from side to side. Plan the lift and size up the load to better reduce spine movement. Keep the patient load as close to the body as possible. Ten pounds at waist height equates to 100 pounds force on the back with arms extended away from the body. Bend at the knees when lifting loads from floor level.

Ten pounds at floor height with bent knees is equal to 100 pounds of force when bending at the waist with legs straight. Avoid any twisting motion and pivot the feet to turn. Always push rather than pull loads. Pushing reduces the force necessary to move an object by 50 percent. Use lifting equipment and devices such as chair lifts, mechanical lifts, transfer boards, and gait belts. Keep beds at proper heights. Keep the back straight and maintain correct posture with head up and stomach tucked in.

BOX 3.16   BACK INJURY PREVENTION TIPS

- Educate nurses on proper back care and use of proper body mechanics
- Provide recurring training on patient transfer techniques
- Implement exercise routines for those involved in lifting
- Establish, educate, and train lifting teams
- Conduct periodic ergonomic evaluations to detect problem areas
- Ensure implementation of effective housekeeping procedures
- Acquire and require the use of patient lift and assist devices and equipment

---

1.   Lateral Transfers

Use lateral transfers or sliding techniques to move patients and residents between two horizontal surfaces such as bed to gurney. Helpful equipment and devices include slide boards, transfer mats, slippery sheets, draw sheets, and incontinence pads.

2.   Ambulating, Repositioning, and Manipulating

For help with these types of activities, use equipment and gait belts, transfer belts with handles, slippery sheets, plastic bags, draw sheets, incontinence pads, pivot discs, range of motion machines, fixtures, etc.

3.   Performing Activities of Daily Living

These activities include showering, bathing, toileting, dressing or undressing, and performing personal hygiene and related activities. Equipment devices include shower toilet combination chairs, extension hand tools, shower carts, gurneys, and pelvic lift devices.

4.   Useful Tips

Encourage healthcare workers to use assist equipment and devices. Some suggestions about assist devices and equipment follow:

- Purchase the proper devices in sufficient quantities
- Store devices in areas visible and readily available
- Involve end users in evaluating and selecting devices
- Ensure the organization accomplishes effective training on device usage
- Equip devices with sufficient replacement accessories such as slings
- Implement a comprehensive maintenance plan for all devices

5.   Lift Teams

Some organizations choose to create a special "lift team" dedicated to performing the majority of the lifting or moving of patients or residents. The lift team should coordinate with the nurses and other medical personnel responsible for the patient or resident. Some organizations train teams to

- Eliminate uncoordinated lifts
- Prevent unprotected personnel from performing lifts

- Reduce weight and height differences between partners
- Prevent untrained personnel from lifting
- Encourage the use of lifting equipment when possible

6. Guiding and Slowing Falls

Review patient or resident assessments and watch for signs of weakness. If falls do occur, attempt to guide, slow, and lower the patient or resident to the floor. Try to maintain a neutral body posture when assisting patients. Regulatory reporting requirements may cause employees to try stopping a fall. Reporting of falls should not lead to fault-finding or negative consequences.

7. Transfer Task Safety

Communicate the plan of action to the patient and other workers to ensure that the transfer takes place using smooth techniques that consider unexpected moves by the patient. Remove any obstacles and focus on maintaining sure footing. Patients should wear slippers that provide good traction. Maintain eye contact, communicate with the patient, and stay alert for trouble signs. Record any problems on the patient's chart so that other shifts will know how to cope with difficult transfers. Also note the need for any special equipment. Implement measures to reduce or prevent back injuries such as

- Developing a return to work or modified duty procedure
- Writing job descriptions that establish the appropriate physical requirements
- Requiring immediate reporting and treatment of injuries

8. Personal Factors

Home and recreational activities involving forceful exertions or awkward postures can also lead to or aggravate back injuries. Some examples include sports and home repair work. Physical fitness, weight, diet, exercise, personal habits, and lifestyle may also affect the development of back injuries. Individuals not in good physical condition tend to have more injuries. Excessive body weight can place added stress on the spine and is often associated with a higher rate of back injuries.

Previous trauma or certain medical conditions involving bones, joints, muscles, tendons, nerves, and blood vessels can also contribute to back-related disorders. Psychological factors, such as stress, may influence the reporting of injuries, pain thresholds, and even the speed or degree of healing. Physically fit individuals tend to have fewer and less severe injuries. Remember to consult with a physician or physical therapist about which aerobic, strength, and flexibility exercises to do. This is especially important for those individuals who have preexisting injuries or medical conditions.

9. Work Evaluation Tools

Involve the employees performing the work in evaluating problems and coming up with potential solutions. Following the simple three-step hazard control process can help reduce lifting-related injuries and complaints. The first step in the process is to identify lifting tasks by observing and evaluating patient/resident needs on the unit. The second step involves analyzing both data and observations. Conduct observations for a period of time to validate the actual tasks. The analysis step should help managers identify causal factors related to lifting or moving tasks. Once the analysis step is completed, the identification and assessment team can consider appropriate control to reduce worker risk of injury. Never select and implement controls if accomplishing the identification and analysis steps incorrectly.

N. Training and Education

Provide training at the level of understanding appropriate for those being trained. Give workers an opportunity to ask questions. Provide an overview of the potential risks, causes, and symptoms of back injury and other injuries. Teach workers how to identify existing ergonomic stressors and methods of control. Explain the use of engineering, administrative, and work practice controls needed to conduct patient or resident handling tasks. Encourage workers and staff to stay physically fit. Provide education and hands-on practice that allows feedback. Review the work task analysis and evaluation information. Implementing improvement options or controls should guide the type of education provided. Training and education must focus on the nature and causal factors of worker injuries. Require that employees demonstrate the skills learned in a competency evaluation. Provide a systematic approach reinforced by retraining. Training is usually most effective when it includes case studies or demonstrations. Answer any questions that may arise during the training. Ensure that charge nurses and supervisors participate in the education and training. They should reinforce safety policies and oversee incident reporting requirements. Supervisors should ensure the implementation of task-specific procedures and adherence by workers to published policies.

XVIII. PATIENT TRANSPORT FUNCTIONS

Some hospitals transport patients using members of a trained team. Theses transporters move patients to various locations in the hospital complex. Patient transportation may involve high-risk patients, such as patients using an oxygen tank. Some transport team members also collect and deliver laboratory specimens. Transporters may transport equipment such as stretchers and wheelchairs. Transporting patients from one location to another includes vehicle to bed, room to procedure area, or building to building. Patient transporters serve as the frontline custodians of patient experience. The patient transportation staff must receive thorough training in all aspects of safe patient handling procedures including lifting protocols and infection control. Transportation functions must stress prompt and efficient services. Patient transport services require strong leadership. Effective transportation services permit nurses and staff members to focus on patient care requirements.

A. Transporting Patients

Establish practices to ensure safe care during the transport of patients. Transporters must learn to recognize hazards during the transport journey. Consider the following issues when planning for a transport: (1) need for IV poles, (2) need for transport oxygen tank, (3) conscious state of patient, and (4) the age and size of patient. Determine physical abilities and the condition of the patient. Consider the following when selecting the type of transportation to use:

- Wheel-locking capability and need for safety straps
- Side rail height sufficiency to prevent falls
- Need to transfer IV poles
- Ability to accommodate patient positioning
- Mattress on gurney is held in place
- Ability to use patient transfer device
- Maneuverability of transportation device
- Transportation device deemed safe

B. Patient Care

The individual who is transporting the patient should introduce and identify herself/himself to lessen patient anxiety. Correctly identify the patient to prevent wrong-patient surgery. If the patient is conscious, explain the transfer procedure prior to implementation to reduce the anxiety of the patient and promote safety. Maintain the patient's dignity during the transfer. This will aid in decreasing the patient's anxiety and ensure personal and moral rights. Adhere to all safety procedures, including the following:

- Elevate the side rails and apply a safety strap
- Confirm IV lines, indwelling catheters, monitoring lines, and drains
- Protect head and arms and make the patient as comfortable as possible
- Transport patient feet first and avoid quick movements
- Verbalize to patient to keep hands and arms inside the safety rails
- Explain all actions to conscious patient
- Maintain dignity by keeping the patient covered at all times

C. TRANSPORT TEAM DEVELOPMENT

Interdisciplinary transport teams can help to reduce patient risk during transport by using standardized protocols and policies. Transport team policies should include the use of sound communication techniques and teamwork with specific roles and responsibilities. The facility must obtain the appropriate equipment to ensure safety. Ensure the curriculum for transport team members focuses on ensuring competency. Education should include lessons on intravenous lines, catheters, and oxygen use. It should also include CPR certification, knowledge of patient safety goals, and handoff communication procedures. Organizations must develop standardized handoff communication checklists to ensure patient safety. Transport teams can help patient safety efforts and reduce the potential for adverse events. Develop a transport team model of care with a clear outline of the specific responsibilities for each team member. Coordinate pre-transport communication between the transporter, nurse, and the destination location. Ensure that patient equipment is functional, fully charged, filled, and in good repair.

BOX 3.17   QUESTIONS THAT ASSESS TRANSPORT ACTIVITIES

- Which patients are being transported?
- What are the most frequent source units and patient types?
- To which locations are most patients transported?
- Are these destinations in the main hospital, adjacent buildings, or across the street?
- Are there special safety hazards in any of the units?
- Does the organization conduct pre-transport patient assessments?
- What criteria determine patient stability, patient risk, and level of monitoring during transport?
- Who is responsible for conducting a transport assessment?
- What is the recommended timing for this assessment?
- How is the assessment communicated to the various care teams?

BOX 3.18   QUESTIONS RELATED TO TRANSPORT PERSONNEL

- Do unlicensed and licensed personnel transport patients?
- What are their specific responsibilities before and during transport?
- What are the competency assessments necessary to ensure patient safety during transport?
- What should minimum basic life support training entail for transport personnel?
- Does training cover how to receive and provide handoff communications?

> **BOX 3.19   QUESTIONS RELATED TO HANDOFF COMMUNICATION**
>
> - How are the patient's condition, potential safety risks, and needs communicated?
> - Is a checklist used? Is patient identification included?
> - What is the responsibility of the sending and receiving providers and/or transporters?
> - What are the necessary supplies and equipment needed for transport?
> - What equipment is required to accompany the acute care patient during transport?
> - What person ensures that therapies are maintained during transport?

1. WHEELCHAIR SAFETY

Wheelchair safety requires making plans to address emergencies such as brake failure on a power chair or a manual chair tipping backwards. Do not rush when assisting an individual in a wheelchair. Lifting appropriately protects the patient and can reduce worker back and arm stress. Always lock the brakes before the patient moves in or out of the chair or when leaving a client unattended. Engage both wheel locks. Never use the wheel locks as a brake when moving. Lift the footplates up before the patient gets in or out of the chair. When adjusting the elevating leg rest, support the frame while lowering or raising to prevent a sudden release of the leg rest.

Take care in wet or icy weather, particularly on sloping pavements, as wheelchairs tend to slide to the lowest point. Lack of maintenance or poor maintenance can lead to the wear or failure of components that may cause the wheelchair or the user to change position unexpectedly. This could lead to the user falling from the wheelchair or tipping over with the wheelchair. Adhere to manufacturer maintenance instructions. Always use a qualified technician to service or repair the wheelchair. If the wheelchair is approved by the manufacturer for transportation of a seated person in a vehicle, make sure that you use the wheelchair tie-down and occupant restraint system specified by the manufacturer. If using large public buses or trains, use the dedicated wheelchair space and any restraint systems provided.

> **BOX 3.20   WHEELCHAIR SAFETY BASICS**
>
> - Maintain wheelchairs in top mechanical condition
> - Disinfect wheelchairs on a regular basis
> - Always back a wheelchair down a ramp or into an elevator
> - Never allow wheelchairs to block hallways or exits
> - Lock the brakes if the patient or resident will remain seated for a length of time
> - Ensure chair wheels have hand-rims for better control and safer use of the wheelchair
> - Use high-density foam rolls that slip easily over the backrest to help distribute weight evenly
> - Use wheelchair lap boards to provide support and encourage use upper extremities
> - Place pillows under the buttocks to raise the hips above the knees
> - Use a wedge cushion to help keep the patient in the seat

REVIEW EXERCISES

1. List at least seven basic safety tips for nurses.
2. List five educational topics for personnel experiencing sleep deprivation.
3. In your own words, why does stress impact the lives of so many nurses?
4. List at least seven ways healthcare organizations can help nurses deal with stress.
5. In your opinion, why do healthcare organizations experience a high rate of slips and trips?

6. Why is electrical safety a top priority in most healthcare facilities?
7. List the six factors that contribute to good indoor air quality.
8. List the three common healthcare hazards listed in the chapter.
9. What are at least four key hazards experienced by surgical nurses?
10. What would be the two primary hazards found in dialysis units?
11. Why should nursing personnel show concern about medical device and equipment safety?
12. How does a healthcare security function impact the well-being of nurses?
13. List the four classes of forensic patients.
14. According to the ENA, what percentage of emergency department nurses experienced violence from patients in 2009?
15. List the five elements of a workplace violence prevention plan.
16. In your own words, define the concept of ergonomics and list at least four risk factors.

4 Patient Safety

I. INTRODUCTION

All healthcare safety personnel should understand the importance of patient safety and how it fits into the organization's total safety system. We need to view patient safety as a discipline within healthcare professions and organizations. Healthcare leaders should consider both the science and practice of safety when addressing patient care issues. Healthcare risk and quality management personnel should learn to view patient safety as a function associated within their disciplines. Patient safety could be defined as preventing patient adverse events and errors while minimizing the harm of those events that do occur. We should approach patient safety effectiveness from two key directions. First, consider organizational issues such as leadership, organizational dynamics, operating cultures, and patient care effectiveness. The coexistence of the organizational operating culture and the overall safety culture may be in conflict in many hospitals. They may attempt to operate in parallel dimensions. Second, make patient safety a function of the organization. Educate organizational members that patient safety is not just another program but a subsystem of the total safety system.

Many healthcare leaders consider patient safety a clinical concern, some view it as a function of risk management or quality improvement processes, and others define patient safety concepts using their own definitions or reference points. Medical specialization, organizational fragmentation, and compartmentalization can hinder safety efforts. Improving patient safety performance at all organizational levels must become a strategic goal of all healthcare organizations. Healthcare organizations should learn to "connect the dots" of the many well-intentioned efforts to improve patient safety. Many organizations spend too much time touting the mere existence of their transactional patient safety endeavors. Connecting the dots in a children's book, if done correctly, produces a recognizable picture. The picture of patient safety "progress" over the past 10 years is not a clear one. Patient safety should focus on more than just methods, tools, surveys, policies, buzzwords, and trendy thematic books. Leaders must learn to consider the patient safety function of every medical or healthcare organization as a recognizable "subsystem" of the total safety system. Effective patient safety requires more than a policy statement signed by senior leaders. Patient safety is much more than an office location, a phone number, an email address, or a slogan strategically placed on an organizational website.

When seeking patient safety information on the Web, one quickly learns that the wealth of information is a challenge to sort through, resulting in a disappointing endeavor. Some recent patient safety innovations have yielded good results, such as preventing central line and surgical wound infections. Many hospitals took actions to reduce medication and surgical adverse events with some positive results. Healthcare organizations should learn to connect the dots, build bridges, and understand that many hidden cultures exist in all organizations. These "covert" or "hidden" cultures operate within the "overt" or "formal" cultures on a daily basis. The informal cultures can and in many instances do operate independently of the recognized organizational culture.

Another hindrance to patient safety success at the local level relates to failure of leaders to understand open and closed systems. High-reliability methods, no matter how appropriate they may seem, will never be effective in all hospital areas, departments, or functions. Healthcare can best be described as an "open system" with some "closed micro-systems." High-reliability methods can work well in highly closed and controlled settings. However, "open systems" should anticipate many uncontrolled elements that impact effective operations. The terms "patient" and "their families" should send the message, loud and clear, that healthcare organizations operate as open systems. Healthcare organizations should implement policies and hazard controls to ensure visitor safety while in the facility. Organizations should implement measures to educate visitors on how they can help protect the safety of patients.

We can base patient safety efforts on three key concepts. First, view patient safety as an organizational function and not a program or department. Never consider patient safety efforts as a function of risk management or quality improvement. Second, view patient safety as an operational "subsystem" of the organization safety system. Implementing a comprehensive safety system requires an organizational culture change. I spoke to a "patient safety coordinator" a few years ago and she quickly pointed out that the patient safety function in her organization operated separately from the hospital's "environmental safety" function. I calmly mentioned to her that patient safety occurs or does not occur in the "environment of care." Patient safety does not compete with the components or subsystems of its own system. Patient safety complements the other safety system components of worker safety, visitor safety, contractor and vendor safety, and community safety. The function of healthcare safety within the organization should derive its value from people. The third aspect relates to the fact that many senior leaders incorrectly believe that patient safety requires a "clinical" professional to lead the efforts. Patient safety can only improve if the organization undergoes transformational changes.

A. WORKER FATIGUE AND PATIENT SAFETY

Fatigue resulting from an inadequate amount of sleep or insufficient quality of sleep over an extended period can lead to a number of problems, including inattention and the inability to stay focused. It can also lead to lack of motivation and trouble in solving problems. It can be demonstrated through a person's memory lapses, poor communication ability, lowered reaction speeds, and even loss of empathy. Shift length and work schedules can impact healthcare provider quantity and quality of sleep. The dangers associated with shifts lasting longer than 12 hours is well documented. Healthcare organizations must

- Assess the organization for fatigue-related risks by reviewing staffing policies
- Evaluate handoff processes and procedures
- Invite staff input into designing work schedules
- Create and implement a fatigue management plan
- Educate staff about sleep hygiene and the effects of fatigue on patient safety
- Provide opportunities for staff to express concerns about fatigue
- Encourage teamwork as a strategy to support staff who work extended work shifts or hours
- Consider fatigue as a potentially contributing factor when reviewing all adverse events
- Assess the environment provided for sleep breaks to ensure that it fully protects sleep

II. LEADING EFFORTS

Leading people can prove more difficult than managing other organizational assets. Leaders should assess how well staff and other healthcare providers support the patient safety function. Education, communication, and feedback must become a top priority. Promote the use of "action plans" to guide patient safety processes and needed innovations. Leaders place the focus on improving processes and not blaming people. Don't "talk the talk" unless you "walk the walk." Frequent walking tours promote safety, improve communication, and provide valuable information to the leadership team. Finally, safety leadership requires allocation of adequate resources to support the ongoing function of patient safety. Leaders can help improve organizational safety functions by focusing on patient-centered strategies. Implement system-centered approaches and look for evidence-based solutions. Create teams with the knowledge and resources to act when appropriate. Develop effective monitoring and measurement tools.

> ## BOX 4.1 CARE ENVIRONMENT SAFETY CHALLENGES
>
> - Establishing a multidisciplinary process or committee to resolve care environment issues
> - Appointing appropriate representation from clinical, administrative, and support areas
> - Ensuring that the multidisciplinary improvement team or committee meets at least bimonthly or as necessary to address environment and quality-of-care issues
> - Identifying and analyzing care and environment issues in a timely manner
> - Developing and approving recommendations for improvement as appropriate
> - Establishing appropriate measurement guidelines with appropriate staff input
> - Communicating issues to organizational leaders and improvement coordinators
> - Providing an annual recommendation for at least one performance improvement activity
> - Coordinating environmental safety issues with leadership of the patient safety program

Peer review of physicians historically focused on a punitive process and not an opportunity to learn. To engage physicians in safety, target the 20 percent of physicians who spend the most time in the hospital and not the 80 percent that rarely enter the hospital. Making physicians partners, not consumers, permits the patients to become the customers. This will require physicians to become responsible not just to their patients but also to the system providing care. Organizational leaders committed to improving patient safety should also create an environment that attracts and retains the best nurses. Create ways to acknowledge the value of nurses and support continuous learning activities. The nursing profession needs leaders that value and support frontline nurses. Leaders should learn to encourage and promote more collegial nurse and physician relationships. Developing a true safety culture will include demanding that nurses be treated with respect. Organizational excellence in patient safety can never occur without implementing policies that address nursing safety, well-being, and job satisfaction.

Healthcare organization oversight and governing boards should lead the patient safety journey. Board members spending less than 25 percent of their time addressing patient safety and quality issues do a disservice to their organization. Oversight boards should develop and publish policies addressing formal quality improvement measures. Boards can promote patient safety only by continuously interacting with the medical staff. The board should require that the CEO lead the way by being the person most identified with patient safety and quality improvement.

The Institute of Healthcare Improvement (IHI) recommends that governing boards establish organizational patient safety goals, listen to *sharp end* stories, implement system-level measures, and monitor important issues such as organization culture change. Organization accountability should focus on full disclosure to patients and their families. Conduct a sequence of event analysis after medical errors and adverse events. Gather facts including the timeline of an event. Use the information collected to assist with a focused analysis. Hold individuals accountable by clearly defining roles and relationships. A just culture dictates a balance between nonpunitive actions and situations requiring accountability and discipline. Humans will and can make mistakes. Hold people accountable when they overestimate their abilities and underestimate their limitations. Humans often fail to recognize fatigue, stress, and work environmental issues such as noise or poor lighting. Illness, boredom, frustration, home situations, and substance abuse can also impair job performance. Create policies that enforce and support accountability. Make an effort to educate frontline staff on the accountability system. Educate all managers, supervisors, and team leaders about expectations. Senior leaders should take actions to publicly embrace the need for more accountability and the development of a culture that learns from errors.

BOX 4.2   KEYS TO CREATING CULTURES OF TRUST

- Promote the value of members and acknowledge contributions
- Allow team members to participate in decisions
- Reduce discipline for errors for valid reasons
- Simplify job and process tasks
- Stress the importance of practicing good human relations
- Forbid senior leaders and physicians from verbally abusing subordinates
- Give staff a chance to express a voice in operational matters
- Solicit suggestions for improving safety or clinical processes
- Practice better oral and written communication
- Redesign processes that contribute to errors
- Provide quality continuing education
- Design clinical processes by using a systems approach
- Decrease reliance on personal vigilance and memory
- Develop good data collection and analysis systems

BOX 4.3   PATIENT SAFETY INFORMATION SOURCES

- Agency for Healthcare Research and Quality (AHRQ) Patient Safety Network
- American Hospital Association
- Canadian Nurses Association Patient Safety Resource Guide
- Institute for Safe Medication Practice (ISMP)
- Institute of Medicine of the National Academics
- Joint Commission International Center for Patient Safety
- Medline Plus Patient Safety
- National Center for Patient Safety
- National Coordinating Council for Medication Error Reporting and Prevention
- National Patient Safety Agency
- World Health Organization Alliance for Patient Safety

III.   IOM REPORTS

The 1999 Institute of Medicine (IOM) Report, *To Err Is Human*, challenged the public and health-care industry to create a *climate* to support change. There has been controversy since the release of the 1999 IOM Report related to mandatory versus voluntary error reporting. The 2001 IOM report, *Crossing the Quality Chasm*, focused on the theme of providing a common purpose for changing healthcare. The 2001 IOM report listed the following six aims for improving healthcare: (1) safe, (2) effective, (3) outcome focused, (4) timely, (5) efficient, and (6) equitable care. The 2003 IOM report, *Patient Safety: Achieving a New Standard of Care*, emphasized the importance of an electronic health record (EHR) with regard to patient safety. It also recommended the development of better definitions for patient safety, including the terms "near misses" and "adverse events." The 2004 IOM report, *Transforming the Work Environment of Nurses*, recommended that healthcare organizations evaluate and improve areas such as nurse management practices, workforce capability, workplace design and organizational culture. The 2005 IOM report, *Quality Through Collaboration: The Future of Rural Healthcare,* proposed a strategy for meeting the health challenges facing rural communities, including providing quality care.

IV. ERRORS AND ADVERSE EVENTS

An adverse event is an injury caused by medical management rather than by some underlying disease or patient condition. Medical errors result from a complex series of system-related issues and not from a single individual. Errors may or may not result in an adverse outcome. IOM defines errors in two ways: (1) an "error of execution" refers to a correct action that did not proceed as intended and (2) an "error of planning" occurs when an intended action was accomplished incorrectly. The National Patient Safety Foundation (NPSF) defines patient safety as the prevention of and elimination or mitigation of patient injury by errors. NPSF defines a healthcare error as an unintended outcome caused by a defect in the delivery of care to a patient The AHRQ defines a medical error as an act of commission (doing the wrong thing), omission (not doing the right thing), or execution (doing the right thing incorrectly). Consider the following examples of error:

- Physician failed to use an indicated diagnostic test or misinterpreted a test
- Emergency room personnel could not use a defibrillator with dead batteries
- Patient developed a post-surgical wound infection, resulting in a longer stay
- Patient received the wrong blood type during a transfusion

A. ACTIVE AND LATENT ERRORS

Characteristics of high-reliability organizations include (1) acknowledging and planning for human variability and fallibility, (2) anticipating the worst and planning for it, and (3) planning for failure to help avoid harm when failures occur. Leaders understand technical, organizational, environmental, and human factors that impact error. Trust pervades the organization so people report safety concerns and errors because they understand what constitutes unsafe practice. Reporting can prove valuable to staff and leaders aware of the importance of accurate data. Organizations should reward reporting of errors and near misses. Flexibility gives frontline personnel responsibility for immediate situations.

B. FACTORS IMPAIRING HUMAN PERFORMANCE

Research and anecdotal evidence indicates that a number of factors can impair human performance. Working in complex surroundings, humans can easily experience limited short-term memory. Running late or being in a hurry can impact task performance. Some individuals find it very difficult to multitask.

Others lose their concentration due to job or task interruption. Healthcare workers and professionals often deal with stress, lack of sleep, and fatigue on the job. Workplace environmental factors, personal or home distractions, and substance abuse can also impair performance. James Reason in his studies developed some questions to address errors committed on the job. Consider the following questions when investigating an error or other adverse event:

- Did the incident involve malicious intent?
- Did someone knowingly work impaired?
- Did someone knowingly do something wrong or unsafe?
- Would a person with identical training make the same mistake?
- Has someone demonstrated a history of adverse event involvement?

C. ERROR REPORTING

Collect data in a proactive but nonpunitive manner. Learn from errors to identify trends that may reveal problems with care. Two incentives for reporting include immunity and confidentiality. Effective reporting procedures help determine educational and training needs. Accurate reporting of errors can help identify policies needing revision. Reporting systems should collect information on healthcare providers. Consider voluntary reporting as a passive form of surveillance for near misses, close calls, and unsafe conditions. Active surveillance involves direct observation of providers or chart review using trigger tools.

An effective incident reporting and evaluation system should support privacy and receive information from a broad range of individuals. Reporting systems should also provide timely feedback and contain mechanisms that ensure evaluation and corrective action plan creation. The reporting process should promote a continuous flow of information into the system. Use reported information to assess and develop appropriate educational or training sessions.

1. Reporting Using Technology

Web-based systems can now receive information from electronic medical records. The advantages of voluntary event reporting systems include relative acceptability and involvement of frontline personnel. Ensure voluntary event-reporting systems remain confidential. Compared with medical record reviews or direct observations, event reports capture only a fraction of events. Physicians generally do not utilize voluntary reporting systems. Failure to receive feedback after reporting an event can create organizational problems. Incident reports should combine direct observations, use of trigger tools, and any chart audits. The Patient Safety and Quality Improvement Act legislation provides for the confidentiality and privilege protections for patient safety information when healthcare providers work with approved Patient Safety Organizations (PSOs). Healthcare providers may choose to work with a PSO and specify the scope and volume of patient safety information to share with a PSO. Hospitals should maintain confidential incident reporting systems.

BOX 4.4   METHODS USED TO IDENTIFY ADVERSE EVENTS

- Anonymous reporting systems
- Information from licensing and accreditation surveys
- Infection control surveillance
- Medication data mining
- Legal complaints and lawsuits
- Performance improvement data
- Retrospective clinical record review
- Information from patients and their families
- Satisfaction surveys

D.   ANALYZING EVENTS AND ERRORS

Individuals responsible for analyzing adverse events should recognize the human component of error. Strive to identify system and latent components of error. Conducting proper event analysis can help any organization plan for change and improves processes. Tracking changes confirms that implemented solutions indeed do work and helps determine the impact on risk reduction efforts. Use an independent process or function to analyze adverse events. Place an emphasis on using multidisciplinary approaches to help avoid the trap known as hindsight bias.

Conduct an analysis to determine which systems or redesign processes contribute to adverse events. Analysis of "near miss" events can help identify any trends or causal patterns that could contribute to future events. Focus on the following areas when analyzing an event: (1) defining objectives for supporting families, (2) understanding what happened, (3) identifying opportunities for improvement, and (4) incorporating learning into daily operations. Publish rules that address issues related to blame, transparency, confidentiality, and innovation. Create processes to document the sequence of events while uncovering opportunity for improvement.

E. COMMON ERROR CAUSAL FACTORS

Research and analysis reveal that several common causal factors contribute to medical errors and adverse events. Some studies reveal that poor communication among providers, caregivers, and other staff members contribute to patient-related incidents. Another important causal factor, often overlooked, relates to the unavailability of critical information during times of decision-making. The simple failure of personnel to follow established policies, guidelines, protocols, and processes also contributes to such events. Other common causal factors documented in reports and investigations include improper patient identification, incomplete patient assessment, failure to obtain informed consent, and lack of proper patient education. Healthcare organizations should place a priority and focus on improving systems to reduce medical errors. Leaders must realize that healthcare professionals can and do make mistakes. Undertaking proactive measures that seek to identify and intervene in potential system failures can reduce errors and adverse events. Implement proven surveillance methods or processes to identify challenges and problems that hinder patient safety. Identify and evaluate best practices for possible use at the facility. A spontaneous reporting system permits anyone in the hospital—clinician, employee, volunteer, patient, or family member—to report an issue, concern, or problem. Assess all information contained in risk management reports, quality improvement assessments, malpractice reports, and patient complaints.

F. MEDICAL EDUCATION AND ERROR ACKNOWLEDGMENT

Most institutions teach medical education in an authoritarian manner. Some healthcare professionals inadvertently and incorrectly promote the infallibility of medicine. Many other educators stress perfection to their medical students. Healthcare personnel sometimes perceive senior clinicians as right because of their experience or reputation. Some clinicians believe there is one right answer to every medical situation. Some others view professional overconfidence as being equal to competence. Errors don't truly happen but can equate to incompetence, negligence, or laziness. Errors can mark medical professionals and can result in shame among peers and other professionals. Medical decision-making always carries some degree of uncertainty. Good clinicians should act decisively but remain flexible to needed changes. Each decision should seek to achieve the goal while minimizing risk. This requires vigilance and alertness to prevent harm. Organizations must learn to report, investigate, and analyze errors and their causes. Stress the integration of patient safety education into daily practice and encourage behavioral change. Patient safety education should begin in medical and professional schools. Educational sessions should present topics on human factors, cultures, system methods, and proactive risk reduction.

V. SAFETY CULTURES

Healthcare management and advances in technology can unknowingly encourage a fragmentation of care. Traditional healthcare methods create hierarchy and confusion that can result in gaps in communication, knowledge, and processes. Sometimes a single individual can't overcome the various forces that exist in organizations. Recurring medical errors make up a large number of the reported adverse events. Healthcare should learn from recurrent errors by creating a genuine safety culture that thrives on transparency. Culture change can only happen when senior leaders communicate the important objective of harm-free care.

The concept of a safety culture originated outside of healthcare. Studies of high-reliability organizations show they attempt to minimize adverse events by maintaining a commitment to safety at all organizational functions and levels. Such a commitment provides a foundation for establishing a safety culture. Healthcare organizations must start acknowledging the high-risk nature of an organization's activities. Achieving consistently safe operations requires a blame-free environment where individuals report errors or near misses without any fear of reprimand or punishment. Established safety cultures promote collaboration across all levels, functions, and disciplines, which helps find solutions to patient safety challenges, risks, and problems. Safety culture development requires senior leaders to commit the necessary resources to address organizational concerns. Studies reveal a wide variation in perceptions of safety culture across organizations and job descriptions. Many issues can undermine healthcare safety culture development. Poor teamwork and lack of communication create a culture of low expectations.

The AHRQ offers Patient Safety Culture Surveys and a Safety Attitudes Questionnaire to assist organizations with evaluating individual units and even the entire organization. Improving safety cultures is a challenge for most organizations. However, teamwork training, executive walkrounds, and the establishment of unit-based safety teams help improve safety culture measurements. The culture of individual blame impairs the advancement of a safety culture. A just culture focuses on identifying and fixing systems issues that contribute to unsafe behaviors. A just culture distinguishes between human error and at-risk or reckless behavior. To improve any safety culture, organizations must find solutions for specific problems. Significant variations in safety cultures may exist within an organization. Perceptions can vary from high to low depending on the unit. Many variables can impact safety culture, including professional relationships and local situations. This can create a changing safety culture at micro-system levels. Finally, some key reasons for lack of a true safety culture are poor leadership, lack of teamwork, ineffective communication, low expectations, and lack of authority.

BOX 4.5 KEY FEATURES OF TRUE SAFETY CULTURES

- Acknowledgment of the high-risk nature of an organization's activities
- Determination to achieve consistently safe operations
- Creation of a blame-free environment where individuals report errors without fear
- Encouragement of collaboration to seek solutions to patient safety problems
- Organizational commitment of resources to address safety concerns

A. SAFETY CULTURE PERCEPTIONS

According to a study in the *Journal for Healthcare Quality*, hospital nursing measures, including staff turnover and workload, are associated with staff's perceptions of a safety culture. Researchers examined the relationship between staff perceptions of a safety culture and nursing-sensitive measures of hospital performance at nine California hospitals and 37 nursing units. The measures of skill mix, staff turnover, and workload intensity accounted for 22 percent to 45 percent of the variance in safety culture perceptions between units.

BOX 4.6 SUGGESTIONS FOR IMPROVING NURSING COMMUNICATION

- Discharge Rounds: Use this opportunity to meet with other team members to discuss the patient's progress.
- Hospitalists: Use these practitioners to help nurses address the patient's immediate needs and communicate information to attending physicians.
- Objective Data: When communicating with other team members, it is helpful if the nurse has objective data in hand to support the claim.
- Interact with Other Healthcare Professionals: Nurses should attempt to be present in the room during visits from other healthcare professionals.
- Computerized Documentation: This immediately available information allows other healthcare professionals to view the nurse's notes and objective data related to the patient's condition.
- Importance of Core Measures: The Joint Commission requires implementation of a series of core measures related to common diseases and conditions.
- Demonstrate Leadership: When working with others, nurses should keep in mind one common goal—improving the health of the patient.

B. Healthcare Bureaucratic Structure

Most hospitals function as a bureaucracy with rules and lines of authority. Leaders should learn to decentralize and promote a trusting culture. Culture comes from the sum of individual or group values, attitudes, perceptions, competencies, and patterns of behavior. Leaders should encourage input (voice) and participation (choice) from all organizational members. Elements necessary to build safety cultures include addressing the perceptions associated with teamwork and identifying safety norms and expected behaviors. Organizations should assess job satisfaction, evaluate senior management effectiveness, and recognize the reality of stress. Healthcare organizations should continually determine the adequacy of supervision, education, and training. Focus on evaluating reporting systems, communication effectiveness, and feedback that could help change wrong assumptions about causes of adverse events.

C. Proactive Organizations

Healthcare organizations undergoing change and transitioning to a proactive safety culture should consider a number of issues as vital to the success of the transition. True safety cultures should focus on establishing nonpunitive incident and close call reporting systems. Healthcare team members should receive education in problem-solving and decision-making techniques. Top leaders promote patient harm as untenable and encourage team members to voice any and all concerns. The organization places a strong focus on improving systems and processes. Never tolerate the blaming of individuals for system failures. Finally, all team members should acknowledge that no best safety model exists to address every issue that could arise in complex healthcare systems. Don't just promote vigilance—expect it.

D. Understanding Change

Many healthcare organizations incorrectly believe that change happens because someone in leadership deems it so. Organizational transition change can only take place by implementing transformational processes and not with transactional instruments. Transactional change instruments come in the form of memos, policies, letters, and directives. First-order change refers to transactional change and second-order change refers to transforming change. Those leading change should place people correctly in the organizational structure to execute a strategy of change. Many times senior leadership should use a *burning platform* to create a desire for change. Use a real event or incident to make the *burning platform* relevant to the need for change. Never forget to articulate how change will impact the staff, patients, families, visitors, and vendors. Ensure educational sessions for everyone focus on real world situations. Require good documentation practices, but place more emphasis on continuous learning. Define and describe in detail data collection, analysis, and dissemination processes. Plan, evaluate, and implement innovative processes using system methodologies. Change can't take place unless leaders integrate individual needs with organizational goals. Finally, leaders should learn to implement the culture change before attempting to promote other needed process or system interventions.

E. Organizational Change

Healthcare organizational leaders should expect hindrances to change processes. The science of medicine sometimes values individual thought processes of physicians over systematic change. Many physicians feel strongly about expressing their own style. Standardization of medical practices and treatment protocols offend many practicing healthcare professionals. Some professionals even reject basic teamwork principles. Some healthcare personnel express open unwillingness to accept imperfection. Education processes that do not address decision-making in the context of uncertainty can hinder any change process. Leaders should encourage some rational risk avoidance to minimize patient harm. Finally, the lack of creativity when encouraging change is often under-valued and frequently overlooked. Creativity refers simply to an organized way of thinking outside of the box. Organizational leaders must take action to help with the journey of change. Create teams that understand the importance of continuous improvement processes. Solicit suggestions from teams and other frontline or sharp-end personnel. Teamwork and change processes need effective communication and feedback. Ensure that all organizational members can participate in the change processes. Value their voice (opinions) and let them make some choices (decisions). Educate organizational members about the basic elements of system thinking and methods.

Educate and involve patients and their families in the change processes underway. Healthcare facilities undergoing change and transition to a true safety culture should consider the three over-lapping organizational levels impacted by change: (1) environmental, (2) organizational, and (3) the clinical interface between clinicians and patients.

Before embarking on the organizational change journey, the organization should assess the cur-rent climate. Do this through the use of interviews, focus groups, personal observations, and per-ception surveys. Surveys help establish a baseline measure of perceptions. In large organizations perceptions can easily become truths. Define in an honest manner the strengths and weaknesses of the organization. Find and use available tools to help with the assessment process. Search for the best tools to use by considering any available reliability or validity evidence. During assess-ment of the organizational climate, track innovations and interventions using the timeline approach. Determine adherence with policies, procedures, and directives during the journey.

F. TEAMWORK

Teamwork in medical systems requires coordination of efforts to achieve the desired outcome. The most successful teams cooperate and communicate effectively. AHRQ and the Department of Defense Military Health System developed TeamSTEPPS to provide evidence-based tools to promote teamwork. Teamwork tools should be designed for high-stress settings such as critical care units, emergency department operating rooms, and obstetrical suites. Tailor teamwork curriculums to specific settings or units. Patient safety depends on individuals with disparate roles and responsi-bilities acting together as a team. Communication barriers across hierarchies, failure to understand human fallibility, and poor situational awareness can result in poor teamwork. Teamwork training should focus on developing effective communication skills and creating an open team member atmosphere. Educate team members to cross-check actions of others, offer assistance as needed, and address errors in a nonjudgmental fashion. Debriefing and feedback remain key components of teamwork training.

VI. UNDERSTANDING SYSTEMS

IOM defines a system as many subsystems connected by a culture/mission. Healthcare should learn to respond to safety issues from a systems view. A good example is that patient safety begins before the patient arrives at the facility. Don't fail to search within the organization for people and resources to solve complex patient safety concerns. Leaders should not tolerate, excuse, or cover up poor decisions in a system. Hospitals function as open systems with some closed subsystems or micro-systems. Closed systems exist in high-reliability organizations that can control most outside threats to system operations. Leaders should never forget that modern healthcare delivery systems could experience stress and overload. Teamwork, communication, feedback, and coordi-nation breed success in any organization. Acknowledging the risk of failure remains an inherent element of complex systems with risk serving as the emerging concern. People cannot always see or know systems-related risks. Humans can fail when putting forth their best efforts and systems can also fail regardless of their design. Trained team members can recognize and compensate for system risks.

**BOX 4.7 COMMON ELEMENTS OF SYSTEM
FAILURES AND MEDICAL ERRORS**

- A complex system can break or fail and contribute to harm.
- A system or process may contain a latent defect before harm results.
- Humans develop behaviors to compensate for chronic system flaws.
- System errors can occur far from the sharp end.
- Current medical culture promotes individual accountability.
- People attempting to fix system errors may not see how it impacts patients.
- Seek to identify all contributing factors that could cause harm.
- Recognize that any part of the system can impact care.
- System problems can eventually cause harm.
- Leaders should understand that system problems and failures can impact care.

A. SYSTEMS APPROACH

The systems approach seeks to identify situations or factors likely to contribute to human error. James Reason's analysis of industrial accidents revealed that catastrophic safety failures almost never result from isolated errors committed by individuals. Most incidents result from smaller and multiple errors in components and environments with underlying system flaws. Reason's "Swiss Cheese Model" describes this phenomenon. Errors made by individuals can result in disastrous consequences due to flawed systems that are represented by the holes in the cheese. Reason believed human error would happen in complex systems. Striving for perfection or punishing individuals who make errors does not appreciably improve safety. A systems approach stresses efforts to catch or anticipate human errors before they occur. Reason used the terms *active errors* and *latent errors* to distinguish individual errors from system errors. Active errors almost always involve frontline personnel. They occur at the point of contact between a human and some element of a larger system. Latent errors occur due to failures of the organization or designs that allow inevitable active errors to cause harm. The terms *sharp end* and *blunt end* correspond to active error and latent error. The systems approach provides a framework for analysis of errors and efforts to improve safety.

B. KEY SYSTEM SAFETY ELEMENTS

Systems need a focus on standardization, simplification, and automation. Minimize fatigue, stress, and boredom of workers. Reduce reliance on human memory but promote vigilance. Encourage teamwork and improve reporting accuracy and timeliness. Complex systems need an enhanced information transfer process within the organization. Design equipment to reduce failures but always consider technology and the human interface during the design processes. Study history to ensure that patient safety continues to improve over time. Statistics can help measure the impact of interventions or innovations. Promote the fact that real continuous improvement processes shift the focus from an individual to the team.

C. SYSTEM RELIABILITY

Reliability refers to the probability that a system or process will consistently perform as designed. We should view this as the opposite of its rate of error or failure. Many reliable systems will not work effectively unless accepted in a just culture that values teamwork, communication, accountability, and learning from mistakes. Consider the following reasons that many organizations struggle with reliability issues: (1) current improvement trends focus on trying harder or paying more attention, (2) focusing on individual outcomes tends to exaggerate reliability, (3) clinical autonomy allows for wide performance margins, (4) failure to respond to errors in ways that prevent recurrence, and (5) processes rarely designed to meet specific goals or objectives.

D. Failure Mode and Effect Analysis

Failure Mode and Effect Analysis (FMEA) is a process that attempts to prospectively identify error-prone situations or failure modes within a specific process or system. FMEA begins with identifying all the steps that should take place for a given process to occur. Once mapping is complete, use FMEA to identify the ways in which each step can go wrong, the probability of detecting an error, and the consequences or impact of an undetected error. The estimates of the likelihood of a particular process failure, the chance of detecting such a failure, and its impact can combine to produce a criticality index. This criticality index provides a rough quantitative estimate of the magnitude of hazard posed by each step in a high-risk process. Assigning a criticality index to each step allows prioritization of those elements targeted for correction/improvement. A FMEA analysis of the medication-dispensing process in a hospital would break down all steps from receipt of orders in the central pharmacy to filling automated dispensing machines by pharmacy technicians. Each step in this process would use an assigned probability of failure and a score or rating value.

E. Technology and Safety

Technology contains four common pitfalls: (1) poor design, (2) lack of technology interface with the patient or environment, (3) inadequate implementation plans, and (4) inadequate maintenance. Nursing intuition now relies on technology to detect physical changes in patient conditions. While technology possesses the potential to improve care, it can also create risks. Technology is part of the problem and part of the solution for safer healthcare. Nurses and other care providers can become so focused on data from monitors that they fail to detect subtle patient changes. Patient care technologies of interest to nurses range from relatively simple devices, such as catheters and syringes, to highly complex devices, such as barcode medication administration systems and electronic health records.

VII. PATIENT-CENTERED HEALTHCARE

Elements of patient-centered care include patient preferences, patient needs, and patient values. Integrated performance consists of clinical performance, operational performance, and financial performance. Restore patient and caregiver trust through organizational transparency.

BOX 4.8   INFORMED CONSENT ISSUES

- Execute the informed consent procedure and place in chart
- Document name and signature of the person who explained the procedure
- Provide informed consent forms written at a fourth-grade reading level
- Provide consent in primary language of patient (use interpreter if needed)
- Ensure the patient/legal surrogate recounts the information presented

---

BOX 4.9   DOCUMENTING INFORMED CONSENT

- Patient name, hospital, and medical procedure
- Practitioners' names
- Risks including alternative procedures and treatments
- Signatures of patient or legal guardian and date/time of consent
- Statements that procedure was explained to patient or guardian
- Signature and designation of person witnessing the consent

A. ERROR DISCLOSURE

Surveys can help to define the components of disclosure that matter most to patients and their families: (1) disclosure of all harmful errors, (2) an explanation as to why the error occurred, (3) how to minimize the error's effects, and (4) steps the physician and organization will take to prevent recurrences. Full disclosure of an error incorporates these components as well as acknowledgement of responsibility and an apology by the physician. Many physicians choose their words carefully by failing to clearly explain the error or its effects on the patient's health. Circumstances surrounding an error can become complex. Physicians may not know how much information to disclose and how to explain the error to the patient. Recently developed guidelines should assist physicians with this process. Since 2001, the Joint Commission requires disclosure of unanticipated outcomes of care. In 2006, the National Quality Forum endorsed full disclosure of *serious unanticipated outcomes* as one of its 30 *safe practices* for healthcare.

BOX 4.10   KEY ERROR DISCLOSURE ISSUES

- Acknowledge what happened, how it happened, and seek the why
- Determine what actions would reduce recurrence
- Make a statement that an immediate analysis of the errors will take place
- Express sympathy and compassion
- Inform the patient, representative, or family about social services
- Basic elements of medical disclosure include moral, ethical, and legal
- Reluctance to admit mistakes provides the greatest barrier to physician disclosure

---

B. MEDICAL ERROR DISCLOSURE GUIDANCE

The American College of Physicians and American Society of Internal Medicine suggests disclosing an error if material to a patient's well-being. The AMA advises error disclosure whenever major medical complications occur. The Joint Commission criterion is for any unanticipated outcome. The NPSF bases the disclosure threshold on any injury occurrence. Healthcare organizations must seek to use a predetermined error threshold. Providing proper disclosure also results in an opportunity for learning for both the organization and the healthcare professional. Disclosure enables the patient to obtain appropriate treatment and also gain an understanding of what happened. Disclosing errors and harm does not result in increased litigation. The risk of litigation increases when patients or families sense deception. Some clinicians believe that medical errors or unanticipated outcomes can appear as a negligent disclosure of error or the error could create liability risks.

BOX 4.11   ADVANCE PATIENT DIRECTIVES

- Honors a patient's wish regarding end of life issues
- Helps avoidance of discomfort with the preservation of individual dignity
- Provides for effective pain relief and appropriate emotional support
- Prevents inappropriate intrusive medical assessments or interventions
- Permits family and significant personal support including spiritual care
- Details procedures for palliative medicine or hospice consultation
- Informs staff about decisions regarding the donation of organs and tissues

C.   EVIDENCE-BASED MEDICINE

Evidence-based medicine (EBM) avoids reliance on instincts and experiences. Specialties many times interpret evidence-based practices in a manner most appropriate to their areas and create inconsistency of systems.

D.   GENERAL PATIENT SAFETY PRACTICES

General patient safety practices can refer to any process or structure that reduce the probability of an adverse event occurring. A patient safety practice can address either clinical or nonclinical issues. Consider the following issues below that can impact patient safety efforts.

BOX 4.12   EXAMPLES OF HIGH-RISK MEDICAL PROCEDURES

- Medication management
- Blood and blood product use
- Restraint and seclusion use
- Behavior management and treatment
- Operative and other invasive procedures
- Resuscitation and its outcomes

BOX 4.13   KEY SENTINEL EVENT CAUSAL FACTORS

- Communication
- Orientation and training
- Patient assessment
- Availability of information

BOX 4.14   PRACTICE GUIDELINES

- Widely used to modify physician behavior
- Systematically developed to assist with decisions for specific clinical issues
- Affect both the process and the outcome of care
- Traditionally focused on ensuring a perceived standard of care
- Now emphasize good patient outcomes and safety

BOX 4.15   CRITICAL AND CLINICAL PATHWAYS

- Functions as a multi-disciplinary method way to provide care
- Targets the specific processes or sequences of care
- Integrated with local or national clinical practices
- Incorporates responsibilities of care providers with those of ancillary services

---

E.   USING PATIENT SAFETY CHECKLISTS

An "algorithmic" checklist in many clinical settings can prevent the overlooking of vital steps of a process. Schematic behavior relates to tasks performed reflexively, while attentive behaviors deal with tasks requiring active planning and problem-solving. Refer to errors associated with failures of schematic behavior as slips. This type of error occurs due to lapses in concentration, distractions, or fatigue. Attentive behavior errors refer to mistakes and frequently result from lack of experience or insufficient training. Checklists can help reduce risks associated with risk of central line bloodstream infections. Using demand response checklists for inducing anesthesia, taking surgical timeouts, and transferring a patient out of surgery can help reduce adverse patient events. View checklists as useful tools that can improve patient safety but work to understand their limitations. Checklist usage may require certain co-interventions by the medical staff and caregivers to maximize their impact. Checklists can help prevent errors during clinical tasks that require attentive behaviors.

VIII. HUMAN FACTORS

Human factors engineering functions as a broad discipline that considers human strengths and limitations when designing interactive processes and systems. Human factors science addresses issues that involve people, tools, technology, and work environments. A human factors professional examines an activity in terms of its component tasks. Important considerations include the following: (1) physical demands, (2) skill requirements, (3) mental workload, (4) team dynamics, (5) elements in the work environment, and (6) device design required to complete the task optimally and safely. Human factor assessments focus on how systems operate with fallible humans at the controls. Human factors engineers test new systems and/or equipment under real-world conditions when possible. This real-world testing helps identify unintended consequences of new technologies.

Some key human factor issues remain overlooked or ignored by healthcare leaders. A continuous improvement process always shifts the focus from individuals to teams, processes, or systems. The concept known as migrating decision-making permits a person with the greatest expertise (regardless of rank) to make an important decision. Finally, managing complexity of patient care can and often does exceed the capabilities of individuals.

Three key elements of human factors science are (1) forcing functions, (2) standardization, and (3) resiliency efforts. The forcing functions of a system prevent an unintended or undesirable action from being performed or allow performance only if another specific action is performed first. Standardization increases system reliability, improves information flow, and minimizes cross-training needs. The use of checklists to ensure the performance of safety steps in the correct order can trace its roots to human factors and system methodologies. Resiliency efforts must also focus on design to preclude error. Resiliency approaches tap into the dynamic aspects of risk management and explore how organizations anticipate, adapt, or recover from system anomalies.

IX. PATIENT SAFETY OFFICERS AND COMMITTEES

The IOM recommends healthcare should establish a comprehensive patient safety function overseen/operated by trained personnel in a culture of safety. Each organization should designate a qualified person to serve as the Patient Safety Officer (PSO). An effective PSO must learn to promote action through the training of staff and implementation of proven error reduction methods. Develop a job description that assigns the duties of promoting a culture of safety. A PSO should contribute to the strategic planning through assessing, organizing, and managing organizational patient safety efforts. The appointed person should inspire others and possess effective communication skills. Patient safety officers should possess the healthcare experience and skills to collect and analyze data. Healthcare organizations could establish a committee or other process structure to help ensure patient safety success. Use an approach and structure for your facility that will be effective. Organizations can establish a patient safety committee or use existing committees depending upon the size and resources of the organization. The primary mission should focus on improving communication and care outcomes while promoting better coordination among other functions or departments.

X. QUALITY IMPROVEMENT

As addressed previously, IOM identified six aims of healthcare: (1) effective, (2) safe, (3) patient-centered, (4) timely, (5) efficient, and (6) equitable. The aims of effectiveness and safety target process-of-care measures. Since errors result from system or process failures, adopting various process-improvement techniques can help identify inefficiencies, ineffective care, and preventable errors. Make the identified changes to the associated processes or systems. The complexity of healthcare delivery, the unpredictable nature of healthcare, and the occupational differentiation and interdependence among clinicians and systems make measuring quality difficult. One of the challenges in using measures in healthcare relates to attribution variability associated with high-level cognitive reasoning, discretionary decision-making, problem-solving, and experiential knowledge. Another measurement challenge relates to reporting and tracking near miss incidents. These types of events, under the right circumstances, could produce harm.

A. BENCHMARKING

Define benchmarking in healthcare as the continual and collaborative discipline of measuring and comparing the results of key work processes with those of the best performers in evaluating organizational performance. We can use two types of benchmarking to evaluate patient safety and quality performance. Use internal benchmarking to identify best practices within an organization, to compare best practices within the organization, and to compare current practice over time. Use competitive or external benchmarking to judge performance and identify improvements proven successful in other organizations.

B. OTHER MEASUREMENTS

Structure measures assess the accessibility, availability, and quality of resources, such as health insurance, bed capacity of a hospital, and number of nurses with advanced training. Process measures assess the delivery of healthcare services by clinicians and providers. Outcome measures indicate the final result of care. Many useful measures can apply to different settings. Without commitment and support of senior-level leadership, even the best intended projects could fail. Champions of quality initiatives and improvement processes need to be visible throughout the organization, but especially in leadership positions and on the team.

C. STAKEHOLDERS

When addressing quality improvement efforts we need to recognize the needs of patients, insurers, regulators, patients, and staff. There exists a need to identify priorities for improvement and meet the competing needs of stakeholders. Determine the threshold of variation needed to produce regular desired results. Using a bottom-up approach to changing clinical practice can prove successful if senior leadership is supportive and the organizational culture supports change. Whatever the term or acronym of the method/tool used, the important component of quality improvement is a dynamic of the process that often employs multiple tools.

D. PLAN-DO-STUDY-ACT

The majority of quality improvement efforts using Plan-Do-Study-Act (PDSA) find greater success by using a series of small and rapid cycles to achieve the goals for the intervention. This helps implement the initiative gradually and allows the team to make changes early in the process. The ability of the team to successfully use PDSA improves greatly after a team receives instruction and training on the process. Using feedback and information from baseline measurements, meetings, and collaborating with others helps achieve goals. Some teams experience difficulty in using rapid-cycle change processes, collecting data, and constructing run charts. However, applying simple rules of the PDSA cycle can generate success when used in complex systems.

E. ROOT CAUSE ANALYSIS

Root cause analysis (RCA) should attempt to discover and analyze causal factors that feed surface problems. After defining the problem or issue, the RCA team attempts to identify, group, and analyze both surface and below the surface causal factors. The interface between humans and complex systems that contain hidden factors results in adverse events. The following categories of causal factors commonly arise during healthcare root cause sessions: (1) organizational cultures, (2) management deficiencies, (3) work environments, (4) team effectiveness, (5) staffing, (6) task-related issues, and (7) patient issues. Many investigations miss root causes and focus on finding fault. Placing blame can hinder investigations or analysis processes. RCA teams need the knowledge and tools to help them systematically organize and analyze collected data. Teams should understand what happened before they attack the why of an event. Sentinel events and other major patient incidents can relate to multiple causes that require systematic analysis. Many healthcare organizations fail to use creative synergistic techniques to identify and correct system root causes. A simple tool that can aid root cause analysis is the "What, Why, Why, Why" process. Another important tool is asking probing or open-ended questions. Never ask questions that require a yes or no response. Use the following basic causal factor grouping scheme when categorizing early in the RCA process: (1) organizational factors, (2) operational factors, and (3) motivation factors.

F. PATIENT ROLES IN PATIENT SAFETY

Efforts to engage should focus on encouraging patients and their families to become actively involved in the care process. Hospital studies reveal that patients often report errors that were not detected through traditional mechanisms such as chart reviews. The AHRQ Fact Sheet "20 Tips to Help Prevent Medical Errors" and the Joint Commission "Speak Up" promotion help educate patients about safety hazards and provide specific questions that patients can ask regarding their safety. The level of patient and family participation remain very difficult to predict. Patients and caregivers already shoulder a significant emotional burden for ensuring safety while hospitalized. Engaging patients in error prevention simply shifts responsibility for safety from providers and institutions to the patients themselves. Patients may also contribute or cause errors. Patient errors occur due to the difficulties inherent in an individual's interaction with a complex medical system. Patient engagement in safety efforts is a strong priority of influential regulatory and governmental organizations.

XI. IMPROVING PATIENT SAFETY

A. WRONG SURGERIES

Few medical errors provide more terror for patients than those experiencing surgery on the wrong body part, undergoing the incorrect procedure, or receiving a procedure intended for another patient. These "wrong-site, wrong-procedure, wrong-patient errors" (WSPEs)—rightly termed "never events"—should never occur. Wrong-site surgery may involve operating on the wrong level of the spine, a common mistake for neurosurgeons. A classic case of wrong-patient surgery involved a patient who underwent a cardiac procedure for another patient with a similar last name. According to AHRQ, a seminal study estimated that these errors occur in approximately 1 of 112,000 surgical procedures. A study using Veterans Affairs data found that fully half of WSPEs occurred during procedures outside of the operating room. Early efforts to prevent WSPEs focused on developing redundant mechanisms for identifying the *correct* site, procedure, and patient. These procedures included "sign your site" initiatives. It soon became clear that even this simple intervention posed problems. Site-marking protocols did increase use of preoperative marking but implementation and adherence differed significantly across surgical specialties and hospitals. In some instances confusion occurred about whether the marked site indicated the area needing surgery or the area to avoid. However, site marking remains a core component of the Joint Commission's Universal Protocol. Root cause analyses of WSPEs reveals that communication-related failure continues as the prominent underlying factor. The use of a surgical timeout or a planned pause to review important aspects of a procedure with all involved personnel improves communication in the operating room and helps prevent WSPEs. The Universal Protocol also specifies use of a timeout prior to all procedures. Although initially designed for operating room procedures, an organization should use timeouts before any invasive procedure. The use of checklists can improve surgical and postoperative safety. The incidence rate makes it difficult to establish any single intervention that can reduce or eliminate WSPEs. Preventing WSPEs depends on the use of system solutions such as strong teamwork, establishing a safety culture, and individual vigilance. The National Quality Forum considers wrong-patient, wrong-site, and wrong-procedure errors as never events. The Joint Commission treats such errors as Sentinel Events. In February 2009, the Centers for Medicare and Medicaid Services (CMS) announced that hospitals lose reimbursement for any costs associated with WSPEs.

B. Diagnostic Errors

Cognitive psychology addresses several types of errors that clinicians make due to anchoring errors or premature closure that relied on initial diagnostic information. Framing errors occur when providers make biased diagnostic decisions based on subtle cues and collateral information. Blind obedience errors result when placing undue reliance on clinical test results or expert opinions, which can delay the proper diagnosis. Biases on the part of individual clinicians do play roles in many diagnostic errors. Underlying healthcare system problems also contribute to many missed or delayed diagnoses. Many diagnostic errors occur in primary care, pediatrics, emergency medicine, and surgery. Poor teamwork and communication among clinicians can contribute to diagnostic errors. Preventing some diagnostic errors would mitigate the effect of these biases and provide physicians with more objective information for their decision-making. Many clinicians remain unaware of diagnostic errors they committed. No hard evidence indicates that computerized diagnostic decision support improves overall diagnostic accuracy. Information technology can improve clinicians' ability to follow up on diagnostic tests in a timely fashion. Structured protocols for telephone triage, teamwork training, communication improvement, and increased supervision of trainees could help improve diagnostic performance. A key goal should encourage clinicians to reflect about their personal thinking processes. A recent commentary termed diagnostic error "the next frontier for patient safety" and called for more research into solutions for individual and systems causes of diagnostic error.

C. Display Confusion

Displays are built into many medical devices to convey a wide range of sometimes critical data.

However, some displays are ambiguous or counterintuitive. This introduces a particularly insidious problem, since treatment decisions will be based on this information—these decisions could be perfectly reasonable and yet wrong.

To reduce this risk, consider the following:

- Device displays should be assessed during pre-purchase trials and demonstrations, and an eye kept open for those that pose a real risk of confusion.
- Alert manufacturers if problems appear after the device has been put into use.
- Users must understand how to interpret the misleading display if the manufacturer cannot provide an immediate solution and the device cannot be removed from service.

D. Air Embolism from Contrast Media Injectors

Angiography—the x-ray imaging of blood vessels—involves injecting contrast media into a patient's vasculature. Power contrast media injectors can replace handheld syringes to improve the control and precision of injection, but the procedure still creates the risk of injecting air, potentially resulting in a fatal embolism. Most power injectors are equipped with safety features to reduce the risk of air embolism, including systems that detect air in the injection line or the presence of a used media syringe (which contains only air). However, none of these features is foolproof. There are still reports of air embolism associated with contrast media injectors. For example, in the United States, a review of reports submitted to the Food and Drug Administration (FDA) over the past 10 years revealed 32 such cases—3 of them fatal. To reduce the risk of embolism, facilities using contrast media injectors should

- Ensure current safety features are installed and enabled on injectors
- Alert users to the fact that, while safety features are built in, vigilance is still the best defense
- Establish standard protocols that define who is responsible for specific tasks—such as tubing checks—so no one mistakenly assumes someone else will perform them
- Ensure those preparing injectors are trained on the specific models being used and only disposable items that are labeled as being compatible with those models are used
- Require a second clinician to verify that injection tubing is tight and leak-free following air purging, and that there is no contrast medium on the outside of the tubing, since this can interfere with air-detection systems
- Ensure clinicians inspect tubing and injectors for air bubbles before injection

E. Serious Reportable Events

The National Quality Forum (NQF) is a nonprofit organization that strives to improve the quality of American healthcare by establishing goals for performance improvement, endorsing national standards for measuring and reporting on performance, and promoting the attainment of national safety goals through education. The NQF Board recently approved a list of 29 serious reportable events (SREs) in healthcare in their 2011 Consensus Report. Of these 29 events, 25 were updated from 2006, and 4 new events were added to the list (NFQ, 2011). This newly expanded list of serious reportable events (never events) provides healthcare professionals with an opportunity to improve patient safety. The term refers to particularly shocking medical errors, such as a wrong-site surgery, that should never occur. The list includes identifiable and measurable events, and is grouped into seven categorical events: surgical, product or device, patient protection, care management, environmental, radiologic, and potential criminal. Refer to the current list of SREs on the NQF website. According to the NQF (2011), over 50 percent of states currently use the NQF-endorsed list of SREs in their public reporting programs.

F. Handoffs

Handoffs happen frequently and studies link them to adverse events occurring in settings ranging from emergency departments to intensive care units. Guidelines for safe handoffs focus on standardizing policies. The components of a safe and effective sign-out include the following: (1) accurate administrative data such as patient name, medical record information/number, and location, (2) update of any new clinical information, (3) covering provider clearly explaining needed tasks, (4) communication of illness severity, and (5) outline of contingency plans for changes in clinical status to assist cross-coverage management of the patient overnight. The Joint Commission requires that all healthcare providers implement a standardized approach to handoff communications including an opportunity to ask and respond to questions.

1. Implementing Structured Handoff and Sign-Out Procedures

Refer to the process of transferring responsibility for care by using the term handoff and use the term sign-out when referring to an act of transmitting information about a patient. Studies reveal a link between sign-outs and adverse event occurrence. Communication failures among providers can contribute to preventable errors, according to studies of closed malpractice claims affecting emergency physicians and trainees. The simple task of communicating an accurate medication list can help prevent errors. Hospitals must reconcile medications across the continuum of care. Ensure the accuracy of the following:

- Administrative data such as patient's name, medical record number, and location
- Updating of new clinical information
- Clear explanation of any tasks performed by the covering provider
- Communication of illness severity
- Outline contingency plans for changes in clinical status to assist cross-coverage management of a patient overnight

The 2006 Joint Commission National Patient Safety Goal 2E required implementation of a standardized approach to handoff communications including an opportunity to ask and respond to questions. The Joint Commission National Patient Safety Goal also contains specific guidelines for the handoff process:

- Use of interactive communications
- Provide up-to-date and accurate information
- Limit the number of interruptions
- Establish a process for verification
- Provide the opportunity to review any relevant historical data

Note: The Accreditation Council for Graduate Medical Education also requires that residency programs maintain formal educational sessions to address handoffs and care transitions.

G. Foreign Bodies

Safety agencies frequently receive reports of foreign bodies left inside patients. A retained device is any item inadvertently left behind in a patient's body and is usually associated with surgery, where objects such as sponges and clamps may become hidden by tissue. The other risk relates to unretrieved fragments of a device that breaks away and remains inside the patient. Many times clinicians fail to notice it and at other times a decision is made to leave it because its location makes retrieval too risky. Accidental retention may lead to serious infection or damage to the surrounding tissue. If a patient subsequently undergoes an MRI scan, the retained metal can heat or migrate, resulting in burns or worse. To reduce the risk of object retention, visually inspect devices before use and if a device appears damaged, immediately remove it from service. Be alert for resistance during device removal, which could indicate that the device is trapped and at risk of breakage. Consider the options—such as repositioning the patient—before continuing. Visually inspect devices as soon as they are removed from the patient and follow accepted surgical count procedures. Systems to locate retained surgical sponges before completing a procedure are available, and similar technologies for other devices and fragments may eventually be introduced.

H. Surgical Fires

Surgical fires do not happen often, but patients can be seriously injured or killed when they do. Most can be avoided if surgical personnel learn to recognize and control the elements that combine to cause fires. Ensure devices that could serve as ignition sources—most commonly electrosurgical units, electro-cautery devices and lasers—are in good condition and being used properly. Also make sure disposable components, which can stay hot for some time after use, are discarded safely. Fires are more likely in the presence of supplemental oxygen. To reduce oxygen enrichment, arrange surgical drapes to prevent the pooling of oxygen. Keep the amount of oxygen used to the minimum necessary and begin limiting the administration of oxygen at least 30 seconds before using an ignition source. Potential fuel sources such as fenestration towels and gowns must be kept as far as possible from the ignition source. If an item cannot be moved, find another way to limit risk posed by it. Allow alcohol-based preparations to evaporate fully before electrosurgery is started, and moisten sponges to reduce their flammability.

In addition, surgical booms housing electrical cables and oxygen hoses should be inspected regularly.

I. Anesthesia Hazards due to Inadequate Inspection

Regular reports occur each year that indicate the discovery of serious problems with anesthesia equipment just before use. These problems can include misconnected breathing circuits, ventilator leaks, and empty gas cylinders. Many problems go unnoticed until the patient becomes seriously or even fatally injured. Ensure inspections of anesthesia systems meet standards as prescribed by the manufacturer for that model. Inspections involve not only the anesthesia unit but also other devices and accessories that may not be specified in the procedure for the unit, such as scavenging equipment and manual resuscitators. Keep documentation easily accessible and if a checklist is required, ensure it is physically attached to the anesthesia unit. Ensure individuals responsible for performing the inspections are familiar with the procedure and understand its importance.

J. Medical Device Alarm Safety

Many medical devices contain alarms including electrocardiogram machines, pulse devices, and blood pressure monitors. Other devices include bedside telemetry, central station monitors, infusion pumps, and ventilators. These alarm-equipped devices help provide safe care to patients. Alarms can produce similar sounds and can cause confusion. Alarm signals per patient per day can reach several hundred, depending on the unit within the hospital. Most alarms do not require clinical intervention. Clinical personnel can become desensitized and can suffer a type of fatigue. Major patient safety events can occur and the following alarm issues can contribute to the incident:

- Absent or inadequate alarm system or improper alarm settings
- Alarm signals not audible in all areas
- Alarm signals inappropriately turned off
- Equipment malfunctions and failures

ECRI Institute continues to report the dangers related to alarm systems. The FDA Manufacturer and User Facility Device Experience database revealed that more than 550 alarm-related patient deaths were reported between January 2005 and June 2010. Consider the following recommendations aimed at reducing adverse events related to alarms:

- Establish a process for safe alarm management and response in high risk areas as identified by the organization.
- Conduct and document an inventory of alarm-equipped medical devices used in high-risk areas and for high-risk clinical conditions.
- Establish guidelines for alarm settings on alarm-equipped medical devices used in high-risk areas and for high-risk clinical conditions.
- Develop procedures for tailoring alarm settings and limits for individual patients.

- Inspect, check, and maintain alarm-equipped devices to provide for accurate and appropriate alarm settings, proper operation, and detectability.
- Provide education and training for all clinical care staff on the processes for safe alarm management and response in high-risk areas.
- Create a cross-disciplinary team that includes representation from clinicians, clinical engineering, information technology, and risk management, to address alarm safety.
- Review trends and patterns in alarm-related events to identify opportunities for improvement.
- Report or share information about alarm-related incidents with appropriate organizations such as ECRI Institute, FDA, the Association for the Advancement of Medical Instrumentation (AAMI), and the Joint Commission.
- Ensure devices handle alarms in a way that is logical, safe, and consistent with their facility's practice, and that they limit false or excessive alarms, which can desensitize staff.
- Ensure that alarm conditions are quickly and consistently conveyed and factors such as volume, floor layout, and physical distance do not prevent them from being heard. Ensure that visual alarm indicators can be seen.
- Ensure staff understand the purpose and significance of alarms and know how to set alarm limits to appropriate, meaningful values. Low-saturation alarms on pulse oximetry monitors and low minute-volume or high peak-pressure alarms on ventilators are regular subjects of this sort of error.

K. DISCHARGED PATIENT EVENTS

Studies suggest that one in five patients experience adverse events within three weeks of discharge from the hospital. Most were drug-related incidents. Minimizing post-discharge adverse events should become a patient safety priority. Patient safety begins before a patient arrives and continues after discharge. The transition of care process contributes to most of the events. Lack of continuity and poor communication among inpatient and outpatient providers occurs too frequently. Inadequate medication reconciliation results in an increased risk of adverse drug events. Low health literacy contributes to the problem. Hospitals should conduct a thorough assessment of discharged patients and their ability to care for themselves after release. The segmentation or fragmentation of care also hinders a hospital's incentive to improve discharge processes. Healthcare organizations should develop systematic discharge processes that focus on medication reconciliation, discharge communication, and education of patients and their families on diagnoses and follow-up requirements. Use trained staff to meet with patients before and sometimes after discharge to address medication reconciliation, self-care instructions, and how to facilitate communication with outpatient physicians.

XII. HEALTHCARE-ASSOCIATED INFECTIONS

According to the CDC, almost 1.7 million hospital-acquired infections (HAIs) occur yearly, contributing to approximately 99,000 deaths. Such infections were long accepted by clinicians as an inevitable hazard. Recent efforts demonstrate that simple measures can prevent the majority of common infections. Hospitals and providers must work to reduce the burden of these infections. Four specific infections account for more than 80 percent of all hospital-related infections. They are surgical site infections, catheter-associated urinary tract infections, central venous catheter–related bloodstream infections, and ventilator-associated pneumonia. Preventing the transmission of antibiotic-resistant bacteria such as methicillin-resistant Staphylococcus aureus (MRSA) remains an important infection control priority. Effective measures exist to prevent the most common healthcare-related infections.

A. CENTRAL VENOUS CATHETER–RELATED BLOODSTREAM INFECTIONS

Employ maximal sterile barrier precautions. Use aseptic techniques including the use of a cap, mask, sterile gown, sterile gloves, and a large sterile sheet for the insertion of all central venous catheters (CVCs). Use 2% chlorhexidine gluconate solution for skin sterilization at the CVC insertion site. Avoid the femoral site for nonemergency CVC insertion and ensure prompt removal of unnecessary catheters.

B. SURGICAL SITE INFECTION

Ensure administration of appropriate prophylactic antibiotic, generally begun within 1 hour before skin incision and discontinued within 24 hours. Avoid shaving of the operative site and use clippers or other methods for hair removal in the area of skin incision(s). Ensure maintenance of blood glucose less than 150 mg/dL during postoperative period. Use tighter controls when needed in specific patient populations.

C. VENTILATOR-ASSOCIATED PNEUMONIA

Ensure elevation of the head of the bed is 30 to 45 degrees for all mechanically ventilated patients. Minimize duration of mechanical ventilation by minimizing sedative administration (including daily "sedation holidays") and/or using protocol-based weaning.

D. CATHETER-ASSOCIATED URINARY TRACT INFECTION

Ensure use of skin antisepsis at insertion and proper aseptic technique for maintenance of catheter and drainage bag, and the use of closed urinary drainage system. Ensure removal of urinary catheter when no longer essential for care.

XIII. MEDICATION SAFETY

Medications include prescriptions, samples, herbal remedies, vitamins, over-the-counter drugs, vaccines, diagnostic drugs, and contrast agents used on/administered to persons to diagnose, treat, or prevent disease. The list includes radioactive medications, respiratory therapy treatments, blood derivatives, intravenous solutions, and any product designated by the FDA as a drug. The definition of medication does not include enteral nutrition solutions, oxygen, and other medical gases. Consider medication management as an important component in the palliative, symptomatic, and curative treatment of many diseases or conditions.

A. MEDICATION ERRORS

Clinicians must deal with more than 10,000 prescription medications. One-third of adults in the United States take five or more medications. Patients admitted to a hospital commonly receive new medications or incur changes to their existing medications. Hospital-based clinicians also may not access a patient's complete medication list or remain unaware of recent medication changes. As a result, the new medication regimen prescribed at the time of discharge may inadvertently omit needed medications, unnecessarily duplicate existing therapies, or contain incorrect dosages. Such unintended inconsistencies in medication regimens may occur at any point of transition in care, such as transfer from an intensive care unit to a general nursing unit. Studies show that unintended medication discrepancies occur in nearly one-third of patients at admission, a similar proportion at the time of transfer from one site of care within a hospital, and in 14 percent of patients at hospital discharge. Medication reconciliation refers to the process of avoiding such inadvertent inconsistencies across transitions in care by reviewing the patient's complete medication regimen at the time of admission, transfer, and discharge and comparing it with the regimen being considered for the new setting of care. Though most often discussed in the hospital context, medication reconciliation can prove equally important in ambulatory care, as many patients receive prescriptions from more than one outpatient provider.

Advances in clinical therapeutics can result in major improvements in the health of patients. These benefits can become overshadowed by increased risks. An adverse drug event (ADE) is defined as harm experienced by a patient as a result of exposure to a medication, and ADEs account for nearly 700,000 emergency department visits and 100,000 hospitalizations each year. ADEs affect nearly 5 percent of hospitalized patients, making them one of the most common types of inpatient errors; ambulatory patients may experience ADEs at even higher rates.

As with the more general term adverse event, the occurrence of an ADE does not necessarily indicate an error or poor quality care. A *medication error* refers to an error (of commission or omission) at any step along the pathway that begins when a clinician prescribes a medication and ends when the patient actually receives the medication. Preventable adverse drug events result from a medication error that reaches the patient and causes any degree of harm. We can characterize medication errors that do not cause any harm because of interception before reaching the patient or by simple luck as potential ADEs. A certain percentage of patients can still experience ADEs from medications correctly prescribed and administered appropriately. Consider these events as adverse drug reactions or nonpreventable events. For example, the safe use of heparin requires weight-based dosing and frequent monitoring of tests of the blood's clotting ability. Taking these actions can help avoid either bleeding complications for high doses or clotting risks for low doses. Prescribing an incorrect dose of a medication would result in an error even if a pharmacist detected the mistake before the patient received the dose. If the incorrect dose gets dispensed and administered, but no clinical consequences occurred, that would still classify as a potential ADE.

B. RISK FACTORS FOR ADVERSE DRUG EVENTS

There exist patient-specific and drug-specific risk factors for adverse events. Older patients take more medications and can prove more vulnerable to specific medication adverse effects. Pediatric patients experience a more elevated risk, particularly when hospitalized due to poor weight dosing. Other well-documented patient-specific risk factors include limited health literacy and math ability. Ambulatory patient factors remain overlooked as an important source of adverse drug events. Studies show that both caregivers and patients can commit medication administration errors at surprisingly high rates. The ISMP maintains a list of high-alert medications—medications that can cause significant patient harm if used in error. These include medications with dangerous adverse effects, but also include lookalike, soundalike medications, with similar names and physical appearance but containing completely different pharmaceutical properties.

C. PREVENTION OF ADVERSE DRUG EVENTS

The pathway between a clinician's decision to prescribe a medication and the patient actually receiving the medication consists of several steps:

- Ordering: The clinician should select the appropriate medication and determine dose and frequency of administration.
- Transcribing: In a paper-based system, an intermediary (a clerk in the hospital setting, or a pharmacist or pharmacy technician in the outpatient setting) should read and interpret the prescription correctly.

- Dispensing: The pharmacist should check for drug–drug interactions and allergies, then release the appropriate quantity of the medication in the correct form.
- Administration: Supply the correct medication for administration to the correct patient at the correct time.

While the majority of errors likely occur at the prescribing and transcribing stages, medication administration errors do occur frequently in both inpatient and outpatient settings. Analysis of serious medication errors invariably reveals other underlying system flaws, such as human factors engineering issues and impaired safety culture, that allowed individual prescribing or administration errors to reach the patient and cause serious harm. Integration of information technology solutions, computerized provider order entry, and barcode medication administration into "closed-loop" medication systems holds great promise for improving medication safety in hospitals. Preventing ADEs remains a key priority for accrediting and regulatory agencies. The Partnership for Patients now includes ADE prevention as a patient safety improvement goal. The Partnership for Patients set a goal of reducing preventable ADEs in hospitalized patients by 50 percent.

D. ACCOMPLISHING MEDICATION RECONCILIATION

A 2012 systematic review of inpatient medication reconciliation studies did find some evidence supporting pharmacist-led medication reconciliation processes. However, the study did not reach any firm conclusions regarding the most effective strategies. As of July 2011, medication reconciliation became incorporated into National Patient Safety Goal 3, "Improving the safety of using medications." This National Patient Safety Goal requires that organizations "maintain and communicate accurate medication information" and "compare the medication information the patient brought to the hospital with the medications ordered for the patient by the hospital in order to identify and resolve discrepancies."

Medication reconciliation processes can help avoid inadvertent inconsistencies across transitions in care. Accomplish this by reviewing the patient's complete medication regimen at the time of admission, at transfer, and upon discharge to compare it with the regimen being considered for any new setting of care. Researchers continue to study a variety of methods including (1) pharmacists performing the entire process, (2) linking medication reconciliation to existing computerized provider order entry systems, and (3) integrating medication reconciliation within the electronic medical record system. In 2009, the Joint Commission announced that they would no longer formally score medication reconciliation during onsite accreditation surveys. This policy change occurred because of the lack of proven strategies for accomplishing medication reconciliation.

BOX 4.16 KEY MEDICATION RECONCILIATION SUGGESTIONS

- Identify all medications of patients being admitted
- Require the patient or a family member to validate the list if possible
- Compare admission orders with the pre-admission medication list
- Make the list readily available to prescribing professionals
- Provide reconciliation information to the next unit or patient care setting.
- Give the complete list of medications to the patient at discharge

---

E. MEDICATION ADMINISTRATION

Develop guidelines for staff members administering medications with or without supervision, consistent with law and regulation and organization policy. Address an individual's qualification to administer by medication, medication class, or route of administration. Provide guidelines for prescribing professional notification in the event of an adverse drug reaction or medication error. Identify the patient by using at least two individual identifiers excluding patient location. Verify the correct medication by reviewing the medication order and product label. Verify stability by conducting a visual examination for particulate matter or discoloration and check the medication expiration date.

Verify that no contraindication exists before administering the medication. Validate the medication administration time, prescribed dose, and correct administration route. Advise the patient or the patient's family about any potential adverse reactions. Discuss any significant concerns about the medication with the patient's physician or prescriber. Provide guidance and training to patients doing self-administration of drugs. Training topics should include how to administer, frequency, route of administration, and dosage. Educate caregivers about any possible side effects of the medications administered.

BOX 4.17   MEDICATION ADMINISTRATION SAFETY SUGGESTIONS

- Ensure guidelines are consistent with laws, regulations, and policies
- Address qualifications to administer by medication, class, or route
- Develop guidance for professional notification for an adverse drug event
- Identify patients using at least two individual identifiers excluding room
- Review orders and product labels, and conduct a visual exam of medications
- Check the medication expiration date
- Ensure no contraindication exists before administering
- Verify medication administration time, dose, and route
- Advise patient or the patient's family about any potential adverse reactions

---

BOX 4.18   HIGH-ALERT MEDICATION SAFETY SUGGESTIONS

- Identify all high-alert drugs available at the facility
- Implement processes to identify new medications for placement on the list
- Develop guidelines, dosing scales, and checklists for all high-alert drugs
- Implement a process to audit compliance with the protocols and guidelines

---

F.   REDUCING MEDICATION ERRORS

Many medication errors occur while communicating or transcribing medication orders. Take steps to reduce the potential for error or misinterpretation of written or verbal orders. The written policy should address the required elements of a complete medication order. Develop and publish a list of unacceptable abbreviations, symbols, acronyms, and dose information. Provide guidance on the use and acceptability of generic versus brand name drugs. Implement detailed policies for ordering drugs with look- or soundalike names. Post procedures for dealing with incomplete, illegible, or unclear orders.

BOX 4.19   MEDICATION ERROR CATEGORIES

- Failure to administer medication when required or as prescribed
- Administration of the medication at the wrong time or using an incorrect route
- Administration of the wrong dosage or concentration of a drug
- Administration of the wrong medication

(Continued)

> ### BOX 4.19   MEDICATION ERROR CATEGORIES (*Continued*)
> - Misunderstanding verbal/written medication orders including transcription
> - Administering medication to the wrong patient
> - Failure to read container labels and using improper injection techniques

Discourage the use of verbal and telephone orders. Implement a verification process for verbal or telephone orders. Create policies for implementing weight-based dosing.

G.   REPORTING MEDICATION ERRORS

Each organization should comply with internal and external reporting requirements. This may include notifying the United States Pharmacopeia (USP), FDA, or ISMP. Errors and adverse events may relate to professional practice, healthcare products, procedures, and systems. This can include prescribing, order communication, product labeling, packaging, nomenclature, compounding, dispensing, distribution, administration, education, monitoring, and usage. The FDA receives medication error reports on marketed human drugs including prescription drugs, generic drugs, and over-the-counter drugs, and nonvaccine biological products/devices. In 1992, the FDA began monitoring medication error reports forwarded to the FDA from the USP and ISMP. The FDA also reviews MedWatch reports for possible medication errors.

H.   INVESTIGATING MEDICATION ERRORS

The organization should designate a qualified person or department to conduct a thorough investigation to document all the facts. Investigations should seek to determine or document all facts surrounding the incident. Document all facts such as unit, time, date, and shift. Evaluate and determine staffing levels at the time of occurrence. Determine what other factors contributed to the event. Assess the legibility and accuracy of physician orders. Gather information on failure to follow safety precautions or other procedures. Ensure evaluation of facts by senior leaders, nurses, and pharmacy personnel. Document trends or patterns and implement corrective actions.

I.   COMPUTERIZED PROVIDER ORDER ENTRY

The basic steps of a Computerized Provider Order Entry (CPOE) system include (1) ordering appropriate medication, dose, and frequency of administration, (2) transcribing the order correctly and communicating accurately to the pharmacist, (3) dispensing, which requires the pharmacist to check for drug-drug interactions and allergies before releasing the appropriate quantity of the medication in the correct form; and (4) administration, which requires that the nurse should receive the medication and check for accuracy before giving it to the correct patient. CPOE refers to any system in which clinicians directly enter medication orders into a computer system. The system transmits the order directly to the pharmacy. A CPOE system does ensure standardized, legible, and complete orders, which can reduce errors at the ordering and transcribing stages. Other advantages include averting problems with similar drug names, drug interactions, and specification errors. Some unanticipated consequences of using CPOE systems include (1) workflow issues, (2) system demands, (3) changes in communication patterns and practices, (4) negative feelings toward the new technology, (5) unexpected changes in organizational power structure or culture, and (6) an overdependence on the technology. AHRQ and the National Quality Forum both recommend CPOE system use as one of the 30 "Safe Practices for Better Healthcare." The Leapfrog Group also recommends CPOE implementation as one of its first three recommended "leaps" for improving patient safety. A 2009 study found that only 17 percent of US hospitals used a CPOE system.

XIV. EMERGENCY DEPARTMENT PATIENT SAFETY

Emergency department errors can easily become less visible due to lack of feedback. Clinicians may never know if their decisions or actions were right or wrong. Errors can occur since responsibility can be diffused across shifts and teams. Patients can present symptoms or problems of varying acuity. Healthcare personnel may not possess all of the information about patients. Emergency department personnel can experience a high degree of uncertainty despite the urgency to provide medical intervention. Emergency departments face many distractions, including the need to multitask. The nature of the job, including shift work and informal communication networks, can hinder patient care.

Diagnostic errors or flawed decisions continue to commonly occur in emergency departments. Mistakes can occur for any of the following reasons: (1) knowledge gap or inexperience, (2) failure to recognize a disease pattern, and (3) misinterpretation or misapplication of diagnostic testing. High-risk times such as shift transitions, staffing issues, and patient transfers can also contribute to errors.

BOX 4.20   COMMON MALPRACTICE SITUATIONS IN EMERGENCY DEPARTMENTS

- Delay in treatment
- Missed fractures
- Wound care complications
- Abdominal or stomach pain
- Missed meningitis
- Spinal cord injury
- Subarachnoid hemorrhage
- Ectopic pregnancy

XV. OTHER KEY PATIENT SAFETY ISSUES

A. INFUSION PUMPS

With the advent of smart infusion pumps in the last decade, nurses are able to employ more advanced technology to manage medication administration. The increasingly sophisticated devices feature drug libraries or databases that hold the drug dosing information, including dosing limits, infusion parameters, and drug-specific advisories. Integrated software can calculate specific dose and delivery rates. Many smart infusion pumps are able to integrate with electronic health records, and some can become network devices and connect to wireless hospital networks. While smart pumps can help reduce medication errors and prevent patient injury, nurses must continue to follow standard safety precautions. The "five rights" of medication administration include the following: the right patient, the right drug, the right dose, the right route, and the right time. The next step in infusion safety, according to ECRI's recommendations, will involve integrating infusion pumps with electronic ordering, administration, and documentation systems, and fostering a shift in mindset from viewing infusion pumps as standalone devices to viewing them as components of an integrated medication delivery system. Following these protocols from the FDA may help prevent infusion pump errors:

- Before starting an infusion or changing an infusion setting, confirm that the infusion pump is programmed correctly.
- When infusing a high-risk medication, require a second clinician to perform an independent double check of infusion pump settings according to your facility's policy.
- If the infusion pumps at your facility contain a drug library feature, use it according to facility policy.

- When a patient is receiving multiple infusions, consider labeling the infusion pump channels and corresponding tubing with the name of the medication or fluid to avoid programming the wrong channel or infusion pump.
- Don't rely solely on the pump to identify problems. Monitor the patient and infusion according to nursing best practices and your facility's policies and procedures.
- Pay attention to displayed alerts and cautions, and investigate them appropriately.
- When tasking multidisciplinary teams to implement and revise smart pump technologies and drug libraries, ensure representation from all interfacing disciplines.
- Continuously reevaluate drug library settings and modify them to align with the standard of care and facility policies and procedures. This should include implementing and altering soft and hard limits (when clinically relevant) as well as standardizing concentrations, dosing configurations, and names of high-risk medications throughout your institution.
- Make sure you're properly educated to manage all infusion pumps used on your unit.
- Facility policies and procedures should be readily available, and clinicians should be promptly notified of changes or updates.

B. Ambulatory Care

The nature of interactions between patients and outpatient providers can also contribute to adverse events. Circumstances can limit face-to-face interaction between a provider and a patient. Patients then should assume a much greater role in and responsibility for managing their own health. This elevates the importance of patient education to ensure patients understand their illnesses and treatments. Medication errors can occur due to a patient's understanding of the indication, dosing schedule, proper administration, and potential side effects of a drug. Low health literacy and poor patient education contribute to increased error risks. Patients should understand how and when to contact their caregivers outside of routine appointments.

C. Work Hours and Patient Safety

In 2003, the Accreditation Council for Graduate Medical Education (ACGME) implemented new rules limiting work hours for all residents. Residents should work no more than 80 hours per week or 24 consecutive hours at a time. Do not schedule residents for on-call duty for more than every third night. Organizations should provide them one day off per week. A 2008 study of pediatric residents found no change in medication errors after implementation of the ACGME regulations.

1. ACGME 2010 Standards

The ACGME issued new standards for duty hours in the fall of 2010, which went into effect in July 2011. The new regulations concurred with the 2008 IOM report entitled *Resident Duty Hours: Enhancing Sleep, Supervision, and Safety*, by recommending the elimination of extended duration shifts. Early experiences with simulation successfully improved a resident's technical, cognitive, and teamwork skills. The new ACGME regulations do not recommend a significant reduction in overall weekly work hours from the present limit of 80.

D. Rapid Response Systems

Many hospitals now use rapid response teams as a patient safety intervention. Patients whose condition deteriorates acutely while hospitalized often exhibit warning signs in the hours before experiencing adverse clinical outcomes. A 2006 consensus conference advocated use of the term "rapid response system" (RRS) as a unifying term. Many physician hospitalists now assume RRS duties, either as the primary responder or to assist nurse-led teams. Many hospitals permit any staff member to call the team if one of the following criteria is met: (1) high or low heart rate, (2) high or low respiratory rate, (3) systolic blood pressure greater than 180 or less than 90, (4) low oxygen saturation, (5) acute change in mental status, (6) low urine output, or (7) a staff member expresses significant concern about the patient's condition.

E. CALL SYSTEM OPERATION

Consider call systems as patient safety tools. To promote legitimate use of call systems, patients and nurses need additional education. Nurses should know that call systems encourage patients not to do things for which they need help. Recent reports indicate that some nursing personnel may ignore calls during overnight shifts. Even though a call for help may seem to be a nuisance, nurses should be encouraged to consider what might happen if a patient attempted a potentially dangerous activity without assistance. Nursing personnel should explain to each patient the proper use of the call system and specify what the patient can do and should not attempt to do without assistance. Responding promptly and courteously to patient calls will encourage patients to use the system rather than attempt a dangerous activity. Locate call buttons within easy reach of the patient's bed. This action can help prevent major patient incidents.

F. HEALTH INSURANCE PORTABILITY AND ACCOUNTABILITY ACT

The Health Insurance Portability and Accountability Act (HIPAA) covers all healthcare organizations. This includes all healthcare providers, even single physician offices, health plans, employers, public health authorities, life insurers, clearinghouses, billing agencies, information systems vendors, service organizations, and universities. The goal of the act is to promote administrative simplification of healthcare transactions and to ensure the privacy and security of patient information. The act's transaction standards call for the use of common electronic claims standards, common code sets, and unique identifiers for all healthcare payers and providers. The security regulations of the act prescribe the administrative procedures and physical safeguards for ensuring the confidentiality and integrity of protected health information. The security regulations will provide a uniform level of protection of all health information housed or transmitted electronically. Patients retain the right to understand and control the use of their health information. Healthcare providers should provide patients with a clear written explanation of how they use, keep, and disclose patient information. Provide patients access to their medical records. The standard restricts the release of certain information without patient consent. The act prohibits coercion of patient consent. Patients retain recourse options when an organization violates their confidentiality. Protected patient information includes the following:

- Name and specific dates of birth, admission, discharge, or death
- Telephone numbers, Social Security number, and medical record number
- Photographs, city, zip code, and other geographic identifiers

G. PATIENT RESTRAINTS

We can define restraints as any manual method, physical device, or mechanical device used to restrict the freedom of movement or normal access to one's body. Due to an increasing number of reports of injury and death associated with the incorrect use of patient restraints, the FDA warns health professionals to ensure the safe use of these devices. Restraints can include safety vests, lap belts, wheelchair belts, and body holders. Incorrect use of these devices can involve using the wrong size for a patient's weight, errors in securing restraints, and inadequate patient monitoring. Such mistakes can result in fractures, burns, and strangulations. We can simply define a restraint as any manual method, physical device, mechanical device, material, or equipment attached or adjacent to a patient or resident's body that restricts freedom of movement or normal access to one's body. Under this functional definition, other devices or facility practices also may meet the definition of a restraint, such as tucking in a bed sheet so tightly that a patient's movement in or out of bed is restricted, or use of a specialty bed that limits a patient from voluntarily exiting from bed. This definition considers side rails, regardless of size, as restraints. The CMS issued regulations in 2006 clarifying under Medicare's Conditions of Participation for Hospitals who may order patient restraint or seclusion as a delegated responsibility.

The use of restraint should comply with the order of a physician or other licensed independent practitioner (LIP) permitted by the State to order a restraint. An LIP is any professional permitted by state law and hospital policy to order restraints and seclusion for patients independently, within the scope of the individual's license and consistent with the individually granted clinical privileges. Physicians must take individual accountability for the care of their patients. The physician can take action to delegate, or to withhold the delegation of, tasks or responsibilities, as he or she deems appropriate. CMS requires consultation with the attending physician as soon as possible if the attending physician did not order the restraint or seclusion. The final regulations clarify that physicians, PAs, and RNs can perform the face-to-face assessment within one hour. When using restraint or seclusion for the management of violent or self-destructive behaviors jeopardizing the physical safety of a patient, staff member, or others, see the patient within one hour after the initiation of the intervention. The regulations also require that if an RN or PA performs the assessment that they contact the attending physician with the results as soon as possible.

1. Most Medical/Surgical Restraints Exempt

Define a restraint as any manual method, physical or mechanical device, material, or equipment that immobilizes or reduces the ability of a patient to move his or her arms, legs, body, or head freely. Drugs also qualify when used as a restriction to manage the patient's behavior or restrict the patient's freedom of movement. Restraints used in medical or surgical care do not fall under the CMS rule. A restraint does not include devices, such as orthopedically prescribed devices, surgical dressings or bandages, protective helmets, or other methods that involve the physical holding of a patient for the purpose of conducting routine physical examinations or tests, or to protect the patient from falling out of bed, or to permit the patient to participate in activities without the risk of physical harm.

BOX 4.21   TYPES OF PATIENT RESTRAINTS

- Safety Bars: Used on wheelchairs to prevent falls.
- Soft Belts: Similar to seat belts; used to prevent falls from beds and wheelchairs.
- Safety Vest: Provides more support in preventing falls from a chair or bed.
- Wrist Restraint: A limb-holding restraint that prevents the patient from removing tubes or bandages. Note: Physically check every 15 minutes to ensure circulation.
- Mitt Restraint: This type of restraint restricts finger movement but permits movement of the arm and wrist.

Use data collection practices to measure and evaluate restraint-related injuries as well as the overall use of restraints. Use continuous quality improvement processes to monitor injury-related falls. Root cause analysis can help catalog factors contributing to serious falls. Provide staff with continuing education on fall risk assessment, interventions to prevent falls, proper application of restraint, monitoring of the restrained patient, methods to reduce restraint use, and documentation of these care practices. The use of physical restraints may occur only if documentation reflects the presence of a specific medical symptom warranting their use. Document the use of such a restraint to treat the symptom and how the restraint assists the patient in attaining or maintaining his or her highest level of physical, mental, and psychosocial well-being.

H. Patient Fall Prevention

Slips, trips, and falls represent the most common patient-related occurrences in healthcare facilities. Many healthcare workers also become slip, trip, and fall victims each year. Determine trends and problem areas by analyzing appropriate safety and risk data. Falls happen because people and conditions do not remain static. Nursing personnel should become actively involved in fall prevention efforts. Leaders must strive to increase awareness of fall hazards located in the unit or department. Care plans should contain information on each patient or long-term care resident to help minimize the risk of a fall. An effective process should identify, evaluate, and correct all hazards contributing to fall events. Educate caregivers about the physical hazards and behavioral aspects of fall prevention. Establish procedures for analyzing the trends and problems within the facility. Implement hazard surveillance strategies to identify and correct physical fall hazards. Place an emphasis on high-risk areas as determined by a statistical analysis of data. Conduct a quarterly assessment of effectiveness and an annual audit of all plan elements.

I. Environmental Hazards

The healthcare organization should maintain a safe environment for patients, visitors, and staff. Switches should be accessible to the patient upon entering the room and from the bed. Lights should shine bright enough to compensate for limited vision and for the activities performed in the room. Night-lights should be available in patient rooms, bathrooms, and hallways. Lighting should be adequate on stairs and hallways. Floor-level lighting should reduce glare. Secure handrails should be provided on both sides of the staircase. Steps can be painted or outlined for increased visibility and covered with nonslip material. Keep stairs clutter free and well maintained. Floors should never produce glare. Use nonskid wax surfaces and throw rugs with nonslip backing. Tape or tack down carpet edges. Some carpet patterns impair perception and may contribute to falls. Use visual warnings to alert patients about flooring changes and hallway turns. Design unobstructed pathways from bed to bathroom. Ensure the proper location and security of chairs, tables, nightstands, and over-the-bed tables. Make call buttons or bells easily accessible from the bed. Ensure bathroom doors permit easy wheelchair or walker passage. Ensure all tubs and showers contain nonskid strips or mats. Install grab bars securely to the walls and at a height to allow easy access. Place grab bars near tubs, showers, and toilets and ensure easy access to toilet seats.

J. Bed Safety

Between 1985 and January 1, 2009, 803 incidents of patients caught, trapped, entangled, or strangled in beds with rails were reported to the FDA. Of these reports, 480 people died, 138 had a non-fatal injury, and 185 were not injured because staff intervened. Most patients were frail, elderly, or confused. Carefully assess all patients experiencing problems with memory, sleeping, incontinence, pain, uncontrolled body movement, or who get out of bed. Assessment by the patient's healthcare team will help to determine how best to keep the patient safe. Never use or permit bed rails to become restraints. Regulatory agencies, healthcare organizations, product manufacturers, and advocacy groups encourage hospitals, nursing homes, and home care providers to assess patients' needs and to provide safe care without restraints.

Inspect all bed frames, side rails, and mattresses as part of the preventive maintenance procedures. Ensure proper bed alignment and that no gap exists wide enough to entrap a patient. Never replace mattresses and side rails with dimensions different than the original equipment supplied by the manufacturer. Check bedside rails for proper installation using manufacturer's instructions. Establish safety rules and procedures for patients considered at high risk for entrapment. Use bedside rail protectors to close off open spaces which could lead to entrapment. Never use bedside rails as a patient protective restraint. Use of restraints requires frequent monitoring and compliance with local, state, and federal regulations. The Safe Medical Device Act requires healthcare organizations to report bed-related incidents resulting in death or injury. Using a bed rail or other device to restrain the patient could place the patient's safety at risk. Any decision regarding bed rail use or removal from use should be made within the framework of an individual patient assessment. When deemed necessary, take steps to reduce known risks associated with the bed rail's use. Consider medical diagnosis, conditions, symptoms, and/or behavioral symptoms.

XVI. ELECTRONIC RECORDS

Many individuals view electronic medical records (EMRs) and electronic health records as identical. However, the EMR is the legal record of source data created in healthcare environments. EMRs create data used by EHRs. The EHR represents the capability to provide medical information among various stakeholders. EHRs rely upon information contained in EMRs. However, EMRs will also rely upon EHRs when compiling the historical patient treatment information. The National Institutes of Health defines an EHR as a longitudinal electronic record of patient health information generated as the result of receiving medical treatment in various settings. EHRs provide information about a patient's demographics, progress notes, problems, medications, vital signs, medical history, immunizations, laboratory data, and even radiology reports. The EHR helps automate and streamline a healthcare professional's workflow. It can help generate a complete record of a clinical patient encounter. An EHR is normally generated and maintained within a hospital, delivery network, clinic, or physician office improves the quality and timeliness of decision-making by providing nurses, physicians, and other clinicians. EHRs can provide comprehensive and up-to-date information and they provide a source of data for error reporting and analysis. Organizations should employ information technology that uses standards to support data interchange, medical terminologies, and knowledge transfers.

A. KEY COMPONENTS OF ELECTRONIC HEALTH RECORDS

Most commercial EHRs combine data from the large ancillary services, such as pharmacy, laboratory, and radiology, with various clinical care components such as nursing plans, medication administration records, and physician orders. The number of integrated components and features involved in any given situation would depend on the data structures and systems used. The EHR may import data from the ancillary systems via a custom interface or may provide interfaces that allow clinicians to access the silo systems through a portal. The EHR may incorporate only a few ancillaries.

B. ELECTRONIC CHARTING

Point-of-care clinical documentation solutions enable nurses to focus more on the important patient care tasks at hand and less on documentation. Access to electronic documentation at the bedside also streamlines the care process and assists all clinicians in making better patient care decisions. Critical patient data such as lab results and vital signs stay at the clinician's fingertips. Vital signs and other clinical data can be recorded electronically through interfaces to patient monitors, further streamlining the documentation process. Communication among caregivers is enhanced since patient data is now available in real time to any caregiver with access to the system. Bedside charting means a change from using a paper-based system accomplished by clerks to using technology for nurses and physicians to do real-time, electronic charting at the bedside. Some clinical studies did find that the use of electronic bedside patient charting could dramatically decrease medical error occurrence while improving outcomes and decreasing administrative costs. Due to high startup costs for implementing EMR systems, a significant number of hospitals, physicians' offices, and care clinics still use paper-based patient charting. Some hospitals with EMR systems still generally do not use mobile bedside charting technologies. Many hospitals now use workstations on wheels.

These electronic devices provide instant access to patient information. Caregivers can electronically assess, monitor, and chart patient care electronically. Some hospitals use coordinated systems that address patient care, medication verification, and computerized provider order entry.

XVII.   INFANT ABDUCTION PREVENTION

As part of contingency planning, every facility should develop a written protocol plan for infant abduction. Communicate this protocol to all staff members within the maternal child care unit. All departments, including plant operations, communications, switchboard operations, plant engineering, accounting, and public relations, should know the protocol. When formulating a protocol, facilities need to consider several items. For instance, the layout or schematics and traffic patterns differ among facilities. Ask law enforcement to use crime code numbers to help prevent the general population from becoming aware of the situation during the initial response. Follow the media response plan to control the release of information. Clear any releases with law enforcement or hospital authorities. Officials making statements should do so in a forthright manner without invading the privacy of the family. The family should know about the media plan and their cooperation sought in working through the official spokespersons. Notify newborn nurseries, pediatric units, emergency rooms, and outpatient clinics for postpartum/pediatric care at other healthcare facilities about the incident. Provide a full description of the baby and the suspected or alleged abductor. Designate a separate area where friends and family of the parents can gather to receive regular updates on the abduction in order to keep them informed about the case and shielded from the press. Provide operators with a written response for use those inquiring about the situation.

A.   PREVENTIVE MEASURES

Implement a policy that requires the mandatory wearing of identification badges by all hospital workers, medical staff, and other designated provider groups. A photo ID is preferable. Consider using further security measures for those working in nurseries, pediatrics, and obstetrics. Issue specific badges to employees assigned to the area on a temporary basis such as housekeeping and volunteer personnel.

Develop a system to identify the infant, mother, father, or a designated other before the baby leaves the birthing area. Develop procedures to address discharge of the mother prior to the infant leaving the hospital. Ensure adherence to documentation procedures to follow when a mother will not receive the baby on discharge.

BOX 4.22   INFANT ABDUCTION OFFENDER PROFILE

- Female, 12 to 50 years of age and often overweight
- Most likely compulsive; most often relies on manipulation, lying, and deception
- Frequently the offender lost a child or remains unable to bear a child
- Often married or involved in cohabitation arrangement
- Companion's desire for a child may provide motivation for the abduction
- Usually lives in the community where the abduction takes place
- Frequently visits nursery and maternity units prior to the abduction
- Asks detailed questions about hospital procedures and the maternity floor layout
- Frequently uses a fire exit stairwell for escape
- Usually plans the abduction and does not target a specific infant
- Frequently impersonates a nurse or other allied health professional
- Often becomes familiar with hospital personnel and even with the victim's parents
- Abductors provide good care to the baby

> ### BOX 4.23 ROOT CAUSES OF INFANT ABDUCTIONS
>
> - Lack of appropriate security equipment
> - Poor line of sight entrances and unmonitored elevators or stairwells
> - Staff-related factors such as insufficient education, lack of competency, and insufficient manpower
> - Delay in notifying security when an abduction is suspected, improper communication of relevant information among caregivers, and improper communication between hospital units
> - Organization cultural factors such as reluctance to confront unidentified visitors or providers

XVIII. QUALITY AND SAFETY EDUCATION FOR NURSES

The Quality and Safety Education for Nurses (QSEN) project began in 2005. Funded by the Robert Wood Johnson Foundation, the project had three phases between 2005 and 2012. The QSEN project addresses the challenge of preparing future nurses with the knowledge, skills, and attitudes (KSAs) necessary to continuously improve the quality and safety of the healthcare systems within which they work. The QSEN Institute is located at the Frances Payne Bolton School of Nursing of Case Western Reserve University in Cleveland, Ohio. Their website is a central repository of information on the core QSEN competencies, KSAs, teaching strategies, and faculty development resources designed to best support this goal. QSEN Initiatives include

- Defining quality and safety competencies for nursing and proposed targets for the knowledge, skills, and attitudes needed in nursing pre-licensure programs for the competencies of patient-centered care, teamwork and collaboration, evidence-based practice, quality improvement, safety, and informatics.
- Completing a national survey of baccalaureate program leaders and a state survey of associate degree educators to assess beliefs about the extent to which the competencies are included in current curricula, the level of satisfaction with student competency achievement, and the level of faculty expertise in teaching the competencies.
- Partnering with representatives of organizations that represent advanced practice nurses and drafting proposed knowledge, skills, and attitude targets for graduate education.
- Funding work with 15 pilot schools committed to active engagement in curricular change to incorporate quality and safety competencies.

REVIEW EXERCISES

1. Why should nursing personnel view patient safety as an organizational function?
2. Define and contrast covert and overt cultures.
3. What role does nursing fatigue play in patient safety?
4. List seven things healthcare organizations can do to address worker fatigue.
5. Describe six care environment safety challenges facing healthcare organizations.
6. List 10 keys to creating a culture of trust within a healthcare organization.
7. List the six aims found in the IOM report entitled *Crossing the Quality Chasm*.
8. Define the following:
   a. Adverse event
   b. IOM definition of error
   c. AHRQ definition of medical errors
9. What are the three fundamental characteristics of high-reliability organizations?

10. List seven methods used to identify adverse events.
11. According to the text what truly causes change to take place in an organization?
12. List the three overlapping organization levels impacted by change.
13. List seven common elements of system failures leading to medical errors.
14. In your own words, define system reliability.
15. What are the four common pitfalls of medical technology?
16. List and describe the three key elements of human factors science.
17. List five common causal factors discovered during healthcare root cause analysis sessions.
18. What role should patients play in their own care and safety?
20. Why has alarm safety become an important patient safety issue?
21. List three common healthcare-associated types of infection.
22. List four key safety suggestions related to high-alert medications.

5 Emergency Management and Fire Safety

I. INTRODUCTION

The International Board for Certification of Safety Managers developed the Certified Healthcare Emergency Professional (CHEP) credential in 2008. The Board believes that CHEP personnel can help to standardize management and system principles in the field of emergency planning, response, mitigation, and recovery. CHEP personnel, including nurses, working in healthcare facilities can lead the way by promoting healthcare emergency management as a "true" profession. Many of the CHEP credentialed individuals working in nonclinical and support areas understand the importance of coordinating healthcare emergency planning efforts. We can define an emergency as an unexpected or sudden event that significantly disrupts a healthcare organization's ability to provide care or significantly changes or increases the demand for services. Emergencies can result from human-made or natural events or a combination of both. A disaster-type emergency, due to its complexity, scope, or duration, can threaten the organization's capabilities. These events can require outside assistance to sustain patient care activities and facility safety or security. Healthcare organizations need to engage in planning activities to prepare a comprehensive Emergency Operations Plan (EOP).

Healthcare facilities must prepare to respond to and recover from events using the all hazards planning approach. Hospitals must plan to maintain a medical surge capacity and capability that will support the community. Use a multidisciplinary process when conducting a healthcare facility risk assessment for emergency response planning. The healthcare hazard vulnerability analysis (HVA) must consider the impact of realistic emergency or disaster incidents. These incidents could include hazardous materials releases, industrial and chemical accidents, transportation accidents, natural disasters, and even bioterrorism events. The HVA process must also assess the probability of each type of event, risks involved, and the organization's level of preparedness to respond. Emergency management consists of mitigation, preparedness, response, and recovery phases. These four phases occur over time, with mitigation and preparedness generally occurring before an emergency. Response and recovery phases usually occur during or after an emergency event.

II. JOINT COMMISSION REQUIREMENTS

The EOP identifies the individuals with the authority to activate the response and recovery phases of the emergency response. The plan must identify alternative sites for care, treatment, and services that meet the needs of the hospital patients during emergencies. The Joint Commission spells out emergency management–related responsibilities in its Emergency Management Standard.

The Joint Commission Standard provides excellent guidance for accomplishing actions to support the four phases of emergency management. The Joint Commission allows hospitals to develop a single HVA that accurately reflects all sites of the hospital, or the organization can develop multiple HVAs. Some remote sites may be significantly different from the main hospital site. Community partners may include other healthcare organizations, public health departments, vendors, community organizations, public safety and public works officials, representatives of local municipalities, and other government agencies. The hospital must communicate its needs and vulnerabilities to community emergency response agencies and identify the community's capability to meet its needs. This communication and identification should occur at the time of the hospital's annual review of the EOP and whenever its needs or vulnerabilities change. The hospital hazard vulnerability analysis provides the basis for defining mitigation activities needed to reduce the risk of damage during an emergency. Hospital leaders, including members of the medical staff, should participate in planning activities prior to developing an EOP. Ensure the plan addresses mass casualty situations, including terrorist events of a chemical, biological, or radiological nature. The plan should consider risks and their potential liabilities. Coordinate plans for maintaining a predictable environment of care during any emergency situation. Develop plans to guide response for any situation. The plan must provide for a command structure to assess situations, coordinate actions, and make decisions. Plan to deal with any situation that significantly disrupts the environment of care or patient treatment.

Planners should reference National Fire Protection Association (NFPA) 1600, NFPA 99, 29 CFR 1910.138, 40 CFR 264, applicable accreditation standards, and Department of Homeland Security (DHS) publications for additional information and guidance. Provide realistic training and education for all emergency personnel. Ensure all staff members understand their roles and responsibilities. Validate their understanding during readiness drills. Educational sessions can help reduce fear or anxiety among hospital personnel responding to terrorism-type events. Train medical and hospital staff to report unexpected illness patterns to appropriate agencies. When possible, ensure a physician meets with the local media to provide updated information about medical issues. Make the public aware of any changes in hospital treatment procedures. Ensure the incident command integrates into the community's command structure. The incident command structure should provide scalable mechanisms to better respond to different types of emergencies. The hospital should maintain an inventory of the resources and assets. These assets should include but never be limited to personal protective equipment, water, fuel, and medical-, surgical-, and medication-related resources and assets.

The local EOP must guide the coordination of communications, resources, assets, safety and security, and staff responsibilities. The plan must address patient, clinical, and support activities during an emergency. Emergencies may vary but the effects on these organization functions may be similar. This "all hazards" approach supports a general response capability sufficiently nimble to address a range of emergencies of different duration, scale, and cause. The EOP permits response procedures to address prioritized emergencies. A comprehensive but flexible EOP can guide decision-making at the onset and as a situation evolves. Response procedures should address the following: (1) maintaining or expanding services, (2) conserving resources, (3) curtailing services, (4) supplementing resources from outside the local community, (5) closing the hospital to new patients, and (6) staged evacuation or total evacuation. The EOP describes the processes for initiating and terminating the hospital's response and recovery phases of an emergency, including under what circumstances these phases are activated. It also identifies the hospital's capabilities and establishes response procedures for when the hospital can't be supported by the local community. Hospitals should plan to stockpile enough supplies to last for 96 hours of operation. The EOP should describe the recovery strategies and actions designed to help restore the systems that are critical to providing care, treatment, and services after an emergency.

Develop contingency plans to ensure the availability of critical supplies. Examples of resources and assets that might be shared include beds, transportation, linens, fuel, personal protective equipment, medical equipment, and supplies. The EOP describes the hospital's arrangements for transporting some or all patients, their medications, supplies, equipment, and staff to an alternative care site when environments can't support care, treatment, and services. The EOP also addresses the arrangements for transferring pertinent information, including essential clinical and medication-related information, with patients moving to alternative care sites. The EOP should describe the hospital's arrangements for internal security and safety. It should also address the roles that community security agencies will play in supporting security activities. The EOP must also describe how the hospital will manage hazardous materials and wastes. Plan to address radioactive, biological, and chemical isolation or decontamination activities. The Joint Commission requires hospitals to provide safe and effective patient care during an emergency. Document staff roles and responsibilities in the EOP. Due to the dynamic nature of emergencies, effective training prepares staff to adjust to changes in patient volume or acuity. The EOP should describe the process for assigning staff to all essential staff functions.

The hospital must communicate, in writing, with each of its licensed independent practitioners regarding his or her role in emergency response and to whom they report. The Joint Commission provides guidance to its accredited facilities on how to grant disaster privileges to volunteer licensed independent practitioners when the EOP has been activated. The medical staff must describe how it will oversee the performance of volunteer licensed independent practitioners granted disaster privileges. Before determining whether a volunteer practitioner is eligible to function as a volunteer licensed independent practitioner, the hospital should obtain his or her valid government-issued photo identification and at least one of the following: (1) current picture identification card from a healthcare organization that clearly identifies professional designation, (2) current license to practice, (3) primary source verification of licensure, (4) identification indicating that the individual is a member of a Disaster Medical Assistance Team (DMAT), the Medical Reserve Corps (MRC), the Emergency System for Advance Registration of Volunteer Health Professionals (ESAR-VHP), or other recognized state or federal response organization or group, (5) identification indicating that the individual possesses granted authority by a government entity to provide patient care, treatment, or services in disaster circumstances, or (6) confirmation by a licensed independent practitioner privileged by the hospital or by a staff member with personal knowledge of the volunteer practitioner's ability to act as a licensed independent practitioner during a disaster. During a disaster, the Joint Commission requires that medical staff oversee the performance of each volunteer licensed independent practitioner. Primary source verification of licensure occurs as soon as a disaster is under control or within 72 hours from the time the volunteer licensed independent practitioner presents him or herself to the hospital, whichever comes first. If primary source verification of a volunteer licensed independent practitioner's licensure cannot be completed within 72 hours of the practitioner's arrival due to extraordinary circumstances, the hospital documents all of the following: (1) reasons it could not be performed within 72 hours of the practitioner's arrival, (2) evidence of the licensed independent practitioner's demonstrated ability to continue to provide adequate care, treatment, and services, or (3) evidence of the hospital's attempt to perform primary source verification as soon as possible.

The Joint Commission requires hospitals to conduct an annual review of its planning activities to identify such changes and support decision-making regarding how the hospital responds to emergencies. The hospital must also conduct an annual review of its risks, hazards, and potential emergencies as defined in its HVA. The hospital must conduct an annual review of the objectives and scope of its EOP. The findings must be documented.

Facilities must conduct exercises to assess EOP appropriateness, adequacy, and effectiveness. Key areas to evaluate include logistics, human resources, training, policies, procedures, and protocols. Exercises should stress the limits of the plan to support assessment of preparedness and performance.

The design of the exercise should reflect likely disasters but should test the organization's ability to respond to the effects of emergencies on its capabilities to provide care, treatment, and services. For each site of the hospital that offers emergency services or a community-designated disaster receiving station, at least one of the two required emergency response exercises or drills must include an escalating event in which the local community is unable to support the hospital. Tabletop sessions are acceptable in meeting the community portion of this exercise. For each site of the hospital with a defined role in its community's response plan, at least one of the two emergency response exercises must include participation in a community-wide exercise. Tabletop sessions meet only community portions of the exercise. Emergency response exercises incorporate likely disaster scenarios that allow the hospital to evaluate its handling of communications, resources and assets, security, staff, utilities, and patients. Staff in freestanding buildings classified as business occupancies that do not offer emergency services nor are designated by the community as disaster-receiving stations need to conduct only one emergency management exercise annually. Tabletop sessions, though useful, are not acceptable substitutes for these exercises.

The hospital should designate individuals to monitor drill or exercise performance and document opportunities for improvement. This person must be knowledgeable in the goals and expectations of the exercise and may be a staff member. Hospitals may use observations of those involved in the command structure as well as the input of those providing services during an actual emergency. The hospital must evaluate all emergency response exercises and all responses to actual events using a multidisciplinary process. Communicate all identified deficiencies and opportunities for improvement to the improvement team responsible for monitoring environment of care issues. The hospital must modify the EOP based on findings of emergency exercises and actual events.

The facility must maintain a written inventory of utility system components considering risks for infection, occupant needs, and systems critical to patient care. The facility must identify in writing inspection and maintenance activities for operating components, and must identify the intervals for inspecting, testing, and maintaining all operating utility systems using manufacturer recommendations, risk levels, or hospital experience. The facility must also take actions to minimize pathogenic biological agents in cooling towers, domestic hot and cold water systems, and other aerosolizing water systems.

It must map and document distribution of its utility systems, label utility system controls to facilitate partial or complete emergency shutdowns, and develop written procedures for responding to utility system disruptions.

III. OTHER EMERGENCY PLANNING ISSUES

In order to maintain daily operations and patient care services, healthcare facilities need to develop an Emergency Water Supply Plan (EWSP) to prepare for, respond to, and recover from a total or partial interruption of the facilities' normal water supply. Water supply interruption can be caused by several types of events such as a natural disaster, a failure of the community water system, construction damage, or even an act of terrorism. Because water supplies can fail, it is imperative to understand and address how patient safety, quality of care, and the operations of your facility will be impacted. Below are a few examples of critical water usage in a healthcare facility that could be impacted by a water outage. Water may not be available for hygiene, drinking, food preparation, laundry, central services, dialysis, hydrotherapy, radiology, fire suppression, water-cooled medical gas, suction compressors, HVAC operation, decontamination, and hazmat response. A healthcare facility must be able to respond to and recover from a water supply interruption. A robust EWSP can provide a road map for response and recovery by providing the guidance to assess water usage, response capabilities, and water alternatives. The EWSP will vary from facility to facility based on site-specific conditions and facility size. Regardless of size, a healthcare facility must develop an effective EWSP to ensure patient safety and quality of care while responding to and recovering from a water emergency.

Healthcare facilities are a critical component to a community's response and recovery following an emergency event. A number of incidents could impact the water supply of a healthcare facility. In the case of some natural disasters, such as a hurricane or flood, a facility may know days ahead of the risks. These events allow more time for preparation, which typically speeds up response. Earthquakes, tornados, or external/internal water contamination can occur with little or no warning. Joint Commission standards address the provision of water as part of the facility's EOP. Centers for Medicare & Medicaid Services (CMS) conditions for participation/coverage also require healthcare facilities to make provisions in their preparedness plans for situations in which utility outages of gas, electric, or water may occur. Incorporate the principles and concepts of the plan into the overall facility EOP. It remains vital that the EWSP receives an annual review. Exercise and revise the plan on a regular basis or at least annually. The process of developing an EWSP for a healthcare facility will depend on the size of the facility and will require the participation and collaboration of both internal and external stakeholders.

For a small facility (less than 50 beds) where one individual performs multiple functions, the process may be relatively simple, with a single individual coordinating development of the EWSP. However, for a large hospital of several hundred beds, the process of developing the plan will be more complex.

A. HOSPITAL EVACUATION PLANNING

Develop plans for evacuating the facility either horizontally and/or vertically. Also plan to identify care providers and other personnel during emergencies. Create a priority listing of institutions or facilities to which the patients or residents will be evacuated. Specify the locations that will serve as a staging area pending further decisions. Develop procedures for obtaining an accurate account of personnel after a completion of the evacuation. Designate assembly areas where personnel should gather after evacuating. Establish a method for accounting for nonemployees such as suppliers and visitors. Establish procedures for further evacuation in case the incident expands. The Americans with Disabilities Act (ADA) defines a disabled person as anyone with a physical or mental impairment that substantially limits one or more major life activities. Emergency planning priorities must consider disabled visitors and employees.

IV. COMMUNITY INVOLVEMENT

FEMA developed a publication titled *A Guide to Citizen Preparedness (FEMA Publication H-34)*. The guide contains facts on disaster survival techniques, disaster-specific information, and how to prepare for and respond to both natural and man-made disasters. Healthcare organizations must adopt a community-wide perspective when planning for mass casualty incidents. Senior leaders must maintain a good relationship with response agencies in the community, including other area healthcare facilities. Clinics and nursing homes may play key roles in large disasters. Public health departments will usually institute appropriate public health interventions, including immunizations and prophylactic antibiotics. Establish working relationships with all responders including local emergency management agencies, law enforcement personnel, and local fire officials. Coordinate the healthcare emergency plan with the official responsible for area wide disaster planning.

A. PARTNERSHIP FOR COMMUNITY SAFETY

The Partnership for Community Safety: *Strengthening America's Readiness* serves as a new coalition formed to advocate for strengthening community readiness for biological, chemical, or nuclear terrorism and other disasters. The Partnership will call on federal policymakers to support and sustain comprehensive readiness efforts in the nation's public health departments, emergency departments, hospitals, fire services, ambulance and emergency medical services (EMS) organizations, medical education institutions, and the nursing profession. While proposals pending in Congress represent important first steps, the Partnership will advocate for a comprehensive and sustained approach to community readiness. Partnership members believe the tragic events of September 11 and the subsequent anthrax incidents demonstrate the urgency for strengthening community preparedness plans to protect the public from acts of terrorism. In addition to working together to help shape national policy, the new alliance will promote collaboration among its members to retool disaster plans and focus on the need to increase capacity for frontline responders to prepare for the new challenges of terrorism. In addition, Partnership members will work to reduce duplication of effort and develop a "bank" of best practices through exchanging ideas and highlighting model plans. The Partnership also plans to educate the public about local readiness issues. The Partnership for Community Safety: *Strengthening America's Readiness* represents firefighters, paramedics and other EMS professionals, emergency physicians, all other physicians, hospital officials, medical education professionals, public health officials, nurses, and state regulatory agencies in the United States.

B.   Hospital Roles in Community Emergencies

Healthcare organizations must work to help assess community health needs and available resources to treat evacuees from other areas. The organization must determine community priorities as identified in the hazard vulnerability analysis. Clarify the organization's role during the annual community-wide emergency exercise. Develop plans for coordinating with the media. Establish a media briefing area with established security procedures. Establish procedures for ensuring the accuracy and completeness of all information approved for public release. Provide for decontamination and treatment for any and all victims. Promote a wider level of preparedness in the community by providing low-cost hazard communication or hazardous waste operations and emergency response (HAZWOPER) training for local government and business emergency response personnel. Provide information and services related to emergency preparedness. Participate actively in community planning and preparedness activities.

Healthcare facilities should focus on the following key areas when conducting planning:

- **Communication**: Assess the ability of the organization to maintain communications within the organization and with all appropriate community disaster resource agencies.
- **Resources and Assets**: Develop plans to access necessary materials, supplies, vendors, community resources, and government support if necessary to sustain operations such as patient care, safety, and medical services.
- **Safety and Security**: Create contingency plans to ensure the safety and welfare of all patients, staff, and visitors during emergencies and disasters.
- **Staff Responsibilities**: Design appropriate curriculums to orient, educate, or train staff members about their changing roles and demands during emergency incidents.
- **Utilities Management**: Establish plans for maintaining key utilities such as drinking water, power sources, ventilation, and fuel supplies.
- **Clinical Support Activities**: Establish clinically coordinated policies and procedures that ensure patient care during extreme emergency conditions when organizational resources are stretched.

V.   INCIDENT COMMAND SYSTEM

There are five basic functional areas of management during a major incident including (1) command, (2) operations, (3) planning, (4) logistics, and (5) finance/administration. An Incident Command System (ICS) coordinates responses involving multiple jurisdictions or agencies. It retains the principle of unified command for coordinating the efforts of many jurisdictions. The system must ensure joint decisions in areas such as objectives, strategies, plans, priorities, and communications. The system focuses on responder readiness to manage and conduct incident actions by coordinating before an event. Some benefits include (1) maintaining a predictable chain of accountability, (2) flexible response to specific incidents, (3) improved documentation, (4) common language to facilitate outside assistance, (5) prioritized response checklists, and (6) cost-effective planning.

BOX 5.1   COMMON ICS PRINCIPLES

- **Common Terminology**: The use of similar terms and definitions for resource descriptions, organizational functions, and incident facilities across disciplines.
- **Integrated Communications**: The ability to send and receive information within an organization and externally to other disciplines.
- **Modular Organization**: Assets within each functional unit may be expanded or contracted based on the requirements of the event.

BOX 5.2   COMMON ICS PRINCIPLES

- **Unified Command Structure**: Disciplines and response organizations work through designated managers to establish common objectives and strategies to reduce conflict or duplication.
- **Span of Control**: The structure permits each supervisory level to oversee an appropriate number of assets based on size and complexity of the event.
- **Span of Control Ratio**: Maintaining effective supervision with an element supervising three to seven entities, with five being the ideal.
- **Consolidated Incident Action Plans**: Goals, objectives, strategies, and major assignments are defined by the incident commander or by unified command.
- **Comprehensive Resource Management**: System processes are in place to describe, maintain, identify, request, and track all resources within the system during an incident.
- **Pre-Designated Incident Facilities**: Assign locations where expected critical incident-related functions will occur and ensure adequate space and technical support for the assigned function.

A.   INCIDENT COMMANDER RESPONSIBILITIES

Assign the duties to certain positions and never to specific individuals. The incident commander must maintain emergency command center effectiveness, ensure communications, and maintain security. Key duties include providing public information and media releases, coordinating facilities, sheltering, feeding, and counseling as needed. The incident commander must oversee establishing the morgue and making EMS available as needed.

VI.   STRATEGIC NATIONAL STOCKPILE

The Centers for Disease Control and Prevention's (CDC's) Strategic National Stockpile (SNS) maintains large quantities of medicine and medical supplies to protect the American public if there is a public health emergency such as a terrorist attack, flu outbreak, or earthquake severe enough to cause local supplies to become diminished. Once federal and local authorities agree on the need for SNS, medicines will be delivered to any state in the United States in time for them to be effective. Each state has a plan to receive and distribute SNS medicine and medical supplies to local communities as quickly as possible. The SNS contains enough medicine to protect people in several large cities at the same time. Federal, state, and local community planners work together to ensure that the SNS medicines will be delivered to the affected areas if there is a terrorist attack. Local communities are prepared to receive SNS medicine and medical supplies from the state to provide them to everyone in the community who needs them. The SNS consists of a national repository of antibiotics, chemical antidotes, antitoxins, life-support medications, IV administration, airway maintenance supplies, and medical/surgical items. SNS can supplement and resupply state and local public health agencies in the event of a national emergency anywhere and at anytime within the United States or its territories.

If the incident requires additional pharmaceuticals and/or medical supplies, vendor managed inventory (VMI) supplies will be shipped to arrive within 24 to 36 hours. If well defined, the agent VMI can be tailored to provide pharmaceuticals, supplies, and/or products specific to the suspected or confirmed agent(s). In this case, the VMI could act as the first option for immediate response from the SNS Program. The first line of support lies within the immediate response 12-hour Push Packages. These are caches of pharmaceuticals, antidotes, and medical supplies designed to provide rapid delivery of a broad spectrum of assets for an ill-defined threat in the early hours of an event. These Push Packages are positioned in strategically located, secure warehouses ready for immediate deployment to a designated site within 12 hours of the federal decision to deploy SNS assets. However, SNS does not function as a first-response tool.

VII.   PLANNING FOR TERRORISM

Hospitals make up a substantial portion of the emergency response system. Educate and train staff about possible events and appropriate response actions. Potential risks associated with nuclear, chemical, biological, or radiological weapon attacks by terrorists call for sound emergency planning procedures. Terrorist events can result in potentially large numbers of casualties. The psychological impact of weapons of mass destruction and the relative ease of their acquisition poses a great threat. Healthcare facilities preparing for a bioterrorism response plan should reference *A Template for Healthcare Facilities*, produced by the Association for Professionals in Infection Control. This resource outlines the steps necessary for responding to biological agents, such as smallpox, botulism toxin, anthrax, and plague, and provides information on the unique characteristics, specific recommendations, management, and follow-up for each of these agents. The CDC *National Public Health Strategy for Terrorism Preparedness & Response Guide* contains information on the following topics: (1) detection, investigation, and laboratory sciences; (2) prevention efforts, worker safety, and communication; (3) emergency response; (4) research and long-term consequence management; and (5) workforce development.

VIII.   PANDEMIC PLANNING

The Department of Health and Humans Services (DHHS) Supplement 3 provides healthcare partners with recommendations for developing plans to respond to a pandemic. Focus on planning during the Interpandemic Period for issues such as surveillance, decision-making structures, communications, education and training, patient triage, clinical evaluations, admission, facility access, occupational health, distribution of vaccines, antiviral drugs, surge capacity, and mortuary issues. The activities suggested in Supplement 3 are intended to be synergistic with those of other pandemic influenza planning efforts, including state preparedness plans. Healthcare facilities must be prepared for the rapid pace and dynamic characteristics of pandemic influenza. All hospitals should be equipped and ready to care for a limited number of patients infected with a pandemic influenza virus, or other novel strains of influenza. Hospitals should prepare for a large number of patients in the event of escalating transmission of pandemic influenza. Healthcare facilities must develop planning and decision-making structures for responding to pandemic. This planning includes developing written plans that address (1) disease surveillance, (2) hospital communications, (3) education and training, (4) triage and clinical evaluation, (5) facility access, (6) employee health, (7) use and administration of vaccines or antiviral drugs, (8) surge capacities, (9) supply chain issues, (10) access to critical inventory needs, and (11) mortuary related issues.

IX.   FIRE SAFETY

Facilities contain many fire-related hazards including medical equipment, combustible gases, chemicals with low flash points, and electrical hazards of all types. Planning should consider proper design, including prevention features and egress safety. Fire response planning remains a key element of any emergency management process.

Fire begins with no visible smoke, flames, or significant heat. However, a large number of combustion particles generate over time. The particles created by chemical decomposition possess both weight and mass but remain too small for the eye to see. They behave according to gas laws and quickly rise to the ceiling. As this incipient stage continues, the combustion particles increase until they become visible and create a condition called "smoke." As the fire continues to develop, ignition occurs and flames begin. The level of visible smoke decreases and heat levels increase. At this point the process produces large amounts of heat, flame, smoke, and toxic gases.

The life safety concept began in 1963 with the publication of the Building Exits Code. The NFPA published its First Edition of the Life Safety Code® in 1966. Building codes provide design criteria but NFPA 101® addresses the general requirements for fire protection and systems safety necessary to assure the safety of building occupants during a fire. The code provides minimum requirements for the design, operation, and maintenance of healthcare organization buildings and structures. NFPA 101 requires that new and existing buildings allow for prompt escape or provide occupants with a reasonable degree of safety through other means. It defines hazards and addresses general requirements for egress and covers fire protection features such as fire doors. The code also addresses building service and fire equipment such as heating, ventilating, air conditioning systems, sprinkler systems, fire detection systems, and localized extinguishers. New editions of the code build on the prior editions.

A. LIFE SAFETY CODE COMPARISONS

The 2012 edition of the Life Safety Code offers new design and compliance options for healthcare facilities. Since all jurisdictions do not use the same edition of the code, CMS and the Joint Commission permit the use of the 2012 edition in its entirety or on a single-element basis. The American Society for Healthcare Engineering (ASHE) developed a monograph in 2013 that provided summary changes in the new code. The monograph also provided a comparison of the 2012 Code with the 2000 and 2009 editions. The monograph presented three different options for upgrading from the 2000 edition to the 2009 or 2012 edition. The monograph provided guidance about waivers and equivalencies for using the newer editions of the code.

BOX 5.3   BASIC FIRE PLAN REQUIREMENTS

- Policies implemented to manage fire safety
- Processes developed to protect humans from fire and smoke
- Procedures for inspecting, testing, and maintaining fire protection systems
- Facility-wide fire response procedures
- Area-specific needs including fire evacuation routes
- Specific roles and responsibilities of staff at the fire's point of origin
- Specific roles and responsibilities of staff in preparing for building evacuation

---

BOX 5.4   OSHA FIRE PLAN ELEMENTS (29 CFR 1910.38)

- Fire department notification and follow-up procedures
- Procedures for announcing the fire location using an appropriate method
- Locations for key personnel to assemble and manage the decisions
- Designated personnel who will meet and direct fire department personnel
- Procedure for holding nonemergency calls and giving an "all-clear" signal

---

BOX 5.5   TOPICS TO CONSIDER WHEN DEVELOPING FIRE RESPONSE PLANNING

- Specific needs related to fire evacuation and egress routes
- Specific roles and responsibilities for all staff
- Specific roles and responsibilities for patient evacuation
- Information about alarm systems and signals
- Information related to the location and use of firefighting equipment
- Information related to fire containment procedures

B. DESIGN CONSIDERATIONS

Design, construct, maintain, and operate a building to minimize the possibility of a fire requiring the evacuation of occupants. Develop procedures to address

- Design, construction, and compartmentalization
- Provision for detection, alarm, and extinguishments systems
- Fire prevention planning and training
- Isolation of fire
- Evacuation of the building or the transfer of occupants to areas of refuge

---

BOX 5.6   GENERAL FIRE DRILL PROCEDURES

- Provide information on location of fire, type of fire, and equipment failures
- Ensure installation of an effective and convenient fire alarm system
- Shut off oxygen and gas valves if possible and disconnect unnecessary equipment
- Reassure patients and visitors about the implementation of emergency plans
- Ensure a realistic implementation of the fire plan and conduct drills at varied times
- Critique drills to identify deficiencies and opportunities for improvement

---

C. FIRE PREVENTION

Buildings must contain a fire alarm or fire detection system that should automatically activate an alarm in the event of a fire. Install air conditioning, ducts, and any related equipment in accordance with NFPA 90A, Standard for Installation of Air Conditioning and Ventilating Systems. Ensure people can hear fire alarms over normal operational noise levels. Locate manual fire alarm stations near each exit. Inspect fire extinguishers at least monthly and ensure regular maintenance. Test fire alarm/detection systems once a quarter. Publish and enforce a Smoking Policy. Implement appropriate electrical safety policies and educate all personnel about fire safety and response plans.

D. INSPECTIONS

Conduct quarterly fire inspections for each fire zone. Accomplish the following:

- Assess all equipment including the testing of alarms, detectors, and pull stations
- Evaluate housekeeping practices and sprinkler pressure inspection procedures
- Check water availability and fire hydrant operation
- Check suppression, detection, and activation systems annually
- Coordinate inspections with the local fire marshal and facility engineering

E. FIRE WARNING AND SAFETY

Ensure all systems meet NFPA standards and local requirements. All manually operated fire systems must be electrically supervised. The system must also automatically transmit an alarm to the fire department. Notify the local fire department by other means when the alarm has been activated.

BOX 5.7   FIRE ALARM TYPES

- Central Station Service (NFPA 71)
- Auxiliary Protective Signaling Systems (NFPA 72)
- Proprietary Protective Signaling System (NFPA 72)
- Remote Station Protective Signaling System (NFPA 72)

> ## BOX 5.8   GENERAL REQUIREMENTS FOR ALARMS
>
> - Fire alarms must be received at a central location within the facility.
> - Continuously man and supervise all locations.
> - Protect supervised locations as a hazardous area.
> - Signals received must transmit at once to the local fire department.
> - Provide a copy of the master fire plan at all supervised locations.

1. Manual Alarm Stations

Locate manual alarm stations throughout the facility. Position the alarms to ensure travel distances of no more than 200 feet when located on the same floor. Design audible alarms to exceed the level of any operational noise. Use audible alarms with visual alarms.

2. Electrically Supervised Systems

Monitor all components to ensure personnel are aware of when the system needs repair. The system should signal trouble when

- A break or ground fault prohibits normal system operation
- The main power source fails
- A break occurs in the circuit wiring

3. Special Requirements for Cooking Areas

Install approved systems to protect cooking surfaces, exhaust hoods, and ducts. Consider the following types of systems:

- Automatic carbon dioxide systems
- Automatic dry chemical systems
- Automatic foam water or wet chemical systems
- Automatic sprinkler systems approved by NFPA 13

4. Fire System Inspections

All systems should receive a visual inspection each quarter. Test or inspect each automatic system on an annual basis. Include all systems in the preventive maintenance plan. Test all supervisory signal devices except valve tamper switches on a quarterly basis. Test valve tamper switches and water flow devices semiannually. Test duct detectors, electromechanical releasing devices, heat detectors, manual fire alarm boxes, and smoke detectors on a semiannual basis. Test occupant alarm notification devices to include audible and visible devices at least annually. Maintain appropriate documentation on all fire-related system testing.

F. FIRE CONFINEMENT

Confinement measures consist of dividing a building into small cells. To assure proper protection of openings, install fire doors in accord with NFPA 80, Standard for Fire Doors and Fire Windows. Evaluate the movement of smoke within a structure by considering many factors such as building and ceiling height, suspended ceilings, ventilation, and external wind force or direction. One method of smoke control uses a physical barrier, such as a door or damper, to block the smoke's movement. Regardless of the type of building construction, stair enclosures must provide a safe exit path for occupants. Stair enclosures also retard the upward spread of fire.

G. EMERGENCY EGRESS

Designing exits involves more than a study of numbers, flow rates, and population densities. Exits must provide alternative pathways to counter potential exit blockage by fire. Each employee should recognize and report fire safety hazards.

Exit doors must withstand fire and smoke for a specified length of time. Provide alternative exits and pathways in case fire blocks an exit. Provide exits with adequate lighting and mark exits with readily visible signs. Develop plans to evacuate disabled or wheelchair employees to meet Occupational Health and Safety Administration (OSHA) requirements. The ADA Title III requires organizations to develop plans to safely evacuate disabled visitors.

1. OSHA Egress Standards

OSHA defines a means of egress as a continuous and unobstructed way of exit that travels from any point in a building or structure to a public way and consists of three parts:

- Exit access: That portion that leads to the entrance of an exit.
- Exit: That portion separated from all other spaces of a building or structure by construction or equipment to provide a protected way of travel to the exit discharge.
- Exit discharge: That portion between the termination of an exit and a public way

H. FIRE EXTINGUISHERS

This subsection covers the basic type of equipment used to fight and control fires. The first type of equipment, known as a fixed system, includes automatic sprinklers, standpipe hoses, and various pipe systems. Supplement fixed systems by providing appropriate types and sizes of portable extinguishers. Train personnel expected to use portable fire extinguishers on their operation and safe use.

BOX 5.9   BASIC "PASS" GUIDELINES FOR EXTINGUISHER USE

- Pull the pin on the extinguisher
- Aim the nozzle at the base of the fire
- Squeeze the handle firmly
- Spray in a sweeping motion

---

1. How Fire Extinguishers Work

Portable fire extinguishers apply an existing agent that will cool burning fuel, displace or remove oxygen, or stop the chemical reaction so a fire cannot continue to burn. When the handle of an extinguisher is compressed, it opens an inner canister of high-pressure gas that forces the extinguishing agent from the main cylinder through a siphon tube and out the nozzle. Fire creates a very rapid chemical reaction between oxygen and a combustible material, which results in the release of heat, light, flame, and smoke.

BOX 5.10   EXTINGUISHER CLASSES

- Class A: For fires involving ordinary combustible materials, such as wood, paper, or clothing, where the quenching and cooling effects of water prove most effective, use a pressurized water extinguisher or ABC type dry powder extinguisher.
- Class B: For fires involving flammable liquids and similar materials, use type BC or ABC dry powder extinguishers. Carbon dioxide (CO_2) extinguishers may also be used.

(Continued)

BOX 5.10   EXTINGUISHER CLASSES (*Continued*)

- Class C: For fires in or near energized electrical equipment where the use of a nonconductive extinguishing agent is of first importance, use CO_2, or dry powder (BC or ABC). NEVER USE WATER.
- Class D: Metal fires—combustible metals such as magnesium and sodium require special extinguishers labeled D.
- Class K: Grease fires—use a portable extinguisher designed especially for cooling these types of fire.

BOX 5.11   FIRE EXTINGUISHER PLACEMENT & TRAVEL DISTANCES

- Class A: Travel distance of 75 feet or less
- Class B: Travel distance of 50 feet or less
- Class C: Travel distance based on appropriate A or B hazard
- Class D: Travel distance of 75 feet

BOX 5.12   MONTHLY EXTINGUISHER INSPECTIONS

- Determine proper location and type
- Ensure accessibility to all extinguisher locations
- Document the proper mounting of each extinguisher
- Check gauges to determine adequate pressure
- Verify proper placement of pins and seals
- Look for evidence of damage or tampering
- Ensure no blockage of nozzles

2. Proper Maintenance

Proper maintenance includes a complete examination, and involves disassembly and inspection of each part and replacement where necessary. Conduct maintenance at least annually or more often if conditions warrant. Perform hydrostatic testing of portable fire extinguishers to protect against unexpected in-service failure. Failure can occur due to internal corrosion, external corrosion, and damage from abuse. Perform hydrostatic testing using trained personnel with proper equipment and facilities. OSHA Standard, 29 CFR 1910.157, Table 1 provides test intervals for extinguishers.

I. SURGICAL FIRES

Fires occurring on or inside a patient rarely occur but can cause grave consequences. They can kill or seriously injure patients, injure surgical staff, and damage critical equipment. Flammable materials present in surgical suites range from alcohol-based prepping agents to drapes, towels, gowns, hoods, and masks. Common ignition sources found in operating rooms include electrosurgical or electrocautery units, fiber-optic light sources and cables, and lasers. High-speed drills can produce incandescent sparks that can fly off the target tissue and ignite some fuels, especially in oxygen-enriched atmospheres. Staff should participate in special drills and training on the use of firefighting equipment. They should know the proper methods for rescue and escape.

Ensure each staff member knows the identification and location of medical gas, ventilation, and electrical systems including controls. Educate staff on how to use the hospital's alarm system and contact the local fire department. Healthcare organizations can prevent fires by

- Informing staff members, including surgeons and anesthesiologists, about the importance of controlling heat sources by following published safety practices.
- Managing fuels by allowing sufficient time for patient prep and establishing guidelines for minimizing oxygen concentration under the drapes.
- Developing, implementing, and testing procedures to ensure appropriate response of all members of the surgical team.
- Reporting any instances of surgical fires as a means of raising awareness and ultimately preventing the occurrence of fires in the future.

1. ASTM Surgical Fire Standard

Refer to the *ASTM Standard Guide to Surgical Fires: Fire Risk Assessment, Prevention, and Extinguishment*. The Guide was developed by ASTM Committee F29 on Anesthetic and Respiratory Equipment. The standard offers instruction on the risks of potentially flammable materials used in surgery and provides some much-needed guidance for doctors, nurses, anesthesiologists, technicians, engineers, risk managers, and health administrators.

BOX 5.13   PREVENTING SURGICAL FIRES

- Minimize ignition risks during use of electrosurgical devices and surgical lasers including the safe and appropriate use of electrosurgical pencils and the use of bipolar electro surgery devices.
- Lessen ignition risks by selective wetting of fuels present at the incision, including gauze, sponges, and towels.
- Implement all general procedures established to minimize ignition risks.
- Specific procedures to minimize ignition risks in oropharyngeal surgery include gas scavenging and using wet gauze, sponges, or pledgets.
- Minimize the oxidizer risks of oxygen and nitrous oxide used in general surgery and specifically during oropharyngeal surgery.
- Reduce fuel risks when using flammable surgical instruments.
- Educate personnel on fire prevention and conduct drills for the operating room setting.

BOX 5.14   RESPONDING TO SURGICAL FIRES

- Extinguish small fires by hand.
- Response to large fires on or in the patient: stop the flow of oxidizers, remove burning materials from the patient, extinguish burning material, and care for the patient.
- Follow correct procedures for using the recommended type of fire extinguisher, such as carbon dioxide.
- Do not use water-based and dry-powder extinguishers.
- Rescue the patient, alert the staff, confine the smoke or fire, and evacuate the area.
- Recommend mounting a five-pound CO_2 extinguisher just inside the entrance of each operating room.

2. Fire Blankets

Never locate wool blankets treated with fire retardants in the operating room and never use them for patient fires. Their use will likely cause more severe injuries to the patient. However, they could help when responding to use on a conscious person, such as a surgical team member.

X.   LIFE SAFETY

Healthcare facilities must design and manage the physical environment to comply with the Life Safety Code. Assign an individual(s) to assess compliance with the Life Safety Code. The facility must complete the electronic Statement of Conditions and manage the resolution of deficiencies. The hospital must maintain a current electronic Statement of Conditions. When the hospital plans to resolve a deficiency through a Plan for Improvement (PFI), the hospital must meet the time frames identified in the PFI accepted by the Joint Commission. Hospitals accredited by the Joint Commission must maintain documentation of any inspections and approvals made by state or local fire control agencies. The hospital must protect occupants during periods when the Life Safety Code is not met or during periods of construction. The hospital must notify the fire department or other emergency response group and initiate a fire watch when a fire alarm or sprinkler system is out of service more than 4 hours in a 24-hour period in an occupied building.

A.   INTERIM LIFE SAFETY

Healthcare facilities must implement written interim life safety measures (ILIMs) to address situations when the organization cannot immediately correct deficiencies. The policy should include criteria for evaluating when and to what extent the hospital follows special measures to compensate for increased life safety risk. The facility must inspect exits in affected areas on a daily basis. It must provide temporary but equivalent fire alarm and detection systems for use during fire system impairment. Use smoke-tight temporary construction partitions made of noncombustible or limited-combustible material that would not contribute to development or spread of a fire. Increase surveillance of buildings, grounds, and equipment, giving special attention to construction areas and storage, excavation, and field offices. Enforce storage, housekeeping, and debris-removal practices that reduce the building's flammable and combustible fire load to the lowest feasible level. Provide additional training to those who work in the hospital on the use of firefighting equipment. Conduct one additional fire drill per shift per quarter. Inspect and test temporary systems monthly. The hospital must train those who work in the hospital to compensate for impaired structural or compartmental fire safety features.

REVIEW EXERCISES

1. What organization developed the CHEP credential?
2. Define the concept of an emergency as related to healthcare organizations.
3. When planning for emergencies and disaster, what approach must be used?
4. Describe the purpose and content of an effective hazard vulnerability analysis.
5. List the four phases of emergency management.
6. List six areas needing coordination that must be addressed in the EOP.
7. How often must Joint Commission–accredited facilities review the objectives and scope of their EOP?
8. Describe the mission of the Partnership for Community Safety.
9. List the six key areas healthcare facilities should focus on when conducting planning sessions.
10. Describe the purpose and process of the strategic national stockpile.
11. What publication should healthcare facility planners reference for pandemic planning?
12. When did the NFPA publish its First Edition of the Life Safety Code®?
13. How frequently must healthcare facilities conduct fire inspections for each zone?
14. In your own words, describe the concept known as fire confinement.

6 Hazardous Materials

I. INTRODUCTION

Healthcare organizations use a wide variety of hazardous substances including disinfectants, sterilizing agents, solvents, chemotherapeutic drugs, compressed gases, and hazardous wastes. The Occupational Safety and Health Administration (OSHA), Environmental Protection Agency (EPA), Department of Transportation (DOT), and accreditation organizations including the Joint Commission require healthcare organizations to properly receive, handle, manage, and dispose of hazardous material in an effective manner. Organizations should develop and implement comprehensive written plans that protect staff, patients, and visitors. Organizations must adhere to the requirements of the OSHA Hazard Communication Standard, the EPA Resource Conservation and Recovery Act (RCRA), and DOT Hazardous Materials Regulations. Healthcare organizations should work to consolidate hazardous material management plans with requirements of accreditation, licensing, and regulatory agencies. Using an integrated approach would improve hazardous material safety and disposal efforts.

The OSHA Hazard Communication Standard requires organizations to maintain a Safety Data Sheet (SDS) for each hazardous substance used in the workplace. Refer to the SDS and in some situations, the container label, for information on special storage requirements. Typical storage considerations may include factors such as temperature, ignition control, ventilation, segregation, and identification. Properly segregate hazardous materials according to compatibility. For example, never store acids with bases or oxidizers with organic materials or reducing agents. Corrosives and acids will corrode most metal surfaces including storage shelves or cabinets. Store flammable and combustible materials in appropriate rooms or approved cabinets.

BOX 6.1   HAZARDOUS MATERIAL MANAGEMENT SUGGESTIONS

- Conduct an inventory and control of all materials used, stored, or generated
- Provide adequate space and equipment for handling and storing hazardous materials
- Monitor and document correct disposal of hazardous gases and vapors
- Develop work area and emergency response procedures to address specific hazards
- Use protective equipment when responding to hazardous materials spills or releases
- Maintain hazardous wastes manifests, permits, and licenses
- Ensure proper labeling of all hazardous materials and wastes

---

II. HAZARDOUS SUBSTANCE SAFETY

Identify and mitigate risks associated with selecting, handling, storing, transporting, using, and disposing of chemicals, dangerous medications, and hazardous gases or vapors. Label hazardous materials and wastes to identify the contents and provide hazard warnings.

BOX 6.2   THE JOINT COMMISSION HAZARDOUS MATERIAL & WASTE CATEGORIES

- Hazardous chemicals
- Hazardous medications and drugs
- Radiation hazards (ionizing and nonionizing)
- Dangerous gases and vapors

BOX 6.3   CHARACTERISTICS OF HAZARDOUS SUBSTANCES

- **Corrosiveness:** Any substance with the ability to degrade the structure or integrity of another substance, object, or material. Examples include acids and alkalis.
- **Ignitability:** Any material that can too readily burn or ignite, including some chemicals that can auto-ignite upon contact with the air.
- **Reactivity:** Any substance with the ability to readily combine with other chemicals to produce a sudden or violent release or heat/energy.
- **Toxicity:** Any material with the capability of causing illness or death in man, animals, fish, or plants, or capable of damaging the environment.

A.   HAZARDOUS SUBSTANCE EXPOSURES

Toxic substances can enter the body through the skin, respiratory system, mouth, and eyes. Some substances can also damage the skin or eyes directly without being absorbed. A person can inhale or swallow inorganic lead, but it does not penetrate the skin. Sometimes a chemical substance can enter through more than one route. Exposures to hazardous materials can cause stress on the body if inhaled, absorbed, or ingested. Exposure effects depend on concentration, duration of exposure, route of exposure, physical properties, and chemical properties. Other chemicals, physical agents, and the general health of the person exposed can influence the effects exerted by a hazardous substance. Train workers how to safely handle, store, use, and segregate hazardous materials and waste products. The OSHA Hazard Communication Standard, 29 CFR 1910.1200, specifies education and training for users of hazardous chemicals.

BOX 6.4   EXPOSURE CONSIDERATIONS

- Concentration of hazardous substance
- Duration of exposure
- Available ventilation
- Temperature of the chemical
- Temperature of the surrounding air

BOX 6.5   HAZARDOUS MATERIAL EXPOSURE TERMS

- **Air Contaminant Standards:** A term used by OSHA to describe hazardous materials regulated by specific substance standards or exposure tables of 29 CFR 1910, Subpart Z.
- **Permissible Exposure Limit (PEL):** The maximum allowed OSHA exposure for workers working 8 hours during a 40-hour week.
- **Short-Term Exposure Limit (STEL):** The exposure allowed for a one-time excursion (normally measured in a 15-minute period).
- **Ceiling:** The maximum amount of an airborne concentration exposure.
- **Threshold Limit Value (TLV):** A voluntary time-weighted average (TWA) exposure limit published by the American Conference of Governmental Industrial Hygienists (ACGIH).
- **Air Contaminants:** Hazardous substances regulated by 29 CFR 1910, Subpart Z.
- **OSHA Additive Formula:** The method described in 29 CFR 1910.1000 for use in determining exposure effects of a substance containing two or more hazardous ingredients.

BOX 6.6   IDENTIFYING AND EVALUATING HAZARDOUS SUBSTANCES

- Determine hazardous properties, including toxicity and health hazards
- Identify purpose, quantities, and locations using the substance
- Implement proper storage procedures, including flammable material locations
- Make SDSs readily available for each substance
- Adhere to compliance and regulatory requirements of OSHA, DOT, and EPA
- Develop written plans as required by compliance agencies and accrediting organizations
- Require the use of personal protective equipment (PPE) when handling hazardous materials
- Evaluate possible use of less hazardous substances
- Create detailed spill containment plans and properly train response teams
- Conduct and document personal and area monitoring as required by OSHA standards
- Provide education and training for all workers with any potential exposures

B.   Hazardous Chemical Determination

Consider a substance as hazardous if regulated by OSHA in 29 CFR 1910, Subpart Z. Treat any substances included in the latest edition of the ACGIH *Documentation of the Threshold Limit Values and Biological Exposure Indices* as hazardous. Any substance confirmed or suspected to be a carcinogen by the National Toxicology Program (NTP) and published in their latest edition of the *Annual Report on Carcinogens* would also be hazardous. Finally, any substance listed by the Annual Report for Research on Cancer and in the latest edition of *IARC Monographs* is also considered to be hazardous.

C.   Reproductive Hazards

Some substances or agents may affect the reproductive health of women or men. These risks may manifest as chemical, physical, or biological hazards. Reproductive hazards can include lead, radiation, and even viruses. Reproductive hazard exposure can occur by inhalation, skin contact, and by ingestion. Potential health effects include infertility, miscarriage, birth defects, and child developmental disorders. Organizations must work to limit exposures by the use of workplace engineering controls, proper work practices, and good hygiene practices. Current scientific evidence suggests that chronic exposure to anesthetic gases increases the risk of congenital abnormalities in offspring among female workers. While more than 1000 workplace chemicals may cause reproductive effects in animals, most physical and biological agents in the workplace that may affect fertility and pregnancy outcomes remain unstudied. The inadequacy of current knowledge coupled with the ever-growing variety of workplace exposures pose a potentially serious public health problem. *The Effects of Workplace Hazards on Female Reproductive Health*, National Institute for Occupational Safety and Health (NIOSH) Publication No. 99-104, addresses exposure, prevention, and reproductive hazards for female workers and their unborn babies. The *Effects of Workplace Hazards on Male Reproductive Health*, NIOSH Publication No. 96-13, identifies steps to reduce or prevent workplace exposure to male reproductive hazards.

BOX 6.7   HEALTHCARE REPRODUCTIVE HAZARDS

- Nitrous oxide
- Ethylene oxide
- Toluene
- Xylene
- Some aerosolized drugs
- Cadmium

(Continued)

D.   THRESHOLD LIMIT VALUES (TLVs)

Published by ACGIH, TLVs represent the opinion of the scientific community for the purpose of encouraging exposure at or below the level of a published TLV. The values serve as guidelines and not standards. TLVs help industrial hygienists make decisions regarding safe levels of exposure to various chemical or physical agents found in the workplace. TLVs serve as health-based values established by committees that review existing published and peer-reviewed literature in various scientific disciplines including industrial hygiene, toxicology, and occupational medicine. ACGIH bases TLVs solely on health factors and not economic issues or any technical feasibility.

E.   CHEMICAL PROPERTIES

The physical properties of a chemical substance include characteristics such as vapor pressure, solubility in water, boiling point, melting point, molecular weight, and specific gravity. Chemical properties describe the reactivity of a substance with other chemicals. Reactive substances can burn, explode, or give off hazardous vapors when mixed with other chemicals or when exposed to air or water. Reactive substances can create fire and explosive hazards. Oxidizing chemicals easily release oxygen that can fuel fires when stored near flammable substances. Oxidizers cause other materials to burn, even though most oxidizers won't burn themselves. Ensure storage is away from heat sources because warming causes oxygen release that can create the perfect environment for a fire. Corrosive chemicals can eat through other materials including human skin. Irritants such as ammonia possess corrosive characteristics that attack mucous membranes in the nose and mouth.

F.   FLASH POINTS

According to NFPA 30, Class I flammable liquids possess a flash point of less than 100 degrees Fahrenheit (38 degrees Celsius), while combustible liquids possess a flash point of 100 degrees Fahrenheit or more. Please note the Global Harmonized System (GHS) may define flammable liquids using different criteria. Vapor is simply the gaseous state of material. We can smell some vapors and some possess no odor. Vapors combine with the oxygen in the air, forming a mixture that will ignite easily and burn rapidly, often with explosive force. Vapor density relates to the ratio of the weight of a volume of vapor or gas to the weight of an equal volume of clean but dry air. SDSs contain vapor densities for the chemical substances. Knowing the vapor density can tell you how a vapor will act. A vapor density less than 1.0 will tend to rise and spread out. This reduces the hazard. A vapor density of 1.0 or more will tend to sink to the lowest point on the ground. These vapors can then travel along the ground, sometimes for long distances, and find ignition sources. This makes chemicals with high vapor densities particularly dangerous. Consider an ignition source as anything that causes something to burn. Common ignition sources include sparks from tools and equipment; open flames such as torches, smoking materials, and pilot lights; hot particles and embers generated while grinding or welding; and hot surfaces such as electric coils and overheated bearings. Flowing liquid chemicals can create static electricity. Grounding ensures that an electrical charge goes to the ground rather than building up on the drum of flammable or combustible material. Bonding refers to a process that equalizes the electrical charge between the drum and the transfer container. This prevents the buildup of electrical charges on one of the containers. Ignition temperature refers to the minimum temperature at which a chemical will burn and continue burning without the need for an ignition source. The main difference between flammable and explosive refers to the rate of combustion or the speed at which a material burns. A fire results from a rapid release of energy. An explosion occurs when an instantaneous release of energy involves an extremely rapid rate of combustion.

G. AIRBORNE EXPOSURE

We can express airborne concentrations in terms of milligrams of substance per cubic meter of air (mg/m^3) or parts of substance per million parts of air (ppm). Asbestos and other airborne fibers can be measured using fibers per cubic centimeter (f/cc) or fibers per cubic meter (f/m^3) of air. OSHA requires consideration of feasible administrative or engineering controls to reduce exposure risks. When these controls prove ineffective, organizations must use PPE or other protective measures to protect employees. Ensure the use of any equipment and/or technical measures receive approval from a competent industrial hygienist or other technically qualified person. 29 CFR 1910, Subpart Z, contains Tables Z-1, Z-2, and Z-3, which cover exposure limits for substances not covered by a specific standard.

H. EMERGENCY SHOWERS AND EYEWASHES

OSHA Standard 29 CFR 1910.151 requires employers to provide suitable facilities for quick drenching of the eyes and body for individuals exposed to corrosive materials. OSHA does not specify minimum operating requirements or installation setup requirements. American National Standards Institute (ANSI) Standard Z358.1 recently underwent revisions led by the efforts of the International Safety Equipment Association (ISEA). Approved by ANSI, the standard became known as ANSI/ISEA Z358.1. Organizations should ensure flushing fluids remain clear and free from foreign particles. For self-contained units, manufacturers provide suggested fluid replacement guidelines. Preservatives can help control bacteria levels in flushing fluids. A preservative's performance depends upon several factors, including the initial bacterial load of the water and a potential bio-film in the station. Self-contained eyewash stations should be drained completely, disinfected, and rinsed prior to refilling. Always inspect and test the unit if you doubt its dependability. Identify problems or concerns and establish regular maintenance procedures. Consult the manufacturer's operating manual and ANSI/ISEA Z358.1 for assistance in performing test procedures, maintenance operations, and training. Personal eyewash bottles can provide immediate flushing when located in hazardous areas. However, personal eyewash equipment does not meet the requirements of plumbed or gravity-feed eyewash equipment. Personal eyewash units can support plumbed or gravity-fed eyewash units but cannot serve as a substitute.

BOX 6.8   BASIC REQUIREMENTS FOR EYEWASH AND SHOWER FACILITIES

- Valves must activate in 1 second or less.
- The facilities must be 10 seconds from the hazard.
- They must be located in a lighted area and identified with a sign.
- Train workers on equipment use and appropriate PPE.
- Activate plumbed units weekly.
- Maintain self-contained units according to manufacturers' specifications.

I. COMPRESSED GAS SAFETY

The Compressed Gas Association (CGA) promotes safe work practices for industrial gases and develops safe handling guidelines. OSHA regulates the use and safety of compressed gases in the workplace. Refer to 29 CFR 1910.101 for complete information on inspecting gas cylinders. The DOT regulates the transportation of compressed gases by rail, highway, aircraft, and waterway. Store compressed gas cylinders in cool and dry areas with good ventilation. Storage areas should meet fire-resistant standards. Never store compressed gas cylinders at temperatures higher than 125°F. Do not store cylinders near heat, open flames, or ignition sources. Properly label all cylinders and never remove valve protection caps until securing the cylinder for use. Comply with OSHA 29 CFR 1910.101–105 and DOT 49 CFR 171–179 standards when handling compressed gases. Refer to ANSI Z48.1 and CGA pamphlet C-7 for marking cylinders. When not in use, close and properly secure valves. Use appropriate lifting devices to transport gas cylinders. Refer to the appropriate SDS for information about cylinder content. Inside of buildings, separate oxygen and flammable gas cylinders by a minimum of 20 feet. Cylinders can be stored in areas with a fire-resistible partition between the oxygen and flammable materials.

III. OSHA HAZARD COMMUNICATION STANDARD (29 CFR 1910.1200)

OSHA requires the development of a written hazard communication plan. The plan must address container labeling, SDS availability, and training requirements. Employers must identify the person responsible for each plan element. Organizations must make the plan available on all shifts. The plan must direct the actions taken to communicate appropriate hazard information to all affected or exposed individuals.

A. Globally Harmonized System

The Globally Harmonized System (GHS) is the international approach to hazard communication. This global system provides criteria for classifying chemical hazards and standardizing labels and SDSs. Development of the GHS required a multiyear endeavor by hazard communication experts from different countries, international organizations, and stakeholder groups. OSHA recently modified the Hazard Communication Standard (HCS) to adopt the GHS approach. Since 1983, OSHA has required employers to communicate hazardous materials information to employees. The original performance-oriented standard allowed chemical manufacturers and importers to convey information on labels and SDSs in a variety of formats. The GHS requires the use of a standardized approach when classifying hazards and conveying the information to individuals with a need to know. It requires providing detailed criteria for determining what hazardous effects a chemical poses. It also requires standardized labels assigned by hazard class and category. This will enhance both employer and worker comprehension of the hazards resulting in safer use and handling. The harmonized format of the SDSs will enable employers, workers, health professionals, and emergency responders to access the information more efficiently and effectively. OSHA will require training on new label requirements and SDS format by December 2013. OSHA will require complete compliance in 2015.

B. Major Hazard Communication Standard Changes

The definitions of hazard will change to provide specific criteria for classification of health and physical hazards, as well as classification of mixtures. These specific criteria will help to ensure that evaluations of hazardous effects remain consistent across manufacturers. This will result in more accurate labels and SDSs. Chemical manufacturers and importers must provide a label that includes a harmonized signal word, pictogram, and hazard statement for each hazard class and category. Precautionary statements must also be provided. Finally, the SDSs will contain a specified 16-section format. The GHS does not address harmonized training provisions. However, the revised Hazard Standard requires retraining of all workers within two years of the publication of the final rule. The parts of the OSHA standard not related to the new system such as the basic framework, scope, and exemptions remain unchanged. OSHA did modify some terms to align the revised standard with language used in the GHS. The term "hazard determination" changed to "hazard classification" and "Material Safety Data Sheet" changed to "Safety Data Sheet." Evaluation of chemical hazards must use available scientific evidence concerning such hazards. The revised standard contains specific criteria for each health and physical hazard with instructions about hazard evaluations and determinations. It also establishes both hazard classes and hazard categories. The standard divides the classes into categories that reflect relative severity of the effect. The original standard did not include categories for most of the health hazards covered. OSHA included general provisions for hazard classification and Appendixes A and B to address criteria for each health or physical effect.

Under the original standard, the label preparer provided the identity of the chemical and the appropriate hazard warnings. The preparer determined the method to convey the information. The revised standard specifies what information to provide for each hazard class and category.

BOX 6.9   NEW LABELING REQUIREMENTS

- **Pictogram:** This method uses a symbol plus other graphic elements, such as a border, background pattern, or color, to convey specific information about the hazards of a chemical. Each pictogram consists of a different symbol on a white background within a red square frame set on a point (a red diamond). The system requires the use of nine pictograms. However, OSHA requires the use of only eight pictograms under the revised standard.
- **Signal Words:** This requirement consists of using a single word to indicate the relative level of severity of hazard to alert the reader of a potential hazard. The signal words used include "danger" and "warning." Use danger for severe hazard and warning for less severe hazards.
- **Hazard Statement:** This requirement consists of a statement assigned to a hazard class and category that describes the nature of the hazards of a chemical. It also includes as appropriate, the degree of hazard.
- **Precautionary Statement:** This phrase describes recommended measures to minimize or prevent adverse effects that could result from exposure to a hazardous chemical. It also applies to the improper storage or handling of a hazardous chemical.

The revised standard requires the printing of all red borders on the label with a symbol printed inside. Chemical manufacturers, importers, distributors, or employers who become aware of any significant information regarding the hazards of a chemical must revise labels within six months of becoming aware of the new information. Employers can label workplace containers with the same label affixed to the shipped containers. Employers can also use label alternatives including those described in NFPA 704, Hazard Rating and the Hazardous Material Information System (HMIS). However, information supplied on alternative labels must meet requirements of the revised standard with no conflicting hazard warnings or pictograms.

C.   SAFETY DATA SHEET CHANGES

The information required on the SDS remains essentially the same as the original standard. The original standard required specific information but did not specify a format for presentation or order of information. The revised standard requires presenting the information on the SDS using

---

BOX 6.10   REQUIRED SDS INFORMATION

- Section 1. Identification
- Section 2. Hazard(s) identification
- Section 3. Composition/information on ingredients
- Section 4. First-aid measures
- Section 5. Firefighting measures
- Section 6. Accidental release measures

(Continued)

BOX 6.10   REQUIRED SDS INFORMATION (*Continued*)

- Section 7. Handling and storage
- Section 8. Exposure controls/personal protection
- Section 9. Physical and chemical properties
- Section 10. Stability and reactivity
- Section 11. Toxicological information
- Section 12. Ecological information
- Section 13. Disposal considerations
- Section 14. Transport information
- Section 15. Regulatory information
- Section 16. Other information, including date of preparation or last revision

Note: OSHA does not mandate inclusion of sections 12 to 15 in the SDS

consistent headings in a specified sequence. Appendix D specifies the information required under each heading. The SDS format remains the same as the ANSI standard format.

OSHA plans to retain the requirement to include the ACGIH TLVs on the SDS. OSHA found that requiring TLVs on the SDS will provide employers and employees with useful information to help them assess the hazards presented by their workplaces. OSHA will also require the inclusion of PELs, and any other exposure limits used or recommended by the chemical manufacturer, importer, or employer preparing the SDS. The revised standard provides classifiers with the option of relying on the classification listings of the IARC and NTP to make classification decisions regarding carcinogenicity, rather than applying the criteria themselves. OSHA also included a nonmandatory Appendix F in the revised standard to provide guidance on hazard classification for carcinogenicity. Part A of Appendix F includes background guidance provided by GHS based on the Preamble of the IARC "Monographs on the Evaluation of Carcinogenic Risks to Humans". Part B provides IARC classification information. Part C provides background guidance from the NTP "Report on Carcinogens."

D.   Managing and Communicating Changes to the Hazard Communication Standard

Consider the GHS as a living document with expectations of relevant updates on a two-year cycle. OSHA anticipates future updates of the Hazard Communication Standard to address minor terminology changes and clarify the final rule text; also, there will likely be additional rule-making efforts to address major changes.

E.   Employee Training

The OSHA Hazard Communication Standard (29 CFR 1910.1200) requires employers to provide employees with information and training on hazardous chemicals used in their work areas. Employers must conduct training at the time of their initial assignment and upon the introduction

BOX 6.11   HAZCOM-MANDATED TRAINING TOPICS

- Existence and requirements of the OSHA Hazard Communication Standard
- Components of the local hazard communication plan
- Work areas and operations using hazardous materials
- Location of the written hazard evaluation procedures and hazard communication plan
- Location of the hazardous materials listing
- Location and accessibility of the SDS file

of a new hazardous substance. Training must address the methods and observations used to detect the presence or release of the chemical. It must also address physical and health hazards, protective measures, labeling, and an explanation of the SDS. Employers must inform employees of the hazards of nonroutine tasks and the hazards associated with chemicals in unlabeled pipes.

IV. HEALTHCARE HAZARDOUS MATERIALS

A. ACETONE

Acetone can be used as a chemical intermediate or as a solvent cleaner in fingernail polish remover, paint-related products, and in the chemical production of ketone substances. It possesses a vapor density twice that of air. Inhalation of acetone can result in slight narcosis to respiratory failure at extremely high concentrations. In the event of accidental contact, skin should be washed and affected clothing removed immediately. In case of eye contact, eyes should be rinsed for 15 minutes. Acetone should be stored in safety cans and cabinets that meet OSHA and NFPA 30 requirements. Workers should wear splash goggles and chemical protective gloves made of butyl. For respiratory protection, use air-purifying respirators equipped with organic vapor cartridges set to the manufacturer's maximum-use concentration.

B. ACRYL AMIDE

Acryl amide, a resin usually found in research labs, is used to make gels for biochemical separations. It can cause eye and skin irritation. Long-term exposure could result in central nervous system disorders. Consider acryl amide as a suspected carcinogen and mutagen.

C. AMMONIA

Ammonia is used as a liquid cleaning agent and as a refrigerant gas. Concentrated solutions of ammonia can cause severe burns. Workers should avoid skin contact with ammonia by wearing protective clothing. Workers handling concentrated solutions should wear rubber gloves and goggles or face shields. Provide adequate ventilation in areas where ammonia gas is released from concentrated solutions. Never store ammonia with deodorizing chemicals because the reaction can produce harmful byproducts such as chlorine gas. The OSHA PEL is 50 ppm based on an eight-hour TWA.

D. CADMIUM (29 CFR 1910.1027)

Cadmium is a soft, blue and white metal or grayish-white powder and is commonly used as an anti-corrosive for electroplated steel. Exposures occur mainly in gas meter refurbishing, aircraft repair, and in shipyard industries. Certain materials and products such as paints, batteries, and phosphate fertilizers also contain cadmium. Healthcare safety personnel must remember that the presence of cadmium can occur in lead molds used in radiation medicine. Treat cadmium as a potential lung carcinogen. Breathing in high levels can cause severe damage to the lungs. Short-term effects of exposure include weakness, fever, headaches, chills, sweating, and muscular pain. Long-term effects can include kidney damage, emphysema, and bone deterioration. Cadmium exposure can also cause anemia, discoloration of teeth, and loss of the sense of smell. OSHA requires establishment and implementation of a written plan for cadmium if exposure levels exceed the PEL.

E. CHLORINE COMPOUNDS

Chlorine is commonly used for sanitizing counter and tabletop surfaces. Household bleach commonly used as a disinfecting solution is a mixture of 1/4 cup chlorine to a gallon of water. Chlorine should be mixed fresh daily and used for disinfection on noncritical surfaces such as those in water tanks and bathrooms. It is also used as a laundry additive, a sanitizing solution for dishwashing, and a disinfectant for floors. Chlorine-based substances should not be mixed with materials containing ammonia because the reaction will produce a toxic gas. Mild irritation of the mucous membranes can occur at exposure concentrations of 0.5 ppm. The OSHA PEL for chlorine is a ceiling of 1 ppm according to 29 CFR 1910.1000, Table Z-1. Chlorine possesses an odor threshold between 0.02 and 0.2 ppm, but a person's sense of smell is dulled by continued exposure.

F. Iodine

Iodine works as a general disinfectant and can be used with alcohol for use as a skin antiseptic or with other substances for general disinfecting purposes. Exposure can include irritation of the eyes and mucous membranes, headaches, and breathing difficulties. Crystalline iodine or strong solutions of iodine may cause severe skin irritation because it is not easily removed and may cause burns. The OSHA PEL sets a ceiling for iodine at 0.1 ppm according to 29 CFR 1901.1000, Table Z-1.

G. Isopropyl Alcohol

Isopropyl alcohol, a widely used antiseptic and disinfectant, is used to disinfect thermometers, needles, anesthesia equipment, and other instruments. The odor of isopropyl alcohol may be detected at concentrations of 40 to 200 ppm. Exposure to isopropyl alcohol can cause irritation of the eyes and mucous membranes. Contact with the liquid may also cause skin rashes. The OSHA PEL is 400 ppm for an eight-hour TWA. Workers should use appropriate protective clothing such as gloves and face shields to prevent repeated or prolonged skin contact with isopropyl alcohol. Splash-proof safety goggles should also be provided and required for use where isopropyl alcohol may contact the eyes.

H. Methyl Methacrylate

Methyl methacrylate is used in the fields of medicine and dentistry to make prosthetic devices and as a ceramic filler or cement. It is an acrylic cement-like substance used to secure prostheses to bone during orthopedic surgery. Exposure usually occurs during mixing, preparation, and in the operating room. Symptoms from overexposure can include coughing, chest pain, headache, drowsiness, nausea, anorexia, irritability, and narcosis. Very high levels may cause pulmonary edema and death. Exposure can cause irritation to skin, which can include redness, itching, and pain. The substance may be absorbed through the skin and can irritate the eyes. Dental technicians using bare hands with methyl methacrylate molding putty developed changes in the nerve impulse transmission in the fingers. Repeated skin exposures may cause tingling or prickling sensation of the skin. Persons with pre-existing skin disorders or eye problems, or impaired liver, kidney, or respiratory function may be more susceptible to the effects of the substance. OSHA recommends mixing in a closed system, if possible.

I. Peracetic Acid

This acid is a powerful sterilant with a sharp, pungent odor. At higher concentrations (1 percent) it can promote tumors in mouse skin. A machine system containing 0.2 percent peracetic acid heated to about 50 degrees Celcius can sterilize rigid and flexible endoscopes within a 45-minute cycle time, including a rinse with water filtered through a 0.2-micrometer membrane to remove bacteria. The system uses the peracetic acid once only and is relatively expensive. Minimize odor and toxicity concerns by containing the peracetic acid within the closed machine. Peracetic acid is used to sterilize the surfaces of medical instruments and may be found in laboratories, central supply, and patient care units. It is a strong skin, eye, and mucous membrane irritant. Currently, no standards exist for regulating exposures to peracetic acid.

J. PESTICIDES

The EPA considers insecticides, herbicides, fungicides, disinfectants, rodenticides, and animal repellents as pesticides. OSHA considers pesticides as hazardous substances under the OSHA Hazard Communication Standard. EPA regulates pesticides under their Federal Insecticide, Fungicide, and Rodenticide Act (FIFRA) regulations. All pesticides sold in the United States must carry an EPA Registration Number. Consider these registered substances safe and effective when used according to directions. Pesticides labeled "DANGER - POISON" indicate highly toxic substances. If inhaled, ingested, or left on the skin, they may be lethal. Responsibility for the safe use of these toxic materials begins with purchase and continues until the empty container is properly discarded. The EPA Worker Protection Standard, 40 CFR 156 and 170, contains regulations that address the handling, loading, mixing, or applying of pesticides and the repair of pesticide-applying equipment. This standard affects all forestry, greenhouse, and nursery workers who perform hand labor in pesticide-treated fields. Some pesticide products require verbal warnings and posted warning signs.

K. PHENOL SUBSTANCES

Phenol solutions can prove effective for a wide range of bacteria. Some phenolic substances may also be used for intermediate-level disinfection when effective against tuberculosis (TB). Phenol may work well for surfaces, equipment, prosthetics, bite registrations, and partial dentures. Avoid skin or mucous membrane exposures. Phenol may be detected by odor at a concentration of about 0.05 ppm. Serious health effects may follow exposure to phenol through skin adsorption, inhalation, or ingestion. The OSHA PEL for phenol is 5 ppm for an eight-hour TWA skin exposure. Workers exposed to phenol should wash their hands thoroughly before eating, smoking, or using toilet facilities.

L. QUATERNARY AMMONIUM COMPOUNDS

These substances, widely used as disinfectants, do not work effectively against tuberculosis and gram-negative bacteria. Central sterile, environmental services, patient care areas, and clinical services use quaternary compounds for general low-level disinfecting tasks. These compounds may cause contact dermatitis and nasal irritation, but are less irritating to hands than other types of substances.

M. SOLVENTS

Most solvents remove the natural fats and oils from the skin and may be absorbed. Organic solvents pose flammability hazards. Safety personnel must properly store solvents in approved safety containers. Local exhaust ventilation and enclosure of solvent vapor sources should be used to control laboratory exposures. When selecting engineering and other controls, safety personnel must consider both toxicity and flammability risks. Toluene and xylene used in laboratories can cause eye and respiratory irritation resulting from exposure to liquid and vapor forms. Other exposure symptoms include abdominal pains, nausea, vomiting, and possible loss of consciousness, if ingested in large amounts. Most individuals can sense the odor of toluene at 8 ppm. Inhaling high levels of toluene in a short time can cause light-headed sensations and drowsiness. Toluene must be stored to avoid contact with strong oxidizers such as chlorine, bromine, and fluorine. Xylene can also be found in some maintenance departments and clinical labs. OSHA and NIOSH set xylene exposure limits at 100 ppm. Store all solvents in accordance with NFPA 30 requirements.

N. ETHYL ALCOHOL

Many healthcare facilities use 70 percent ethyl alcohol as a topical application in local skin disinfection. Consider ethyl alcohol flammable in all dilutions where vapor may come in contact with an ignition source. The flash point of a 70 percent solution is approximately 70 degrees Fahrenheit. Ethyl alcohol can enhance the drying of the skin. Take care when using to avoid dermatitis. Dispose of ethyl alcohol after thoroughly diluting with water and only in an area with adequate ventilation. Maintain ethyl alcohol in volumes over 70 percent in a flammable storage cabinet away from patient care areas. In case of fire, use a type BC fire extinguisher.

O. GLUTARALDEHYDE

Glutaraldehyde is used to disinfect and clean heat-sensitive medical, surgical, and dental equipment. Glutaraldehyde solutions serve as a tissue fixative in histology and pathology labs. Absorption may occur by inhalation, dermal contact, or ingestion. Use ventilation controls to prevent overexposure, which can cause allergic eczema and mucous membrane irritation. Date all solutions to ensure effectiveness against bactericidal contamination. Glutaraldehyde is used to disinfect and clean heat-sensitive equipment such as dialysis instruments, surgical instruments, suction bottles, bronchoscopes, and endoscopes. It works well to disinfect ear, nose, and throat instruments. The colorless and oily substance gives off a pungent odor. Hospital workers use it most often in a diluted form mixed with water or in a commercially prepared product. OSHA does not currently publish a PEL for gluturaldehyde. ACGIH recommends a ceiling TLV of 0.05 ppm (parts per million). NIOSH publishes a recommended exposure limit (REL) of 0.2 ppm for gluturaldehyde vapor from either activated or inactivated solutions. Refer to the NIOSH Publication No. 2001 -115, May 2001, *Glutaraldehyde: Occupational Hazards in Hospitals.*

P. ORTHOPTHALALDEHYDE

Orthopthalaldehyde (OPA), which was introduced as a less-toxic alternative to ethylene oxide and glutaraldehyde, is less extensively studied than other commonly used medical sterilizing agents. However, a 2005 study of risks to healthcare workers from these exposures noted that OPA has similar structural and reactive properties to glutaraldehyde and acted similarly as a chemical sensitizer. OPA does not currently have a recommended exposure limit. A 2010 Johnson & Johnson SDS for an OPA product reflected reproductive effects as unknown.

Q. ETHYLENE OXIDE (29 CFR 1910.1047)

Ethylene oxide (EtO) exposure most commonly occurs by dermal absorption or inhalation. EtO exists as a colorless liquid at temperatures below 51.7 degrees Fahrenheit. As a gas, it produces an ether-like odor at concentrations above 700 ppm. It is both flammable and highly reactive. The current OSHA PEL for EtO is 1 ppm for an eight-hour TWA with a 5 ppm excursion level. EtO is used within central supply to sterilize items that can't be exposed to steam sterilization. Exposure usually results from improper aeration of the ethylene oxide chamber after the sterilizing process. Exposure can also occur during off-gassing of sterilized items or poor gas-line connections. You can find EtO in outpatient surgery clinics, cardiac cath labs, operating rooms, dental labs, and autopsy labs. OSHA mandates an action level of 0.5 ppm and a STEL of 5.0 ppm averaged over a sampling period of 15 minutes, and prohibits worker rotation as a way of compliance with the excursion limit. OSHA also requires employers to develop a "written emergency plan" to protect employees during any potential release. Employers must also provide information through signs and labels that clearly indicate carcinogenic and reproductive hazards. Initial and annual training should be given to workers who may be exposed at the action level.

R. FORMALDEHYDE (29 CFR 1910.1048)

Formaldehyde is used as a fixative and is commonly found in most laboratories and the morgue. It can cause eye and respiratory irritation, abdominal pains, nausea, and vomiting. A high concentration of vapor inhaled for long periods can cause laryngitis, bronchitis, or bronchial pneumonia. Prolonged exposure may cause conjunctivitis. OSHA considers formaldehyde a suspected carcinogen and other studies indicate it as a potential carcinogen as well. Exposure risk areas include autopsy rooms, pathology laboratories, and dialysis units. Exposure also commonly occurs in endoscopy and surgical facilities. Pre-placement and periodic examinations should include baseline and periodic pulmonary, dermal, and hepatic evaluations. Require the use of PPE, including appropriate gloves, when responding to spills. Odor is not a reliable warning for the presence of formaldehyde because a worker's ability to smell formaldehyde is quickly extinguished. Airborne concentrations above 0.1 ppm can irritate the eyes, nose, and throat. Formaldehyde is often combined with methanol and water to make formalin. In 1992 OSHA lowered the PEL for formaldehyde to 0.75 ppm as an eight-hour TWA.

V. HAZARDOUS DRUGS

Studies indicate that workplace exposures to hazardous drugs can result in health problems such as skin rashes, infertility, spontaneous abortions, congenital malformations, and possibly leukemia or other cancers. Health risks vary by the extent of the exposure and potency and/or toxicity of hazardous drugs. Potential health effects can be minimized through sound procedures for handling hazardous drugs, engineering controls, and proper use of protective equipment to protect workers to the greatest degree possible. The NIOSH Working Group on Hazardous Drugs in 2004 defined as hazardous any drug exhibiting at least one of the following characteristics in humans or animals: (1) carcinogenic or other developmental toxicity, (2) reproductive toxicity, (3) organ toxicity at low doses, and (4) structure and toxicity profiles of new drugs that mimic existing drugs determined hazardous by the above criteria. Workers must receive standardized training on the hazardous drugs and equipment/procedures used to prevent exposure. Prepare all agents within a ventilated cabinet designed to protect workers and adjacent personnel from exposure. Develop procedures to provide product protection for all drugs that require aseptic handling. Use two pairs of powder-free, disposable chemotherapy gloves with the outer one covering the gown cuff when mixing drugs. Avoid skin contact by using a disposable gown made of a low lint and low permeability fabric. A face shield is required to avoid splash incidents involving the eyes, nose, or mouth when engineering controls do not provide adequate protection. Wash hands with soap and water immediately before using and after removing personal protective clothing, such as disposable gloves and gowns. Use syringes and IV sets with lock-type fittings when preparing and administering hazardous drugs. Decontaminate work areas before and after each mixing activity. Clean up small spills immediately using appropriate safety precautions and PPE. Implement the facility spill response plan for large spills.

A. HAZARDOUS PHARMACEUTICAL WASTES

The EPA defines hazardous wastes in the RCRA. This waste also includes pharmaceutical wastes that contain toxic chemicals or exhibit properties that make them hazardous to the environment and/or humans. RCRA wastes include broken or spilled vials, partial vials, expired products, and patient personal medications. EPA limits hazardous waste maximum storage time to 90 days or 180 days based on generator status. All wastes must be stored in a separate and locked area clearly marked to avoid becoming a food source or breeding place for insects or animals. Like infectious wastes, there is no time limit to fill the container. Not all states mandate the same storage requirements. Contact local and state authorities for additional information. The EPA's P-listed chemicals (40 CFR 261.33) include such pharmaceuticals as epinephrine, nicotine, chloroform, and warfarin over 0.3%. The U-listed chemicals (40 CFR 261.33) include many used in chemotherapy, such as paraldehyde, mercury, phenol, and warfarin under 0.3%

VI. MEDICAL GAS SYSTEMS

Refer to NFPA 99-2012, Chapters 1 to 5, for the medical gas information previously contained in NFPA 99C. Medical gas personnel can also access the NFPA 99-2012 handbook for all of the former NFPA 99C medical gas and vacuum systems content. Refer to the CGA pamphlet *Characteristics and Safe Handling of Medical Gases* (No. P2). Also refer to the OSHA publication *Anesthetic Gases: Guidelines for Workplace Exposures* for occupational safety information as well as *Controlling Exposures to Nitrous Oxide during Anesthetic Administration*, NIOSH Publication No. 94-100 for additional information and guidance. Bulk medical gas systems involving oxygen and nitrous oxide should meet the requirements of the CGA pamphlet *Standard for the Installation of Nitrous Oxide Systems* (No. 8.1) or NFPA 50, Standard for Bulk Oxygen Systems. For additional information refer to ANSI/Z 79.11-1982, Anesthesia Gas Scavenging Devices and Disposal Systems. The NIOSH publication *Development and Evaluation of Methods for the Elimination of Waste Anesthetic Gases and Vapors in Hospitals* contains information about control methods to establish and maintain low concentrations of waste anesthetic gas in operating rooms. The ASSE 6040 Medical Gas Systems Maintenance Personnel certification requires that individuals qualified under the provisions of ASSE 6040 perform all maintenance of medical gas and vacuum systems. The candidates shall be employed or contracted by a healthcare facility, or actively engaged in working with medical gas systems, and document one year of minimum experience in the maintenance of medical gas systems. A minimum score of 75 percent is required to pass the exam.

Inspect, test, and maintain critical components of piped medical gas systems, including master signal panels, area alarms, automatic pressure switches, shutoff valves, flexible connectors, and outlets. Test piped medical gas and vacuum systems upon installation, modification, or repair, including cross-connection testing, piping purity testing, and pressure testing. Maintain the main supply valve and area shutoff valves of piped medical gas and vacuum systems to assure they are accessible and clearly labeled. These systems provide oxygen, nitrous oxide, and compressed air throughout the facility. When using such systems, heavy and bulky bottles or tanks do not require physical transport throughout a building. A plan addressing preventive maintenance and periodic inspection helps ensure that medical gas systems operate safely and reliably. As part of an effective management and maintenance plan, inspections and corrective actions should be documented and any faulty fittings should be repaired or replaced immediately. In the event of a system outage, notify nursing personnel and medical staff. In addition, labeling shutoff controls and providing signs at outlet locations, as well as identifying piping, will help ensure safety in a facility. Keep system piping free from contamination and protect cylinders from weather extremes. Some other requirements include the following:

- Doors or gates to enclosures for the gas supply systems must be locked.
- Enclosures for gas supply systems must not be used for storage purposes other than for cylinders containing the nonflammable gases that are to be distributed through the pipeline.
- Storage of empty cylinders disconnected from the supply equipment is permissible.
- Empty cylinders must be segregated and identified.
- Cylinders not in use must be capped and secured in a vertical position by a chain or similar device.
- Cylinders connected to a manifold must also be secured. Plumbing (tubing, etc.) to the manifold will not suffice for this purpose.
- Smoking is prohibited in the gas supply system enclosure. "No Smoking" signs should be posted.

A. ANESTHETIC GAS HAZARDS

Healthcare worker exposures to anesthetic gases can result in risk of occupational illnesses. Healthcare facilities can now better control anesthetic gases through the use of improved scavenging systems, installation of more effective general ventilation systems, and increased attention to equipment maintenance and leak detection as well as careful anesthetic practices. Exposure can occur in the operating room, recovery room, or post-anesthesia care unit. Some potential health effects of exposure to waste anesthetic gases include nausea, dizziness, headaches, fatigue, irritability, drowsiness, and problems with coordination and judgment, as well as sterility, miscarriages, birth defects, cancer, and liver and kidney disease. Use appropriate anesthetic gas scavenging systems in operating rooms. Appropriate waste gas evacuation involves collecting and removing waste gases, detecting and correcting leaks, considering work practices, and effectively ventilating the room. The American Institute of Architects recommend an air exchange rate of 15 air changes per hour with a minimum of three air changes of fresh air per hour. Never recirculate operating room air containing waste anesthetic gases to the operating room or other hospital locations. Exposure measurements taken in operating rooms during the clinical administration of inhaled anesthetics indicate that waste gases can escape into the room air from various components of the anesthesia delivery system. Potential leak sources include tank valves; high- and low-pressure machine connections; connections in the breathing circuit; defects in rubber and plastic tubing, hoses, reservoir bags, and ventilator bellows; and the connector. In addition, selected anesthesia techniques and improper practices such as leaving gas flow control valves open and vaporizers on after use, spillage of liquid inhaled anesthetics, and poorly fitted face masks or improperly inflated tracheal tube and laryngeal mask airway cuffs also can contribute to the escape of waste anesthetic gases into the surgical area atmosphere. OSHA does not publish exposure limits regulating halogenated agents. NIOSH issues RELs for both nitrous oxide and halogenated agents. The NIOSH REL for nitrous oxide, when nitrous oxide is used as the sole inhaled anesthetic agent, is 25 ppm measured as a TWA during the period of anesthetic administration. NIOSH also recommends that no worker should be exposed at ceiling concentrations greater than 2 ppm of any halogenated anesthetic agent over a sampling period not to exceed one hour. NIOSH does currently publish RELs for isoflurane, desflurane, and sevoflurane.

B. SCAVENGING

Scavenging is the process of collecting and disposing of waste anesthetic gases and vapors from breathing systems at the site of overflow. It is carried out to protect operating room personnel by preventing the dispersal of anesthetic gases into the room air. A scavenging system consists of two key components including a collecting device or scavenging adapter to collect waste gases. The system also has a disposal method to carry gases from the room. Organizations should create a management plan to address techniques for scavenging, maintaining equipment, monitoring air, and minimizing leakage while administering anesthesia. Persons responsible for health and safety in the hospital surgical department should be aware of the availability of new products and new information on familiar products.

BOX 6.12   SCAVENGING SYSTEM COMPONENTS (ASTM F 1343–91)

- A gas collection assembly such as a collection manifold or a distensible bag captures excess anesthetic gases at the site of emission and delivers it to the transfer tubing.
- The transfer tubing then conveys the excess anesthetic gases to the interface.
- The interface provides positive and sometimes negative pressure relief and may provide reservoir capacity. It is designed to protect the patient's lungs from excessive positive or negative scavenging system pressure.

(Continued)

BOX 6.12   SCAVENGING SYSTEM COMPONENTS (ASTM F 1343–91) (*Continued*)

- Gas disposal assembly tubing conducts the excess anesthetic gases from the interface to the gas disposal assembly.
- The gas disposal assembly conveys the excess gases to a point where they can be discharged safely into the atmosphere. Several methods in use include a nonrecirculating or recirculating ventilation system, a central vacuum system, a dedicated waste gas exhaust system, or a passive duct system.

C.   NITRIC OXIDE

Nitric oxide was approved by the FDA in 1999 for use as a vasodilator in the treatment of hypoxic respiratory failure in full- and near-term infants. It is a colorless and essentially odorless gas with a very narrow therapeutic window for patients. Acute exposure effects include mucous membrane irritation and drowsiness. More serious effects include delayed pulmonary toxicity and damage to the central nervous system effects. Exposed employees may seem relatively asymptomatic at the time of exposure. It can take as long as 72 hours to manifest clinical symptoms. OSHA classifies nitric oxide as a highly hazardous substance.

D.   NITROUS OXIDE

Nitrous oxide (N_2O), a clear, colorless, and oxidizing liquefied gas, possesses a slightly sweet odor. The product remains stable and inert at room temperature. While classified by the DOT as a nonflammable gas, nitrous oxide will support combustion and can deteriorate at temperatures in excess of 1202 degrees Fahrenheit. Nitrous oxide is blended with oxygen when used in anesthesia applications. Pure nitrous oxide will cause asphyxiation. The painkilling and numbing qualities of inhaled nitrous oxide begin to take effect at concentrations of 10 percent. The CGA and National Welding Supply Association (NWSA) identify initiatives to address N_2O abuse issues.

VII.   MANAGING WASTE

Virtually all healthcare and industrial facilities generate hazardous waste as defined by the RCRA. Each facility must develop effective waste management plans. Effectively managing inventory provides the next best opportunity to reduce hazardous waste generation. Never discard any hazardous chemical down the drain, in a toilet, or on the ground outside. Never attempt to burn chemical waste under any circumstances. Never place hazardous chemicals in trashcans or garbage containers destined for landfills. Always read the label, check the SDS, and follow established facility procedures. Even a small amount of some chemicals when left in a container can pose a danger. Dispose of all waste containers according to required procedures. Wastes can react with one another and burn, release toxic vapors, or explode.

BOX 6.13   KEY ELEMENTS OF WASTE MANAGEMENT

- Inventory and categorization of wastes
- Identification of hazards during daily activities
- Knowledge of storage requirements
- Distribution and special handling issues
- Special cleanup procedures
- Disposal procedures including identification, transport, pickup, and end point

---

The EPA requires generators to track all materials using the "cradle to grave" management approach. Develop policies and procedures for identifying, handling, storing, using, and disposing of hazardous wastes from generation to final disposal. Provide training for all exposed personnel. Monitor personnel who manage or regularly come into contact with hazardous materials and/or wastes. Evaluate the effectiveness of the planning documents and provide reports to senior leaders. RCRA Subtitle C regulations focus on the management of waste with hazardous properties. The regulations help protect human health and the environment from mismanagement of hazardous waste. Generators normally cannot store hazardous waste for more than 90 days under EPA regulations. RCRA Subtitle C established four characteristics of hazardous waste. The four categories include corrosiveness, ignitability, reactivity, and toxicity. The EPA considers solid waste as hazardous waste if it meets the following criteria: (1) listed as a hazardous waste in the regulations, (2) substance contains a listed hazardous waste, and (3) waste was derived from the treatment, storage, or disposal of a listed hazardous waste. The EPA and state environmental agencies possess the authority to inspect facilities and their records at any reasonable time. When an EPA inspector finds that the facility is in violation of RCRA or its permit, enforcement action in the form of a compliance order, including administratively imposed injunctions or court actions, may follow.

A. Medical Waste

Healthcare facilities, hospitals, clinics, physician offices, dental practices, blood banks, veterinary clinics, medical research facilities, and laboratories all generate medical waste. The Medical Waste Tracking Act of 1988 defines medical waste as "any solid waste that is generated in the diagnosis, treatment, or immunization of human beings or animals, in research pertaining thereto, or in the production or testing of biologicals." This definition includes, but is not limited to, blood-soaked bandages, culture dishes and other glassware, surgical gloves and instruments, sharps and needles, cultures, and pathological waste. The EPA also defines medical waste in 40 CFR 259.10 and 40 CFR 22 as any solid waste generated in the diagnosis, treatment, or immunization of human beings or animals. Most states regulate medical waste within their jurisdictions.

VIII. RESPIRATORY PROTECTION (29 CFR 1910.134)

Respirators prevent the inhalation of harmful airborne substances and provide fresh air in oxygen-deficient environments. An effective respiratory protection plan must address the following: (1) hazards encountered, (2) type and degree of protection needed, (3) medical evaluation for respirator usage, (4) selection and fit requirements, (5) training on use and care, and (6) methods to ensure continued effectiveness.

A. Types of Respirators

Air-purifying respirators come in either full face or half mask versions. These types of respirators use a mechanical or chemical cartridge to filter dust, mists, fumes, vapors, or gaseous substances. Only use disposable air-purifying respirators once or until the cartridge expires. These respirators contain permanent cartridges with no replaceable parts. Reusable air-purifying respirators use both replaceable cartridges and parts. The replaceable cartridges and parts must come from the same manufacturer to retain NIOSH approval. Disposable or reusable air-purifying respirators contain no replaceable parts except cartridges. Gas masks, designed for slightly higher concentrations of organic vapors, gases, dusts, mists, or fumes, use a volume of sorbent much higher than a chemical cartridge. Powered air-purifying respirators use a blower to pass the contaminated air through a filter. The purified air then enters into a mask or hood. They filter dusts, mists, fumes, vapors, or gases like other air-purifying respirators. Never use air-purifying respirators in any oxygen-deficient atmosphere.

Oxygen levels below 19.5 percent require either a source of supplied air or a supplied-air respirator. Consider levels below 16 percent as unsafe, as death could result. Supplied-air respirators provide the highest level of protection against highly toxic and unknown materials. "Supplied air" refers to self-contained breathing apparatuses and airline respirators. Airline respirators contain an air hose connected to a fresh air supply from a central source. The source comes from a compressed air cylinder or air compressor that provides breathable air. Emergency Escape Breathing Apparatuses provide oxygen for short periods of times, such as 5 or 10 minutes, depending on the unit. Only permit these devices for emergency situations such as escaping from environments immediately dangerous to life or health (IDLH). Determine the correct cartridge for air-purifying respirators by contacting a respirator professional or referring to the SDS of the substance needing filtering. Cartridges use a color scheme designating the contaminant needing filtering. Replace the cartridge any time a wearer detects odor, irritation, or taste of a contaminant. The proper selection and use of a respirator depends upon an initial determination of the concentration of the hazard or hazards present in the workplace or area with an oxygen-deficient atmosphere. IDLH atmospheres pose the most danger to workers. Use a full face piece pressure demand self-contained breathing apparatus (SCBA) or a combination full face piece pressure demand supplied-air respirator (SAR) for dangerous atmospheres. Respirator selection requires matching the respirator with the degree of hazard and needs of the user. Choose only devices that fully protect the worker and permit job accomplishment with minimum physical stress.

BOX 6.14   RESPIRATOR SELECTION CONSIDERATIONS

- Physical and chemical properties of the air contaminant
- Concentration of the contaminant
- PELs
- Nature of the work operation or process
- Length of time respirator worn
- Work activities and physical/psychological stress
- Fit testing, functional capabilities, and limitations of the respirator

Persons assigned tasks requiring use of a respirator must possess the physical ability to work while using the device. OSHA requires employers to ensure the medical fitness of individuals that must wear respirators. The fitness evaluation considers the physical and psychological stress imposed by the respirator. It must also evaluate the stress originating from job performance. Employers must ensure that employees pass the evaluation prior to fit testing or permitting use of the respirator for the first time. A physician or other licensed healthcare professional must determine medical eligibility for respirator wear. A qualified healthcare provider includes physicians, occupational health nurses, nurse practitioners, and physician assistants if licensed to do so in the state in which they practice.

The OSHA standard requires the fit testing of all tight-fitting respirators. OSHA does not exclude disposable particulate respirators from fit testing requirements. Some employees may not achieve an adequate fit with certain respirator models or a particular type of respirator. Provide alternative respirator choices to ensure worker protection. Employers must provide a sufficient number of respirator models and sizes from which employees can choose an acceptable respirator with a correct fit. A Quantitative Fit Test (QNFT) refers to the assessment of the adequacy of respirator fit by numerically measuring the amount of leakage into the respirator. A QNFT uses an instrument to take samples from a wearer's breathing zone. Adhere to the OSHA protocol for a QNFT as detailed in Appendix A of 29 CFR 1910.134. A Qualitative Fit Test (QLFT) refers to a pass/fail test that assesses adequacy of respirator fit that relies on the individual's response to a test agent. A QLFT, according to 29 CFR 1910.134, applies only to negative pressure air-purifying respirators that must achieve a fit factor of 100 or less. Since the QLFT relies upon the subjective response of the wearer, accuracy may vary.

The proper fit, usage, and maintenance of respirators remain the key elements that help ensure employee protection. Train employees about the proper use of respirators, and the general requirements of the Respiratory Protection Standard. Training must address employer obligations such as written plans, respirator selection procedures, respirator use evaluation, and medical evaluations. Employers must ensure proper maintenance, storage, and cleaning of all respirators. They must also retain and provide access to specific records as required by OSHA. Employees must know basic employer obligations as related to their protection. OSHA requires annual training of all workers expected to wear a respirator. New employees must attend respirator training prior to using a respirator in the workplace.

Employers must conduct workplace evaluations to ensure the scope of the written respirator plan protects those required to use respirators. Employers must also evaluate the continued effectiveness of the written respirator plan. Proper evaluations help to determine if workers use and wear respirators correctly. The evaluations can also indicate the effectiveness of respirator training. Employers must solicit employee views about respirator plan effectiveness and determine any problem areas.

OSHA requires employers retain written information regarding medical evaluations, fit testing, and respirator plan effectiveness. Maintaining this information promotes greater employee involvement and provides compliance documentation. Employers must retain a record for each employee subject to medical evaluation. This record includes results of the medical questionnaire and, if applicable, a copy of the healthcare professional's written opinion. Maintain records related to recommendations, including the results of relevant examinations and tests. Retain the records of medical evaluations and make them available as required by 29 CFR 1910.1020, Access to Employee Exposure and Medical Records. Retain fit test records for users until the administration of the next test.

REVIEW EXERCISES

1. What three federal agencies regulate some aspect of hazardous materials?
2. List the hazardous materials and waste categories defined by the Joint Commission.
3. Describe the four primary characteristics of hazardous substances.
4. List the five key exposure considerations of hazardous materials.
5. Define the following terms:
   - PEL
   - STEL
   - TLV
   - OSHA Additive Formula
6. List six reproductive hazards found in healthcare facilities.
7. Define the concept of vapor density.
8. Describe the basic requirements for emergency showers and eyewashes.
9. Describe the basic OSHA requirements for hazard communication plans.
10. List the OSHA Hazard Communication mandated training topics.
11. Describe the hazards of the following chemical substances:
    - Cadmium
    - Methyl Methacrylate
    - Solvents
    - Glutaraldehyde
    - Ethylene Oxide
12. How did the NIOSH Working Group in 2004 define hazardous drugs?
13. What voluntary standard contains information about scavenging system components?

7 Infection Control and Prevention

I. INTRODUCTION

Healthcare-associated infections (HAIs) pose significant risks to both patients and healthcare personnel. Healthcare organizations must adhere to current Centers for Disease Control and Prevention (CDC) guidelines concerning infection prevention and control. The Occupational Safety and Health Administration (OSHA) mandates organizations take specific actions to minimize employee exposures to bloodborne pathogens. Healthcare safety personnel should work closely with the infection control and prevention staff on issues of joint concern such as compliance with the OSHA Bloodborne Pathogens Standard (29 CFR 1910.1030). Healthcare personnel must strive to minimize their exposure risks to blood, sputum, aerosols, and other body fluids by following proper work practices and using personal protective equipment (PPE)/clothing. The use of appropriate PPE, along with relevant education and training, provides the foundation for infection prevention. Any effective infection control and prevention function should stress personal hygiene, individual responsibility, surveillance monitoring, and investigation of infectious diseases and pathogens. Efforts should emphasize identifying infection risks, instituting preventive measures, eliminating unnecessary procedures, and preventing the spread of infectious diseases through the healthcare facility. Facilities should develop and implement plans to prevent and control infections by focusing on

- Integrating infection control into safety and performance improvement efforts
- Assessing risks for the acquisition or transmission of infectious agents within the facility
- Using an epidemiological approach to focus on surveillance and data collection data
- Implementing infection prevention and control processes based on sound data
- Coordinating plan design and implementation with key leaders
- Establishing coordinated processes to reduce the risks of organization-acquired infections
- Appointing one or more qualified individuals to lead facility efforts
- Reporting information about infections both internally and to public health agencies

---

BOX 7.1 INFECTION CONTROL PLAN DEVELOPMENT CONSIDERATIONS

- Device-related, intravascular devices, ventilators, and tube-feeding infections
- Surgical-site infections and HAIs in special care units
- Infections caused by organisms that are antibiotic resistant
- Tuberculosis and other communicable diseases
- Infections in the neonate population
- Geographic location of the facility
- Volume of patient or resident encounters
- Patient populations served
- Clinical focus of the facility
- Number of employees and staff

- Designing processes to reduce rates or trends of epidemiologically significant infections
- Implementing strategies to reduce risks and prevent transmission of infections
- Adopting strategies that consider scientific knowledge, practice guidelines, laws, or regulations
- Considering endemic rates and epidemic rates when analyzing data
- Ensuring management systems support infection control objectives
- Stressing risk identification using surveillance, prevention, and control activities
- Coordinating with external organizations to reduce infections from the environment

II.   HEALTHCARE IMMUNIZATIONS

Healthcare organizations should establish a comprehensive written policy regarding immunizing personnel, develop a listing of all required and recommended immunizations, and refer all staff to the employee health function to receive education about needed immunizations appropriate for their positions. The employee health function should consider medical history and job position exposure risk to determine needed vaccinations.

BOX 7.2   HEALTHCARE VACCINE CATEGORIES

- **Strongly Recommended:** Diseases posing special risks including hepatitis B, influenza, measles, mumps, rubella, and varicella.
- **Recommended in Some Situations:** Active and/or passive immunizations as indicated by job circumstances to prevent occurrences of tuberculosis, hepatitis A, meningitis, and typhoid fever.
- **Recommended for All Adults:** Immunization for tetanus, diphtheria, and pneumonia disease.

A.   GUIDELINES OF THE ADVISORY COMMITTEE FOR IMMUNIZATION PRACTICE

Healthcare personnel should meet the Advisory Committee for Immunization Practice (ACIP) guidelines for immunization against mumps, rubella, diphtheria, and measles. Consider the need for the following vaccinations:

- **Rubella:** Require for individuals considered at risk, including those with direct contact with pregnant patients.
- **Hepatitis B:** Offer all staff members with any potential exposure to bloodborne pathogens the vaccine within 10 days of their job assignment as mandated by OSHA.
- **Measles:** Consider immunization for persons susceptible by history or serology.
- **Influenza:** Healthcare personnel should receive flu immunization to help prevent spread of influenza.

B.   OTHER VACCINATION CONSIDERATIONS

Healthcare organizations must develop comprehensive policies and protocols for management and control of outbreaks of vaccine preventable diseases as described in the ACIP Guidelines. Healthcare employees working abroad should consider vaccinations for diseases such as hepatitis A, poliomyelitis, encephalitis, meningitis, plague, rabies, typhoid, and yellow fever. Healthcare organizations should develop written policies regarding work restrictions or exclusion from duty for immunization and infection control reasons. Require healthcare staff members to report any illnesses, medical conditions, or treatments that could make them susceptible to opportunistic infections.

III.   CENTERS FOR DISEASE CONTROL AND PREVENTION

The CDC publishes guidelines, advisories, and recommendations that do not carry the force of law. The CDC bases their guidance and recommendations on scientific studies. However, some infection control practices applicable to one setting may not apply in all healthcare situations. The guidance offered by the CDC gives healthcare infection control personnel the information necessary to make informed decisions. Organizations must provide proper education and training on current infection control and prevention practices including the latest OSHA requirements and CDC recommendations.

The continuous evaluation of care practices under the supervision of the infection control staff can help ensure continued adherence to correct practices. The Association for Professionals in Infection Control and Epidemiology (APIC) publishes up-to-date articles and guidelines on healthcare infection control.

A.   CDC GUIDELINES FOR HAND HYGIENE IN HEALTHCARE SETTINGS

CDC guidelines highly recommend the placement of alcohol-based hand-rub solutions in convenient locations of patient care areas of healthcare organizations. Clinical studies indicate that the frequency of handwashing relates to the accessibility of hand-hygiene facilities. Installing hand-rub dispensers immediately outside patient or resident rooms or within suites of rooms improves the overall efficacy of staff use by over 20%.

B.   GUIDELINES FOR ENVIRONMENTAL INFECTION CONTROL IN HEALTHCARE FACILITIES

The guidelines provide excellent information on maintaining a safe healthcare environment and include infection control tips to follow during inspection, construction, or renovation activities in patient care and treatment areas. The guidelines provide a comprehensive review of the relevant literature with a focus on conducting a risk assessment before undertaking any activities that could generate dust or water aerosols. The guidelines also review infection control measures for catastrophic events such as flooding, sewage spills, and loss of utilities, including ventilation. Environmental infection control procedures must consider disease transmission via surfaces, laundry, plants, animals, medical wastes, cloth furnishings, and carpeting. These guidelines do not apply to sick buildings, terrorism, or food safety. Key suggestions include

- Evaluating the impact of activities on ventilation and water systems
- Creating a multidisciplinary team to conduct infection control risk assessment
- Using dust-control procedures and barriers during construction activities
- Implementing special control measures in any areas with patients at high risk
- Using air sampling to monitor air filtration and dust-control measures
- Controlling tuberculosis risks in operating rooms when infectious patients require surgery
- Culturing water as part of a control plan for Legionella if appropriate
- Recovering from water system disruptions, leaks, and natural disasters
- Disinfecting surfaces to control antibiotic-resistant microorganisms
- Developing specific infection-control procedures for laundries
- Establishing control procedures for using animals in activities and therapy
- Managing the use of all service animals in healthcare facilities
- Developing strategies for animals receiving treatment in human facilities
- Measuring water use from main lines for dialysis, ice machines, hydrotherapy, dental water lines, and automated endoscope reprocessing equipment

Other CDC Infection Control Guidelines include

- CDC Position Statement on Reuse of Single Dose Vials (2012)
- Basic Infection Control and Prevention Plan for Outpatient Oncology Settings (October 2011)

- Guide to Infection Prevention in Outpatient Settings: Minimum Expectations for Safe Care (July 2011)
- CDC Issues Checklist for Infection Prevention in Out-Patient Settings to Accompany New Guide (July 2011)
- Guideline for the Prevention and Control of Norovirus Gastroenteritis Outbreaks in Healthcare Settings (2011)
- Guideline for Disinfection and Sterilization in Healthcare Facilities (2008)
- Guideline for Isolation Precautions: Preventing Transmission of Infectious Agents in Healthcare Settings (2007)
- Guideline–Management of Multidrug-Resistant Organisms in Healthcare Settings (2006)
- Public Reporting of Healthcare-Associated Infections (2005)
- Bloodstream Infection: Guideline for the Prevention of Intravascular Catheter-Related Infections (2011)
- Dialysis–Multi-Dose Vials Infection Control (2008)
- Environmental Infection Control (2003)
- Hand Hygiene (2002)
- Infection Control–Healthcare Personnel (1998)
- Occupational Exposures (2005)
- Pneumonia (2003)
- Surgical Site Infection (1999)
- Guidelines for Preventing Transmission of Mycobacterium Tuberculosis in Healthcare Facilities (2005)
- Urinary Tract Infection (2009)

C. CDC STANDARD PRECAUTIONS

Consider handwashing as the first line of defense in preventing exposures to diseases, bloodborne pathogens, and infections. The CDC's Standard Precautions provide the major features of blood and body fluid precautions designed to reduce the risk of transmission of bloodborne pathogens. Use the Standard Precautions to reduce the risk of transmission of microorganisms from both recognized and unrecognized sources of infection in hospitals. After using the Standard Precautions, facilities should then apply the appropriate Tiered Precautions for airborne, droplet, and contact routes of infection. Facilities should learn to recognize the pathogenic risk of body fluids, secretions, and excretions and should take precautions against the various routes of transmission by designing processes that eliminate confusion with regard to infection control or isolation requirements. Facilities should refer to the guidelines for information on clinical syndromes and empiric precautions. They should also adhere to specific transmission precautions for patients colonized with pathogens. CDC's tier precautions provide guidelines with regard to

- Handwashing, glove use, and patient placement procedures, including transport
- Use of masks, gowns, and other protective apparel
- Procedures for patient care equipment, linen, and laundry
- Cleaning dishes, glasses, cups, and eating utensils

D. CDC ISOLATION PRECAUTIONS

Federal, state, and local health agencies publish rules and guidelines that define isolation procedures. Healthcare organizations should follow these guidelines because several routes can transmit infectious agents:

- **Contact:** Contamination due to close proximity with persons with a contagious disease
- **Indirect contact:** Contamination by contacting an object contaminated by an infected person

- **Droplet:** Contamination caused by a person sneezing, coughing, or talking
- **Common vehicle:** Disease spread by food, water, drugs, devices, or equipment
- **Airborne:** Air-suspended infectious nuclei or dust that could be inhaled or digested
- **Vector-borne:** Organisms carried by animals or insects

1. Airborne Precautions

Airborne precautions reduce the risk of airborne transmission of infectious agents disseminated by airborne droplet nuclei (small-particle residue 5 micrometers or smaller), evaporated droplets that may remain suspended in the air for long periods of time, or dust particles containing the infectious agent. Microorganisms disperse widely by air currents and may become inhaled or deposited on a susceptible host within the same room. Depending on environmental factors, use special air handling and ventilation to prevent airborne transmission. Airborne precautions apply to patients with known or suspected pathogens such as measles, varicella, and tuberculosis.

2. Droplet Precautions

Droplet precautions reduce the risk of droplet transmission of infectious agents. Droplet transmission involves contact of the conjunctive or mucous membranes of the nose or mouth of a susceptible person with large-particle droplets (larger than 5 micrometers). The droplets contain microorganisms generated from a person with a clinical disease or who serves as a carrier of the microorganism. Droplets generated from the source person during coughing, sneezing, or talking and during performance of certain procedures such as suctioning and bronchoscopy can result in exposure risks. Transmission of a disease via large-particle droplets requires close contact between the source and recipient. Droplets do not remain suspended in the air and generally travel only short distances (3 feet or less). Special air handling and ventilation is not required to prevent droplet transmission. Droplet precautions apply to any patient with known or suspected infections such as meningitis, pneumonia, sepsis, pharyngeal diphtheria, Mycoplasma pneumonia, pertussis, streptococcal group A pharyngitis, scarlet fever, Rubella, and pneumonic plague.

3. Contact Precautions

Contact precautions reduce the risk of transmission of epidemiologically important microorganisms by direct or indirect contact. Direct contact transmission involves skin-to-skin contact or the physical transfer of microorganisms to a susceptible host. This can occur when caregivers turn patients, bathe patients, or perform other patient-related activities with physical contact. Direct contact transmission can also occur between patients. Indirect contact transmission involves contact of a susceptible host with a contaminated object or surface in the patient environment. Contact precautions apply to specific patients with a known or suspected infection with epidemiologically important microorganisms transmitted by contact via gastrointestinal tracts, respiratory systems, skin surfaces, wounds, and multidrug-resistant bacteria. Use contact precautions for hepatitis A, contagious skin infections, herpes simplex viruses, scabies, and viral or hemorrhagic infections.

E. NEW INFECTION RISK

The CDC recently released information that four percent of U.S. hospitals and 18 percent of nursing homes had treated at least one patient with the bacteria called carbapenem-resistant Enterobacteriaceae (CRE) during the first six months of 2012. CRE are in a family of more than 70 bacteria called enterobacteriaceae, including *Klebsiella pneumoniae* and *E. coli*, that normally live in the digestive system. In recent years, some of these bacteria became resistant to last-resort antibiotics known as carbapenems. Most CRE infections occur in patients with prolonged stays in hospitals, long-term facilities, and nursing homes. These bacteria can kill up to 50 percent of infected patients. According to the CDC, these bacteria can easily spread from patient to patient on the hands of caregivers. CRE bacteria can transfer their antibiotic resistance to other bacteria of the same type. To reduce spread of these bacteria, the CDC wants hospitals and other healthcare facilities to take the following steps:

- Enforce infection-control precautions
- Group together patients with CRE
- Segregate staff, rooms, and equipment with CRE patients
- Inform facilities about the transfer of patients with CRE
- Use antibiotics carefully

IV. CENTERS FOR MEDICARE & MEDICAID SERVICES, HOSPITAL-ACQUIRED CONDITIONS, AND PRESENT ON ADMISSION INDICATORS

On February 8, 2006 the President signed the Deficit Reduction Act (DRA) of 2005, which required there be an adjustment in Medicare Diagnosis Related Group (DRG) payments for certain hospital-acquired conditions (HACs) with a component that addresses new Present on Admission (POA) coding. Section 5001(c) of the DRA required the Secretary to identify, by October 1, 2007, at least two conditions for which hospitals under the Inpatient Prospective Payment System (IPPS) would not receive additional payment beginning on October 1, 2008, if the condition was not POA. The conditions must involve high cost or high volume or both and result in the assignment of a case to a DRG with a higher payment when present as a secondary diagnosis. Preventing the incident could reasonably occur through the use of evidence-based guidelines. For discharges occurring on or after October 1, 2008, hospitals would not receive additional payment for cases in which one of the selected HACs was not POA. HACs for potential reduced payment include the following infections:

- Catheter-associated urinary tract infections. Note: ICD-9 coding does not distinguish between catheter-associated infection and inflammation.
- Vascular catheter-associated blood stream infection (BSI). The Centers for Medicare & Medicaid Services (CMS) developed a specific code for central-line vascular catheters (CVC) and CVC-BSI is not limited to the intensive care unit (ICU).
- Surgical-site infection mediastinitis occurring after coronary artery bypass grating (CABG) surgery.
- Orthopedic surgeries involving spinal fusion and other surgeries of the shoulder and elbow
- Bariatric surgery for morbid obesity including laparoscopic gastric bypass and gastro-enterostomy

V. DISINFECTANTS, STERILANTS, AND ANTISEPTICS

We can divide chemical germicides into three general categories:

- Sterilizing agents, used to eliminate all microbial life on objects or surfaces, including bacterial spores that can survive other germicides
- Disinfectants, classified as high, medium, or low, depending on the strength required, and which can destroy nearly all microbial life on objects or surfaces except for bacterial spores
- Antiseptics, used to inactivate or destroy organisms on skin or living tissue

A. GERMICIDAL EFFECTIVENESS

Bacterial spores exhibit the most resistance to germicides, followed by myco-bacteria, non-lipid viruses, fungi, and vegetative bacteria. Lipid viruses exhibit the least resistance. Facilities should use Food and Drug Administration (FDA)- or EPA-approved cleaning agents and should read and follow the manufacturer's instructions to ensure proper use. Their effectiveness depends on

- Shape and texture of surface
- Amount of contamination on the surface
- Resistance of contaminants to the germicide
- Amount of soil buildup, including blood, mucous, or tissue
- Chemical composition of the germicide
- Time of exposure to the germicide
- Temperature of the germicide

B. REGULATORY APPROVAL OF DISINFECTANTS

The EPA oversees the manufacture, distribution, and use of disinfectants. Manufacturers must use pre-established test procedures to ensure product stability, determine toxicity to humans, and assess microbial activity. If the product passes these requirements, the EPA registers the substance for use. The EPA regulates disinfectants under the authority of the Federal Insecticide, Fungicide, and Rodenticide Act (FIFRA). The FDA regulates liquid chemical sterilants and high-level disinfectants such as hydrogen peroxide and peracetic acid under the authority of the Medical Devices Amendment to the Food, Drug, and Cosmetic Act of 1976. The FDA regulates the chemical germicides if marketed for use on specific medical devices. Regulatory authority requires the manufacturer to provide instructions for the safe and effective use of substances with that device. The FDA uses the same basic terminology and classification scheme as does the CDC, which categorizes medical devices as critical, semicritical, and noncritical. The scheme classifies antimicrobial effectiveness or sterilization as high, intermediate, and low level.

The EPA registers environmental surface disinfectants based on the manufacturer's microbiological activity claims. The EPA does not use the terms "intermediate level" and "low level" when classifying disinfectants. The CDC designates any EPA-registered hospital disinfectant without a tuberculocidal claim as a low-level disinfectant. Consider an EPA-registered hospital disinfectant effective against tuberculosis as an intermediate-level disinfectant. The EPA also lists disinfectant products according to their labeled use against certain organisms. OSHA requires the use of EPA-registered hospital tuberculocidal disinfectants or EPA-registered hospital disinfectants labeled effective against human immunodeficiency virus (HIV) and hepatitis B virus (HBV) for decontaminating work surfaces. Hospitals can use disinfectants with HIV and HBV claims if surfaces contain no contamination requiring the use of a higher-level disinfectant.

C. EPA's REGISTERED STERILIZERS AND TUBERCULOCIDAL AND ANTIMICROBIAL PRODUCTS

All of EPA's registered pesticides must possess an assigned EPA Product Registration Number. Alternative brand names possess the same EPA Product Registration Number as the primary product name. The EPA Product Registration Number remains the key way to identity the substance. An EPA Establishment Number refers to the production location. A formulation or device uses a set of codes that consist of the registrant's ID Number followed by the state where produced and Facility Number. The EPA updates their registered disinfectant lists periodically to reflect label changes, cancellations, and transfers of product registrations. Information on the following list does not constitute a label replacement. Inclusion of products in these lists does not constitute an endorsement of one product over another. The EPA organizes the lists alphabetically by product names and by numerical order of their EPA Registration Numbers.

BOX 7.3   EPA REGISTRATION CATEGORIES

- List A: EPA's Registered Antimicrobial Products as Sterilizers
- List B: EPA's Registered Tuberculocide Products Effective Against Mycobacterium tuberculosis (TB)
- List C: EPA's Registered Antimicrobial Products Effective Against Human HIV-1 Virus
- List D: EPA's Registered Antimicrobial Products Effective Against Human HIV-1 and HBV
- List E: EPA's Registered Antimicrobial Products Effective Against TB, HIV-1, and HBV
- List F: EPA's Registered Antimicrobial Products Effective Against Hepatitis C Virus (HCV)
- List G: EPA's Registered Antimicrobial Products Effective Against Norovirus
- List H: EPA's Registered Antimicrobial Products Effective Against Methicillin-Resistant Staphylococcus aureus (MRSA) and Vancomycin-Resistant Enterococcus faecalis or faecium (VRE)
- List J: EPA's Registered Antimicrobial Products for Medical Waste Treatment
- List K: EPA's Registered Antimicrobial Products Effective Against Clostridium difficile Spores

D.   CDC RECOMMENDATIONS

The CDC does not test, evaluate, or otherwise recommend specific brand-name products of chemical germicides. The CDC recommends disinfecting environmental surfaces or sterilizing or disinfecting medical equipment with products approved by the EPA and FDA. When no registered or approved products are available for a specific pathogen or use situation, the CDC suggests following specific guidance regarding unregistered uses for various chemical germicides. No antimicrobial products hold registered status for use against severe acute respiratory syndrome (SARS), Norwalk virus, or Creutzfeldt–Jakob disease agents.

BOX 7.4   CDC DISINFECTING LEVELS

- High-level disinfection processes can expect to destroy all microorganisms with the exception of high numbers of bacterial spores.
- Intermediate-level disinfection inactivates Mycobacterium tuberculosis, vegetative bacteria, most viruses, and most fungi but does not necessarily kill bacterial spores.
- Low-level disinfection can kill most bacteria, some viruses, and some fungi but does not kill resistant microorganisms such as tubercle bacilli or bacterial spores.

E.   SELECTING A DISINFECTANT

Healthcare facilities use a number of disinfectants, including alcohol, chlorine, chlorine compounds, hydrogen peroxide, phenolic substances, and quaternary ammonium compounds. Never routinely interchange disinfectants. Proper selection and use of disinfectants provide the key to effective safety and quality control. Alcohols demonstrate variable effectiveness against some bacteria and fungi. Alcohols act fast, leave no residue, and can compatibly combine with other disinfectants such as quaternaries, phenolic substances, and iodine to form tinctures. Aldehydes can prove effective against a wide spectrum of bacteria and viruses, including spores, when used properly. They also demonstrate activity against other pathogens, including vegetative bacteria and viruses. Chlorine works very well for cleaning up blood or body-fluid spills. Chlorine compounds work as effective biocides on tuberculosis and vegetative bacteria. Chlorine compounds prove effective against HIV after 10 to 20 minutes and demonstrate effectiveness at a 1:5 dilution against bacterial spores and mycobacteria. Diluted chlorine neutralizes rapidly in the presence of organic matter. Chlorine compounds work very well for the decontamination of HBV, HCV, and cleanup of biohazardous spills.

VI. OSHA BLOODBORNE PATHOGENS STANDARD (29 CFR 1910.1030)

On November 6, 2000, the President signed the Needlestick Safety and Prevention Act (Pub. L. 106-430). The Act required OSHA to revise the OSHA Bloodborne Pathogens standard within six months of enactment of the Act. To facilitate expeditious completion of this directive, Congress explicitly exempted OSHA from procedural requirements generally required under the rule-making provision of the act (paragraph 6(b)) and from the procedural requirements of the Administrative Procedure Act (5 U.S.C. 500 et seq.). The Bloodborne Pathogens standard sets forth requirements for employers with employees exposed to blood or other potentially infectious materials. In order to reduce or eliminate the hazards of occupational exposure, an employer must implement an exposure control plan for the worksite with details on employee protection measures. The plan must also describe how an employer will use a combination of engineering and work practice controls, ensure the use of personal protective clothing and equipment, and provide training, medical surveillance, hepatitis B vaccinations, and signs and labels, among other provisions.

Engineering controls provide the primary means of eliminating or minimizing employee exposure and include the use of safer medical devices, such as needleless devices, shielded needle devices, and plastic capillary tubes. Many different medical devices can now reduce the risk of needle sticks and other sharps injuries. These devices replace sharps with nonneedle devices or incorporate safety features designed to reduce injury. Despite advances in technology, needle sticks and other sharps injuries continue to occur at high rates. The revised OSHA standard became effective April 18, 2001, adding new requirements for employers, including additions to the exposure control plan and keeping a sharps injury log. It did not impose any new requirements for employers to protect employees from sharps injuries. The original standard already required employers to adopt engineering and work practice controls that would eliminate or minimize employee exposure from hazards associated with bloodborne pathogens. The revision requires organizations to implement engineering controls such as using safer medical devices to reduce or eliminate exposure risks. Exposure control plan requirements must make clear that employers must implement safer medical devices proven appropriate, commercially available, and effective. Organizations must get input on selecting devices from those responsible for direct patient care. The updated standard also requires employers to maintain a log of injuries occurring from contaminated sharps.

A. EXPOSURE CONTROL PLAN

The revision included new requirements regarding the employer's exposure control plan, including an annual review and update to reflect changes in technology that eliminate or reduce exposures to bloodborne pathogens. The employer must

- Consider new innovations in medical procedures and technology that reduce the risk of exposure to needle sticks
- Consider and document use of appropriate, commercially available, and effective safer needles
- Realize that no single medical device can prove effective for all circumstances
- Identify devices used, the method in place to evaluate those devices, and justification for the eventual selection

- Select devices based on reasonable judgment, but never jeopardize patient or employee safety
- Select devices that will make an exposure incident involving a contaminated sharp less likely to occur

B.   OSHA Hand Hygiene Requirements

The OSHA Bloodborne Pathogen Standard requires that personnel wash their hands immediately or as soon as feasible after removal of gloves or other PPE. OSHA requires that employees wash their hands and any other skin with soap and water or flush mucous membranes with water immediately or as soon as feasible following contact of such body areas with blood or other potentially infectious materials. Personnel removing gloves after exposure to blood or other potentially infectious materials must wash their hands using an appropriate soap and running water. Staff members with no access to a readily available sink after an exposure may decontaminate hands with a hand cleanser or towelette. However, staff must wash their hands with soap and running water as soon as feasible. If no exposure or contact occurs with blood or other potentially infectious materials, consider use of alcohol-based hand cleansers as appropriate. Use alcohol based sanitizing solutions when the location does not support handwashing facilities. Use a sanitizer with an alcohol concentration of 62% or greater. It is important to note the hand sanitizers are effective against common diseases, but they are ineffective against certain organisms such as bacterial spores. Never use sanitizers as a substitute for soap and water. Research shows that with just three applications of an alcohol-based sanitizer, the effectiveness of the sanitizer decreases. The reason for the decreased effectiveness is alcohol in the sanitizers can remove natural oils from your hands, which will cause your hands to dry out and crack.

C.   Employee Involvement

Employers must solicit input from nonmanagerial employees responsible for direct patient care regarding the identification, evaluation, and selection of effective engineering controls, including safer medical devices. Employees selected should represent the range of exposure situations encountered in the workplace, such as those in geriatric, pediatric, or nuclear medicine and others involved in the direct care of patients. OSHA will check for compliance with this provision during inspections by questioning a representative number of employees to determine if and how their input was requested. Employers must document in the exposure control plan how they received input from employees. Organizations can meet this obligation by listing the employees involved and describing the process used to obtain their input. Employers can also present other documentation, including references to the minutes of meetings, copies of documents used to request employee participation, or records of responses received from employees.

D.   Recordkeeping

Employers with employees experiencing occupational exposure to blood or other potentially infectious materials must maintain a log of occupational injuries and illnesses under existing recordkeeping rules, but must also maintain a sharps injury log. Maintain this log in a manner that protects the privacy of employees. The sharps injury log may include additional information that the employer must protect due to privacy requirements. Employers can determine the format of the log. At a minimum, the log must contain the following:

- Type and brand of device involved in the incident
- Location of the incident
- Description of the incident

E. Engineering Controls

Engineering controls include all control measures that isolate or remove a hazard from the workplace, such as sharps disposal containers and self-sheathing needles. The original Bloodborne Pathogens standard was not specific regarding the applicability of various engineering controls (other than the above examples) in the healthcare setting. The revision now specifies that safer medical devices, such as sharps with engineered sharps injury protection and needleless systems, constitute an effective engineering control. The phrase "sharps with engineered sharps injury protection" describes nonneedle sharps or needle devices containing built-in safety features used for collecting fluids and administering medications or other fluids. It can also describe other procedures involving the risk of sharps injury. This description covers a broad array of devices including

- Syringes with a sliding sheath that shields the attached needle after use
- Needles that retract into a syringe after use
- Shielded or retracting catheters
- Intravenous medication (IV) delivery systems that use a catheter port with a needle housed in a protective covering

F. Needleless Systems

The term "needleless systems" refers to any devices that provide an alternative to needles for procedures with a risk of injury involving contaminated sharps. Examples include intravenous medication systems that administer medication or fluids through a catheter port using nonneedle connections. Consider jet injection systems, which deliver liquid medication beneath the skin or through a muscle, as another example.

G. Exposure Determination

Exposure determination involves listing all job classifications in which employees could encounter potential exposure. This includes physicians, nurses, and other clinical personnel. Maintenance, environmental services, and laundry personnel can work in situations that could pose exposure risks. List any specific procedures or tasks in which exposure could occur without regard to the use of PPE.

H. Control Measures

Employers should take appropriate preventative measures against occupational exposure. These include engineering controls and work practice controls. Examples of engineering controls include biohazard hoods, puncture-resistant sharps containers, mechanical pipette devices, and other devices that permanently remove the hazard or isolate individuals from exposure. Organizations must evaluate and incorporate new safer devices, including needleless devices, needles with sheaths, and blunt suture needles. Work practice controls must include handwashing policies, sharps handling procedures, proper waste disposal techniques, and other actions that would reduce the likelihood of exposure.

I. Personal Protective Equipment

Employers must provide PPE to all personnel with occupational exposure. Select personal protective equipment that does not permit blood or other potentially infectious materials to pass through or reach a person's outer clothing, undergarments, skin, eyes, mouth, or other mucous membranes. Ensure personnel wear gloves when hand contact occurs with blood or other potentially infectious materials. Replace disposable gloves as soon as possible when contaminated or no longer in condition to provide barrier protection. Decontaminate reusable utility gloves and discard immediately if cracked, discolored, or punctured, or if they show signs of deterioration. Require personnel to wear masks, eye protection, and face protection when exposed to potentially infectious splashes, spray, or droplets. Require the use of gowns, aprons, and other clothing to protect against anticipated exposure to the body, head, and feet.

J.   HOUSEKEEPING, LAUNDRY, AND WASTE PRACTICES

Employers should create a schedule for periodic cleaning and appropriate disinfecting to ensure that the worksite remains clean and sanitary. Personnel should place and transport contaminated laundry in properly labeled or color-coded bags and containers. They should disinfect contaminated work surfaces after completing the task. Clean surfaces contaminated by splashes or spills and when they come into contact with blood or other potentially infectious materials. Clean the area at the end of the work shift. Place all blood or infectious materials, contaminated items that could release infectious materials, or contaminated sharps in appropriate sharps containers or closable, color-coded, or properly labeled leak-proof containers or bags. Dispose of infectious waste in accordance with federal, state, and local regulations. Attach warning labels to all containers used for the storage or transport of potentially infectious materials. Use labels of orange or red-orange color with the biohazard symbol in a contrasting color. Employers can substitute red containers or bags for warning labels.

K.   HEPATITIS B VIRUS

HBV causes an estimated two million deaths annually worldwide, establishes a carrier state in many victims, and generally produces some jaundice along with many acute symptoms such as (1) painful joint aches, (2) significant skin rashes, and (3) serve liver damage mediated by host immune reactions to the presence of HBV particles. If jaundice appears, it can persist for two to six weeks. HBV infection can cause severe fatigue and weakness, brown urine, and pale stools. The virus that causes HBV is found in blood and other body fluids, including semen, vaginal secretions, urine, and even saliva. Most people recover, but up to 10 percent become chronic carriers. These chronic carriers can spread the disease to others for an indefinite period of time and create a high risk for other diseases including cirrhosis of the liver and primary liver cancer. Although the blood and blood products provide the key transmission vehicle, viral antigen can also appear in tears, saliva, breast milk, urine, semen, and vaginal secretions. The virus can survive for seven days or more on environmental surfaces exposed to body fluids containing the virus. Infection may occur when the virus transmitted by infected body fluids or implanted via mucous surfaces becomes introduced through breaks in the skin.

1.   Hepatitis B Vaccination

All healthcare personnel with potential exposure to blood, blood-contaminated body fluids, other body fluids, or sharps should receive a vaccination. Administer the hepatitis B vaccine using the intramuscular route in the deltoid muscle. The OSHA Bloodborne Pathogens standard requires employers to offer the hepatitis B vaccine free of charge to all potentially exposed employees within 10 days of hire. Administer post-exposure prophylaxis with hepatitis B immunoglobulin (passive immunization) and/or vaccine (active immunization) when indicated after per-cutaneous or mucous membrane exposure to blood known or suspected to contain hepatitis B. Needle-stick or other percutaneous exposures of unvaccinated persons should lead to initiation of the hepatitis B vaccine series. HBV vaccination requirements are as follows:

- OSHA requires employers to offer the HBV vaccination series to all personnel with potential occupational exposure to blood or other potentially infectious material within 10 days of hire.

- Employers should always follow U.S. Public Health Service and CDC recommendations for hepatitis B vaccination, serologic testing, follow-up, and booster dosing.
- Employers should test personnel for anti-HBs 12 months after completion of the three doses.
- Healthcare staff members should complete a second three-dose vaccine series or receive evaluation to determine if HBV-positive (if no antibody response occurs to the primary vaccine series).
- Retest personnel for anti-HBs at the completion of the second vaccine dose. If no response to the second three-dose series occurs, retest nonresponders for HBV.
- Before vaccinating HBV negative nonresponders, counsel them regarding their susceptibility to HBV infection and precautions.
- Employers should provide employees with appropriate education regarding the risks of HBV transmission and availability of the vaccine. Employees who decline the vaccination should sign a declination form. Employers must maintain the form.
- Make the vaccination available without cost to the employee, at a reasonable time and place for the employee, by a licensed healthcare professional and according to recommendations of the U.S. Public Health Service, including routine booster doses.
- Provide the healthcare professional designated by the employer to implement this part of the standard with a copy of the Bloodborne Pathogens standard.
- The healthcare professional must provide the employer with a written opinion stating whether the hepatitis B vaccination is indicated for the employee and whether the employee received the vaccination.

BOX 7.5   WHEN THE HEPATITIS B VACCINATION IS NOT REQUIRED

- Employees previously completed the hepatitis B vaccination series
- Immunity confirmed through antibody testing
- Vaccine contraindicated for medical reasons
- Following participation in a prescreening plan
- Employees who decline the vaccination

Note: Employees who decline to accept the hepatitis B vaccination must sign a declination form indicating the employer offered the vaccination.

L.   HEPATITIS C

Hepatitis C (HCV), a contagious liver disease, results from infection with the hepatitis C virus. It can range in severity from a mild illness lasting a few weeks to a serious, lifelong illness that damages the liver. Hepatitis C can occur in "acute" or "chronic" forms. Consider acute hepatitis C infection as a short-term illness that occurs within the first six months after exposure. Approximately 75 to 85 percent of people who become infected with the hepatitis C virus develop chronic infection. For reasons unknown, 15 to 25 percent of people "clear" the virus without treatment and do not develop chronic infection. Chronic hepatitis C virus infection can progress into a long-term illness that occurs when the virus remains in a person's body. Over time, it can lead to serious liver problems, including liver damage, cirrhosis, liver failure, or liver cancer. Before widespread screening of the blood supply began in 1992, hepatitis was also commonly spread through blood transfusions and organ transplants. Although rare, outbreaks of hepatitis C do occur from blood contamination in medical settings. Most people with hepatitis C present no symptoms. Symptoms can appear two weeks to six months after exposure. Symptoms can include fever, fatigue, no appetite, nausea, vomiting, abdominal pain and dark urine, clay colored bowel movements, joint pain, and jaundice.

M. HUMAN IMMUNODEFICIENCY VIRUS

HIV affects the immune system, rendering the infected individual vulnerable to a wide range of disorders. Infections typically lead to the death of the patient. Symptoms can occur within a month and can include fever, diarrhea, fatigue, and rash. Exposed persons may develop antibodies and not present symptoms for months to years. The infected person may finally develop a wide range of symptoms depending on the opportunistic infections against which the body's immune system cannot defend.

BOX 7.6   HIV EXPOSURE AND TRANSMISSION ROUTES

- Contact with blood, semen, vaginal secretions, and breast milk
- Sexual intercourse
- Using needles contaminated with the virus
- Contact with HIV-infected blood under the skin, mucous membranes, or broken skin
- Mother-to-child contact at the time of birth
- Blood transfusions or organ transplants

BOX 7.7   WORKPLACE TRANSMISSION OF HIV

- Body fluids such as saliva, semen, vaginal secretions, cerebrospinal fluid, synovial fluid, pleural fluid, peritoneal fluid, pericardial fluid, amniotic fluid, and any other body fluids visibly contaminated with blood
- Saliva and blood contacted during dental procedures
- Unfixed tissue or organs other than intact skin from living or dead humans
- Organ cultures, culture media, or similar solutions
- Blood, organs, and tissues from experimental animals infected with HIV or HBV

BOX 7.8   MEANS OF TRANSMISSION

- Accidental injury with a sharp object contaminated with infectious material, such as needles, scalpels, broken glass, and anything that can pierce the skin
- Open cuts, nicks, skin abrasions, dermatitis, acne, and mucous membranes
- Indirect transmission, such as touching a contaminated object or surface and transferring the infectious material to the mouth, eyes, nose, or open skin

N. OTHER KEY TOPICS

Employees should know what to do when confronted with an emergency involving blood or other potentially infectious materials, post-exposure evaluations, the HBV vaccine, and the use of signs and labels. After training, make vaccinations available to those who run the risk of exposure. Employers should establish a medical record for each employee with occupational exposure. Keep this record confidential and keep it separate from other personnel records. Employers can keep these records onsite or healthcare professionals providing services to the employees can retain the records. The medical record contains the employee's name, Social Security number, HBV vaccination status, date of the HBV vaccination (if applicable), and the written opinion of the healthcare professional regarding the hepatitis B vaccination. Note any occupational exposure in the medical record to document the incident, and include the results of testing following the incident. The post-evaluation written opinion of the healthcare professional becomes a part of the medical record. The medical record must document what information was provided to the healthcare provider. Maintain medical records for 30 years past the last date of employment of the employee. Ensure confidentiality of medical records. Never disclose a medical record or part of a medical record without direct written consent of the employee or as required by law. Keep training records for three years. Training records must include the date, content outline, trainer's name and qualifications, and names and job titles of all persons attending the training sessions. Employers who cease to do business should transfer the medical and training records to the successor employer. Upon request, make both medical and training records available to the Assistant Secretary of Labor for Occupational Safety and Health. Make training records available to employee upon request. The employee or anyone given the employee's written consent may obtain medical records.

O.   LATEX ALLERGIES

OSHA's Bloodborne Pathogens standard requires handwashing after removal of gloves or other PPE to help minimize the amount of powder or latex remaining in contact with the skin. Employees can develop latex sensitivity or latex allergy from exposure to latex in products such as latex gloves. The National Institute of Occupational Safety and Health (NIOSH) estimates that 8 to 12 percent of healthcare personnel may experience latex reactions ranging from contact dermatitis to possibly life-threatening sensitivity. Among the alternatives are synthetic, low-protein, and powder-free gloves. Powder-free gloves may reduce systemic allergic responses. Employees should never wear latex gloves when no risk of exposure to blood or other potentially infectious materials exists. Never assume hypoallergenic gloves, glove liners, or powder-free gloves are latex-free. Use good housekeeping practices to remove latex-containing dust from the workplace.

Frequently clean areas contaminated with latex dust such as upholstery, carpets, and ventilation ducts. Frequently change ventilation filters and vacuum bags used in latex-contaminated areas. Employ appropriate work practices to reduce the chance of reactions to latex such as not using oil-based hand creams or lotions, which can cause glove deterioration, unless shown to reduce latex-related problems and maintain glove barrier protection. After removing latex gloves, wash hands with a mild soap and dry thoroughly.

P.   INFORMATION AND TRAINING

All employees with occupational exposure must receive initial and annual training on the hazards associated with exposure to bloodborne pathogens. The training must also address the protective measures taken to minimize the risk of occupational exposure. Employers must conduct retraining when changes in procedures or tasks occur. Consider OSHA employee-training requirements as performance oriented. However, employers may tailor their presentations to the employees' backgrounds and responsibilities. Ensure the training addresses the topics listed in paragraph (g)(2)(vii) of 29 CFR 1910.1030.

Employers must provide training at the time of initial employment and at least annually thereafter. Provide annual retraining for employees within one year of their original training date. Refresher training must cover topics listed in the standard to the extent needed and must emphasize new information or procedures. Employers must train part-time employees, temporary employees, and those referred to as "agency" or "per diem" employees. OSHA requires training to include an explanation of the use and limitations of methods that will prevent or reduce exposure, including appropriate engineering controls, work practices, and PPE. Training must include instruction in any new techniques and practices. Use hands-on training if possible.

1.   Training Methods and Interactive Question Opportunities

Training employees solely by means of a film or video without the opportunity for a discussion period would constitute a violation of 29 CFR 1910.1030(g)(2). Never consider a computer program, even an interactive one, as appropriate unless the employer supplements such training with the required site-specific information. Provide trainees with direct access to a qualified trainer during all training sessions. OSHA permits employers to meet the requirement if personnel can directly access a trainer by way of a telephone hotline. OSHA does not consider use of electronic mail systems to answer employee questions unless a trainer answers e-mailed questions at the time the questions arise.

2. Trainer Qualifications

OSHA requires persons conducting training to possess knowledge in the subject matter covered by the elements contained in the training plan. The trainer must demonstrate expertise in the area of the occupational hazard of bloodborne pathogens and know local procedures. Trainers, such as infection control practitioners, registered nurses, occupational health professionals, physician's assistants, emergency medical technicians, industrial hygienists, and professional trainers, may conduct the training, provided they know the subject matter covered in the training plan as it relates to the workplace. In dentist and physician offices, individual employers may conduct the training, provided they understand bloodborne pathogen exposure control and the subject matter required by the standard.

Q. MEDICAL RECORDKEEPING

Medical recordkeeping, covered by 29 CFR 1910.1020(h), requires employers to keep medical and training records for each employee. OSHA permits employers not to retain medical records of employees working for less than a year if given to the employee upon termination of employment. Keep medical records confidential except for disclosures permitted by the standard or by other federal, state, or local laws. Make all medical records required by the standard available to OSHA. The Compliance Officer must protect the confidentiality of these records. If copied for the case file, follow the provisions of 29 CFR 1913.10. Consider records about employee exposure to bloodborne pathogens and documenting HIV/HBV status as medical records.

1. Training Recordkeeping

OSHA requires accurate recordkeeping of training sessions, including titles of the employees who attend. The records assist the employer and OSHA in determining whether the training plan adequately addresses the risks involved in each job. Additionally, this information can prove helpful in tracking the relationship between exposure incidents and the corresponding levels of training. Store training records onsite to permit easy access. Do not consider training records as confidential. Retain training records for three years from the training date.

R. HAZARDOUS WASTE OPERATIONS AND EMERGENCY RESPONSE (29 CFR 1910.120)

The standard covers all personnel expected to respond to emergencies caused by the uncontrolled release of a hazardous substance. The definition of hazardous substance includes any biological agent or infectious material that may cause disease or death. Potential scenarios where the Bloodborne Pathogens and Hazardous Waste Operations and Emergency Response (HAZWOPER) standards may interface include healthcare staff members responding to an emergency caused by the uncontrolled release of infectious material. Employers of employees engaged in these types of activity must comply with the requirements in 29 CFR 1910.120 as well as the Bloodborne Pathogens standard. If there is a conflict or overlap, the provision that is more protective of employee safety and health applies.

S. POST-EXPOSURE EVALUATION AND FOLLOW-UP

Employers should provide a confidential medical evaluation for any employees involved in an exposure incident. The evaluation documents the exposure route and all circumstances related to the incident, including blood testing, HIV/HBV status of source, and appropriate medical/psychological treatment. An exposure incident can occur to a specific eye, the mouth, mucous membranes, nonintact skin, or any other contact with potentially infectious material that results from the performance of an employee's duties. Employees should immediately report exposure incidents to permit timely medical evaluation and follow-up by a healthcare professional. The employer can request testing of the source individual's blood for HIV and HBV. Consider a source individual as any patient whose blood or body fluids provide the source of an exposure incident to an employee. At the time of the exposure, the exposed employee must report to a healthcare professional. The employer must provide the healthcare professional with a copy of the Bloodborne Pathogens standard and a description of the employee's job duties as they relate to the incident. The employer also must provide a report of the specific exposure, including route of exposure, relevant employee medical records (including hepatitis B vaccination status), and results of the source individual's blood tests, if available. Draw a baseline blood sample if the employee consents. If the employee elects to delay HIV testing of the sample, the healthcare professional must preserve the employee's blood sample for at least 90 days. Never repeat testing for known HIV- or HBV-positive individuals. Most states require written consent before testing. Treat the results of the source individual's blood test as confidential. Make the results available to the exposed employee through consultation with the healthcare professional. The healthcare professional will provide a written opinion to the employer. The opinion can only provide a statement that the employee received the results of the evaluation. The employer must provide a copy of the written opinion to the employee within 15 days. This requirement remains the only information shared with the employer following an exposure incident. Treat all other employee medical records as confidential. Provide all evaluations and follow-up visits at no cost to the employee. They must take place at a reasonable time and place. Perform evaluations and follow-up visits under the supervision of a licensed physician or another licensed healthcare professional. All evaluations must follow the U.S. Public Health Service guidelines current at the time. Conduct all laboratory tests by using an accredited laboratory and at no cost to the employee.

**BOX 7.9   EVALUATING THE CIRCUMSTANCES
SURROUNDING AN EXPOSURE INCIDENT**

- Engineering controls in use at the time
- Work practices followed
- Description of the device being used
- Protective equipment or clothing used at the time of the exposure incident
- Where the incident occurred
- Procedure being performed when the incident occurred
- Employee's training

VII.   TUBERCULOSIS

Every healthcare setting should conduct initial and ongoing evaluations for risk of TB transmission. The TB risk assessment determines the types of administrative, environmental, and respiratory-protection controls needed for a setting and serves as an ongoing evaluation tool of the quality of TB infection control and for the identification of needed improvements in infection-control measures. Review the community profile of TB disease in collaboration with the state or local health department. Consult the local or state TB-control plan to obtain epidemiologic surveillance data necessary to conduct a TB risk assessment for the healthcare setting. Review the number of patients with suspected or confirmed TB disease encountered in the setting during at least the previous five years. The screening plan should consist of four major components: (1) baseline testing for TB infection, (2) serial testing for TB, (3) serial screening for symptoms or signs of TB disease, and

(4) TB training and education. Surveillance data from healthcare workers (HCWs) can protect both HCWs and patients. Screening can prevent future transmission by identifying lapses in infection control and expediting treatment for persons with latent TB infection (LTBI) or TB disease. Test for TB and document TB screening according to procedures in the 2005 CDC Guidelines. Maintain the protection of privacy and confidentiality of all test results.

A.   TB Screening Procedures for Settings Classified as Low Risk

All employees should receive baseline TB screening upon hire, using a two-step tuberculin skin test (TST) or a single blood assay Mycobacterium tuberculosis (BAMT) to test for infection. After baseline testing, do not conduct additional screening if an exposure to TB occurs. Individuals with a baseline positive or newly positive test result for TB or documentation of treatment for LTBI or TB disease should receive one chest radiograph result to exclude TB disease or provide an interpretable copy within a reasonable time frame, such as six months. Do not repeat radiographs unless symptoms or signs of TB disease develop or as recommended by a clinician.

B.   TB Screening Procedures for Settings Classified as Medium Risk

All healthcare personnel should receive baseline TB screening upon hire, using two-step TST or a single BAMT to test for infection. After baseline testing, personnel should receive TB screening annually. Personnel with a baseline positive or newly positive test result or documentation of previous treatment for LTBI or TB disease should receive one chest radiograph result to exclude TB disease. Instead of participating in serial testing, personnel should receive a symptom screen annually. This screen should be accomplished by educating the employees about symptoms of TB disease and instructing them to report any such symptoms immediately. Treatment for LTBI should be in accordance with CDC guidelines.

C.   TB Screening Procedures for Settings Classified as Potential Ongoing Transmission

Perform testing for TB infection every eight to ten weeks until lapses in infection control are corrected and no additional evidence of ongoing transmission exists. Consider the classification of potential ongoing transmission as a temporary classification only. It warrants immediate investigation and

---

BOX 7.10   TB TRAINING AND EDUCATION TOPICS

- Basic concepts of transmission, pathogenesis, and diagnosis
- Explanation of the difference between latent and active tuberculosis
- Signs and symptoms of active tuberculosis
- Increased risk for those infected with HIV
- Potential for occupational exposure
- Information about prevalence of tuberculosis in the community
- Situations that increase the risk of exposure
- Principles of infection control
- Importance of skin testing and significance of a positive test
- Principles of preventive therapy for latent tuberculosis
- Drug therapy procedures for active tuberculosis
- Importance of notifying the facility
- Information about medical evaluation for symptoms of active tuberculosis

corrective steps. After a determination that the ongoing transmission has ceased, reclassify the setting as medium risk. Recommend maintaining the classification of medium risk for at least one year.

D.   OSHA Tuberculosis Exposure Enforcement Guidelines

These guidelines address patient and healthcare staff testing, source control methods, decontamination techniques, and prevention of tuberculosis-contaminated air. This enforcement policy refers to CDC guidelines and the OSHA general-duty clause. OSHA conducts inspections in response to complaints and during routine compliance visits in the following workplaces:

- Healthcare settings
- Correctional institutions
- Homeless shelters
- Long-term care facilities
- Drug treatment centers

1.   OSHA Citations for TB Exposures

OSHA can issue citations to employers as a result of exposure or potential exposure to the exhaled air of a suspected or confirmed case of tuberculosis. Exposure can occur during a high-hazard procedure performed on an individual with suspected or confirmed tuberculosis. OSHA can issue citations under the respirator standard (29 CFR 1910.134) when employers fail to provide respirators and fit testing to potentially exposed employees.

2.   OSHA Abatement Methods

Strive to identify persons with active tuberculosis. Provide medical surveillance at no cost to the employee, including pre-placement evaluation, tuberculosis skin tests, annual evaluations, and twice yearly exams for those exposed. Evaluate and manage individuals with a positive skin test. Use acid-fast bacilli isolation rooms for those with active or suspected TB infection. Maintain such rooms under negative pressure and use outside exhaust or high efficiency particulate air (HEPA)-filtered ventilation. Develop an employee information and training plan.

3.   OSHA Tuberculosis Respirator Requirements

OSHA requires healthcare organizations to meet the provisions of 29 CFR 1910.134, which covers respiratory protection for general industry. OSHA enforces all provisions, including annual fit testing, with regard to TB exposures. When using disposable respirators, never permit reuse unless maintaining the functional and structural integrity of the respirator. Maintaining functionality depends on adherence to the manufacturer's instructions. Facilities should address the conditions under which a disposable respirator is considered contaminated. Whenever using reusable or disposable respirators, employers must implement a respiratory protection plan to meet the requirements of 29 CFR 1910.134. This entails creating a written respiratory protection plan for managing respirator selection and use, employee instruction and training, surveillance of work area conditions, and respirator fit testing. 29 CFR 1910.134 provides specific guidance on appropriate fit testing procedures. CDC guidelines, NIOSH recommendations, and selection criteria in the OSHA standard indicate that most facilities should use half-mask, N95, air purifying, filtering face piece respirators for TB protection. This type of respirator contains a securely fitting face piece. Effective protection requires a good seal between the face and the face piece to ensure individual protection. The OSHA Respiratory Protection standard does not require annual medical evaluations. However, employers must conduct annual fit tests for all respirator users. Never consider respirator fit testing as a hazard-specific or industry-specific activity. Annual fit testing provides the opportunity for employees to receive feedback on how well they don their respirator. Employees should receive fit testing annually as part of training.

E. TB EXPOSURE CONTROL PLAN

Every healthcare setting should implement a TB infection control plan as a part of their overall infection control and prevention efforts. The specific details of the TB infection control plan can differ depending on patients encountered. Administrators making this distinction should obtain medical and epidemiologic consultation from state and local health departments. The TB infection-control plan should consist of administrative controls, environmental controls, and a respiratory-protection plan. Every setting providing services to persons with suspected or confirmed infectious TB disease, including laboratories and nontraditional facility-based settings, should develop a TB infection control plan. Take the following steps to establish a TB exposure-control plan: (1) assign supervisory responsibility for the TB infection-control plan to a qualified person or group; (2) delegate authority to conduct a TB risk assessment and implement and enforce TB infection-control policies; (3) ensure the completion of education and training; (4) develop a written TB exposure control plan that outlines control procedures; and (5) update the plan annually. Employers must develop procedures for evaluating suspected or confirmed TB disease when not promptly recognized or appropriate precautions or controls fail. Collaborate with the local or state health department to develop administrative controls consisting of the risk assessment, the written TB infection-control plan, management of patients with suspected or confirmed TB disease, training and education, screening and evaluation, problem evaluation, and coordination. Create a plan for accepting patients with suspected or confirmed TB disease when transferred from another setting.

1. Administrative Controls

The first and most important level of TB controls is the use of administrative measures to reduce the risk for exposure to persons with TB disease. Administrative controls consist of the following activities: (1) assigning responsibility for TB exposure control in the setting, (2) conducting a TB risk assessment of the setting, (3) developing and instituting a written TB exposure control plan to ensure prompt detection, (4) use of airborne precautions, (5) treatment of persons with suspected or confirmed TB disease, and (6) ensuring the timely availability of recommended laboratory processing, testing, and reporting of results to the ordering physician and infection-control team. Other administration controls include implementing effective work practices for the management of patients with suspected or confirmed TB disease and ensuring proper cleaning and sterilization or disinfection of potentially contaminated equipment such as endoscopes. Provide training and education with specific focus on prevention, transmission, and symptoms. Conducting screening and evaluation of those at risk for TB disease remains a key administrative action.

2. Environmental Controls

Environmental controls provide the second line of defense in the TB infection-control plan, after administrative controls. Environmental controls include technologies for the removal or inactivation of airborne TB. These technologies include local exhaust ventilation, general ventilation, HEPA filtration, and ultraviolet germicidal irradiation (UVGI). These controls help to prevent the spread and reduce the concentration of infectious droplet nuclei in the air. CDC provides a summary of environmental controls and their use in prevention in the 2005 CDC TB Guidelines. Primary environmental controls consist of controlling the source of infection by using local exhaust ventilation (hoods, tents, or booths) and diluting and removing contaminated air by using general ventilation. Secondary environmental controls consist of controlling the airflow to prevent contamination of air in areas adjacent to the source (airborne infection isolation [AII] rooms) and cleaning the air by using HEPA filtration or UVGI.

3. Respiratory-Protection Controls

The first two control levels minimize the number of areas in which exposure to M. tuberculosis might occur and, therefore, minimize the number of persons exposed. These control levels also reduce, but do not eliminate, the risk for exposure in the limited areas in which exposure can still occur.

Because persons entering these areas might be exposed to M. tuberculosis, the third level of the hierarchy is the use of respiratory protective equipment in situations that pose a high risk for exposure. Use of respiratory protection can further reduce risk for exposure from droplet nuclei expelled into the air from a patient with infectious TB disease. Take the following measures to reduce the risk of exposure: (1) implement a respiratory-protection plan, (2) train employees on respiratory protection, and (3) educate patients about respiratory hygiene and cough etiquette procedures.

4. Engineering Controls

Engineering controls prove critical in preventing the spread of tuberculosis within a facility. The CDC guidelines recommend exhausting air from possibly infected areas to the outside. Healthcare facilities can develop isolation rooms with negative pressure. Recommend a rate of six air changes per hour. New construction requires 12 air changes per hour. Some facilities use UVGI to supplement ventilation and isolation efforts.

VIII. HEALTHCARE OPPORTUNISTIC INFECTIONS

A. BACTERIA

Once classified as a member of the plant kingdom, we now classify bacteria as a totally separate kingdom. Bacteria adapt remarkably and survive in diverse environmental conditions. They exist in the bodies of all living organisms and in all parts of the world, even in hot springs and the stratosphere. Bacteria normally exhibit one of three typical shapes: (1) rod shaped (bacillus), (2) round (cocci), and (3) spiral (spirillum). An additional group (vibrios) appears as incomplete spirals. We can characterize bacteria by growth patterns such as the chains formed by streptococci. Bacillus and spirillum exhibit motile or swimming motions similar to the whip-like movements of flagella. Other bacteria possess rod-like appearances (called pili) that serve as tethers. Aerobic forms of bacteria function only in the presence of free or atmospheric oxygen. Anaerobic bacteria cannot grow in the presence of free oxygen but obtain oxygen from other compounds. Bacteria do not make their own food and must live in the presence of other plant or life. Bacteria grow when they find food and favorable conditions. A cough or sneeze releases millions of bacteria from the body.

B. METHICILLIN-RESISTANT STAPHYLOCOCCUS AUREUS

Staphylococcus aureus can live on the skin or in the nose of healthy people. During the past 50 years, these infections became resistant to various antibiotics, including penicillin-related antibiotics. Infection control personnel refer to these resistant bacteria as methicillin-resistant MRSA. Colonization can occur when the staph bacteria survive on or in the body without causing illness. Staph bacteria causes a variety of illnesses, including skin infections, bone infections, pneumonia, and severe bloodstream infections. MRSA occurs more commonly among persons in hospitals and healthcare facilities. The infection usually develops in elderly hospitalized patients with serious illnesses or in an open wound such as a bedsore. Factors that place some patients at risk include long hospital stays, receiving broad-spectrum antibiotics, and being kept in an intensive care or burn unit. Keep cuts and abrasions clean and covered with a proper dressing or bandages and avoid contact with wounds or material contaminated by wounds. Implement contact precautions when identifying MRSA and the Infection Control and Prevention Function believes that a special clinical or epidemiological significance exists.

C. VIRUSES

Smaller than bacteria, viruses contain a chemical compound with protein. They must infect a host to survive for long periods. Viruses depend on the host cells to reproduce. Outside of a host cell, a virus exists as a protein coat or capsid that can enclose within a membrane. While outside the cell, a virus remains metabolically inert. A virus can insert genetic material to take over the functions of the host. An infected cell begins to produce more viral protein and genetic material instead of its usual products.

Some viruses may remain dormant inside host cells for long periods and cause no obvious change in the host cells. When stimulated, a dormant virus enters a phase that results in new viruses bursting and infecting other cells. Viruses cause a number of diseases, including smallpox, colds, chickenpox, influenza, shingles, hepatitis, polio, rabies, and AIDS. Disinfectants destroy viruses very easily.

D. ASPERGILLUS

The mold spore produced by Aspergillus can create pathogenic infection opportunities. Aspergillus exists worldwide and can thrive at elevated temperatures. Ideal growth conditions include damp areas with decaying vegetation. Aspergillus appears initially as a flat thread-like white growth that soon becomes a powdery blue-green mold spore. Most infections result from inhaling these spores. Most people possess natural immunity and do not develop any disease. Patients with serious ailments tend to experience a greater risk of infection. The severity of Aspergillus depends on the individual's immune system. Aspergillus infection can range from sinusitis conditions to pulmonary infections, including pneumonia. Ventilation plays a key role in maintaining an Aspergillus-free environment in healthcare settings. Establish procedures to control dust during renovations that occur near patient areas.

E. ANTHRAX

Exposure to this spore-forming bacterium results in black, coal-like skin lesions. In the naturally occurring forms, anthrax passes on by contact with anthrax-infected or anthrax-contaminated animals and animal products. Anthrax does not spread from one person to another person. Humans can host three forms of anthrax: inhalation, cutaneous, and gastrointestinal. Inhalation anthrax occurs when the anthrax spore is inhaled. Cutaneous anthrax, the most common naturally occurring form, is contracted by handling contaminated hair, wool, hides, flesh, blood, or excreta of infected animals and from manufactured products such as bone meal. It is introduced through scratches or abrasions of the skin. Gastrointestinal anthrax occurs as a result of ingesting insufficiently cooked infected meat or from flies. The spores enter the lungs, migrate to the lymph nodes, change to the bacterial form, multiply, and produce toxins.

F. SEVERE ACUTE RESPIRATORY SYNDROME

Severe Acute Respiratory Syndrome (SARS) is an emerging, sometimes fatal, respiratory illness. The first identified cases occurred in China in 2002. Some experts believe a virus causes SARS; however, the specific agent remains unidentified. No laboratory or other test can definitively identify cases. Most suspected SARS cases occurring in the United States involve individuals returning from travel to Asia and healthcare staff members in contact with patients. Casual contact does not appear to cause SARS. Transmission appears to occur primarily through close contact with a symptomatic patient. Signs of illness include a decreased white blood cell count in most patients as well as below-normal blood platelet counts, increased liver enzymes, and electrolyte disturbances in a number of patients.

G. PSEUDOMONAS

Pseudomonas, a motile rod-shaped organism, uses glucose in an oxidative manner. These bacteria pose a clinically important risk because of their resistance to most antibiotics. They can survive conditions that few other organisms can tolerate, aided by their production of a protective slime layer. The key targets include immune-suppressed individuals, burn victims, and individuals on respirators or with indwelling catheters. Additionally, these pathogens colonize the lungs of cystic fibrosis patients, increasing the mortality rate of individuals with the disease. Infections can occur at many sites and can lead to urinary tract infections, sepsis, pneumonia, and pharyngitis.

Rarely does Pseudomonas cause infection in healthy individuals. Its noninvasive nature limits its pathogenic capabilities. Pseudomonas prefers to inhabit moist environments but it can survive in a medium as deficient as distilled water.

H. LEGIONELLA

Studies link a majority of outbreaks of Legionella to cooling towers and domestic water systems. Other sources include evaporative condensers, respiratory equipment, showers, faucets, whirlpool baths, humidifiers, and decorative fountains. Hot-water systems provide a perfect breeding habitat, as Legionella grows best in temperatures ranging from 95 to 115 degrees Fahrenheit. Uncontrollable incidents that can cause Legionella problems include surges in water pressure that may disperse dirt into the water system or dislodge Legionella-laden scale and sediment from the walls of water pipes. Legionella can enter cooling towers, air intakes, or water pipes. In addition, new or renovated water lines not properly flushed prior to opening may contain Legionella. Idle plumbing can hold heavy contamination due to stagnant water. Current human treatment includes the antibiotics erythromycin and rifampin for severe cases.

I. INFECTION CONTROL RISK ASSESSMENT

Infection Control Risk Assessment (ICRA) functions best as a multidisciplinary process that focuses on reducing risk from infection throughout facility planning, design, construction, and renovation activities. A multidisciplinary team considers environment, infectious agents, human factors, and impact of a proposed project on controlling infections. The team includes, at a minimum, experts in infectious disease, infection control, patient care, epidemiology, facility design, engineering, construction, and safety, as circumstances dictate.

Educate both the construction team and healthcare staff about high-risk patient areas with regard to the airborne infection risks associated with construction projects, dispersal of fungal spores during such activities, and methods to control the dissemination of fungal spores. Incorporate mandatory adherence agreements for infection control into construction contracts, with penalties for noncompliance and mechanisms to ensure timely correction of problems. Establish and maintain surveillance for airborne environmental disease such as Aspergillus during construction, renovation, repair, and demolition activities to ensure the health and safety of high-risk patients.

IX. MEDICAL WASTE

Medical waste is defined in 40 CFR 259.10 and 40 CFR 22 as any solid waste generated in the diagnosis, treatment, or immunization of human beings or animals, in related research, biological production, or testing. Currently medical waste is regulated by most states. Segregate infectious waste from other waste at the point of generation within the facility. Ensure packaging contains the infectious waste from the point of generation up to the point of proper treatment or disposal. Place contaminated sharps immediately in rigid, leak-proof, puncture-resistant containers. Conspicuously identify all containers used for disposal of infectious waste. Handle and transport infectious waste to maintain integrity of the packaging. Do not transport plastic bags containing infectious waste using chutes or dumbwaiters. Establish procedures for storing infectious waste that would inhibit rapid microbial growth and minimize exposure potential. Isolate the area after a spill, discovery of ruptured packaging, or other incident involving potentially infectious materials. When practicable, repackage all spilled waste and containment debris. Disinfect all containment equipment and surfaces appropriately. Complete the appropriate incident report.

Regulated medical waste, a subset of all medical wastes, includes several categories, which can vary depending on the state or locality.

BOX 7.11 SOME CATEGORIES OF REGULATED BIOMEDICAL WASTES

- Cultures and stocks of infectious agents
- Human pathological wastes such as tissues and body parts
- Human blood and blood products
- Sharps (needles and syringes used in care areas)
- Certain animal wastes
- Certain isolation wastes (e.g., wastes from patients with highly communicable diseases)
- Unused sharps (suture needles, scalpel blades, hypodermic needles)

A. SHARPS CONTAINERS

Red or clear biohazard sharps containers can also contain regulated medical or infectious waste, but will more specifically hold items that could potentially puncture the skin and transmit an infectious disease. Needles and syringes make up the first type of sharps people correlate with sharps containers. Other items can become sharps hazards including pipettes, scalpels, and lancets. Contents can consist of used or unused sharps that could potentially contain pourable, squeezable, or dried flaky blood. Since the primary components placed in sharps containers include needles and syringes, sharps containers in some states may require incineration to prevent misuse of these items.

B. SEPARATING MEDICAL AND HAZARDOUS WASTES

Medical facilities and pharmaceutical companies bear the enormous task of segregating and disposing of the wastes that accumulate during the development of medicines and treatment of patients. These wastes pose very different hazards and impacts on the environment and human beings.

Many types of containers can assist with proper segregation of wastes. Proper training on the use of containers protects the environment and humans but also reduces disposal costs by identifying what should and should not go into specified containers. Use red biohazard bags only for Regulated Medical Waste consisting of solid, liquid, or semi-liquid blood, or other potentially infectious material and contaminated items that, if compressed, could release blood or infectious material during normal handling. This includes waste generated during the treatment or diagnosis of humans or animals or in the testing of biologicals. Common contents could include empty IV bags and tubing, gloves, gowns, or other materials containing pourable, squeezable, flaking, or dripping blood, including bodily fluids. Use red bags or bags clearly identified with the international biohazard symbols to contain materials requiring disinfection. Rendering materials noninfectious could permit disposal in a nonhazardous landfill. When red bag waste requires incineration, send the materials to a Regulated Medical Waste Incinerator permitted by the EPA. Consider incineration for all pathological waste.

C. INFECTIOUS WASTES

Blood and body fluids include bulk laboratory specimens of blood tissue, semen, vaginal secretions, cerebrospinal fluid, synovial fluid, pleural fluid, peritoneal fluid, pericardial fluid, and amniotic fluid. Precautions do not apply to feces, nasal secretions, sputum, sweat, tears, urine, or vomit unless they contain visible blood. Handle free-flowing materials or items saturated to the point of dripping liquids containing visible blood or blood components. Pathological waste includes all discarded waste from renal dialysis contaminated with peritoneal fluid or blood visible to the human eye. Consider solid renal dialysis waste as medical waste if it is saturated and demonstrates the potential to drip/splash blood or other regulated body fluids. Waste sharps include any used or unused discarded article that may cause punctures or cuts.

D. Medical Waste Best Practices

Provide medical waste spill response and cleanup training. Educate staff on how to properly identify all nonregulated waste and dispose of it. Inform local regulators and involve them in the effort to reduce waste and properly regulate medical waste. Establish and post guides for identifying regulated medical waste at every site of generation and disposal. Weigh each container, log them, and compare totals from each generator; retrain departments with high utilization. Provide PPE to facilitate alternative disposal methods. Date sharps containers when first used and remove only when 75 percent full or when meeting an established time limit of use. Place medical waste receptacles in such a way as to avoid indiscriminate use.

E. Waste Handling for Offsite Transfer

The outermost layer of packaging for medical containers, excluding sharps, should be of a red background color or use red lettering with a contrasting color. Clearly label the container with warnings such as "Infectious Waste," "Medical Wastes," or "Biohazardous." The container should also bear the universal biohazard symbol. Print wording on the container or securely attach a label on two or more sides. Ensure the symbol measures at least six inches in diameter and contains print with a contrasting color to the background. Ensure the use of impermeable containers of sufficient strength to resist ripping, tearing, or bursting under normal conditions of use. Place sharps in rigid, leak-proof, puncture-resistant, sealed containers to prevent loss during handling. Clearly label all containers. Place small containers used to collect untreated medical waste inside larger containers during storage, transportation, and disposal activities.

F. Containers

During collection, storage, and transportation, use containers constructed of materials compatible with the treatment methods utilized. Ensure the use of burnable single-use containers for waste destined for incinerators. Containers destined for steam sterilizers should allow proper treatment of the waste. Decontaminate reusable containers after each use using only approved methods. Never reuse containers unless decontaminated and before use remove all medical waste labeling.

G. Medical Waste Disposal

The EPA promulgated new source performance standards and emission guidelines in 1997. Refer to 40 CFR 60, Subparts E and C, for standards on new and existing incinerators. Existing sources include those for which construction was commenced on or before June 20, 1996. Hospitals, clinics, and clinical research laboratories operating a medical incinerator must comply with the New Source Performance Standards (NSPS) as defined in 40 CFR 60, Subpart C. Steam sterilization uses saturated steam within a pressure vessel at temperatures high enough to kill infectious agents. Steam sterilization works effectively with low-density materials such as plastics. Containers effectively treated can include plastic bags, metal pans, bottles, and flasks.

Never sterilize materials with high-density polyethylene and polypropylene because they do not allow steam penetration. Refer to 40 CFR 60.51 for classification information on waste streams such as hospital waste, medical and infectious waste, and pathological waste. Consider other methods such as (1) shredding or compacting with steam autoclaving, (2) microwave irradiation, (3) thermal treatment, and (4) chemical/biological treatment. Discharge into the sewer system suctioned fluids, waste in liquid apparatus, bodily discharges, and dialysis waste. Sterilize substances such as cultures, etiologic agents, and other laboratory wastes. Use chemical disinfection methods in dialysis equipment and for specimen spills prior to cleaning. Never use compactors or grinders to process infectious wastes. When shipping waste offsite, collect, transport, and store in a manner prescribed by the contractor. Require contractors to only pick up wastes in leak-proof and fully enclosed containers. Require contractors to maintain all required permits relevant to medical waste disposal. Contractors must also maintain regulatory documentation and records relevant to disposal actions including incinerated waste.

H. DOT Infectious Shipping Requirements

The Pipeline and Hazardous Materials Safety Administration (PHMSA) develops and issues hazardous materials regulations (49 CFR 100–180). The regulations govern the classification, hazard communication, and packaging of hazardous materials for transportation. The DOT classified infectious substances, including regulated medical waste, in Division 6.2 of the hazardous materials rules. Shippers of infectious substances must meet all standards related to commerce by rail, water, air, or highway. Regulated Medical Waste Packaging (49 CFR 173.197) must meet certain requirements. Ensure the use of rigid, leak-resistant, impervious to moisture, and strong packing materials to prevent tearing or bursting under normal conditions of use and handling. Properly seal all waste containers to prevent leakage during transport. Ensure the use of puncture-resistant sharps containers. Ship waste materials in break-resistant containers, tightly lidded, or stopper closed for fluids. Package and mark with the "Biohazard" marking in accordance with OSHA standard, 29 CFR 1910.1030. Diagnostic specimens and biological products do not qualify for regulation except when transported as regulated medical waste. An infectious substance consists of a viable microorganism, or its toxin, that causes or may cause disease in humans or animals. Consider regulated medical waste as any waste or reusable material that contains an infectious substance and is generated in the diagnosis, treatment, or research of humans or animals. This definition does not include discarded cultures or stocks.

A biological product refers to any material prepared and manufactured in accordance with certain regulations of the Department of Agriculture or the Department of Health and Human Services. A diagnostic specimen refers to any human or animal material being shipped for purposes of diagnosis. It includes but is not limited to excreta, secreta, blood, blood components, tissue, and tissue fluids.

BOX 7.12   DOT TRAINING AND REPORTING REQUIREMENTS

- All hazardous material employees must be trained in accordance with 49 CFR 172
- Training must address general awareness, function-specific, and safety rules
- Test each employee and maintain record of testing
- Retrain hazardous materials employees every two years
- Report any infectious substance spill to the CDC according to 49 CFR 171.15
- For other reportable hazardous material spills, contact the DOT National Response Center

REVIEW EXERCISES

1. List at least seven infection control plan development considerations.
2. List the three healthcare vaccine categories.
3. Healthcare employees should meet the ACIP guidelines for immunization of which four diseases?
4. The Guidelines for Environmental Infection Control in Healthcare Facilities would not apply to which three broad categories of biological risks?
5. List the six types of disease and infection transmission routes.
6. What three diseases or pathogens can transmit by airborne routes in healthcare facilities?
7. List three hospital-acquired conditions that could qualify for potential reduced payments.
8. List the three types of chemical germicides.
9. Which federal agency regulates liquid chemical sterilants and high-level disinfectants?
10. Describe in your own words the three CDC-defined disinfecting levels.
11. What three types of information does OSHA require on sharps injury logs?
12. Describe the CDC-recommended TB screening procedures for settings classified as low risk.
13. Describe the process known as an ICRA.

8 Radiation, Lab, and Drug Hazards

I. RADIATION SAFETY

Ionizing radiation occurs naturally from the decay of radioactive materials or by the operation of x-ray emitting devices. A radioactive element spontaneously changes to a lower energy state and emits particles and gamma rays from its nucleus. X-rays result when highly energized electrons strike the nuclei of a targeted material. The electrons deflected from their path release energy as electromagnetic radiation or x-rays. Ionizing radiation occurs anytime an electron dislodges from its parent atom or molecule. Ionizing radiation's ability to penetrate the body depends on the wavelength, frequency, and energy of the material. Alpha particles do not penetrate the skin. An alpha particle cannot penetrate a thin layer of paper or clothing. Highly energized beta particles require shielding using some type of low-density material such as plastic, wood, water, or acrylic glass. Beta particles can travel a few centimeters into living tissue. Beta contaminants that remain on the skin for a prolonged period can cause injury. Beta emitting contaminants can also cause harm if deposited internally. Gamma and x-rays can penetrate human tissue. Radioactive materials that emit gamma radiation constitute both an external and internal hazard to humans. Gamma radiation frequently accompanies the emission of alpha and beta radiation. X-rays also possess longer wavelengths, lower frequencies, and lower energies than alpha or beta particles. The international community measures radiation using the International System of Units (SI). The United States uses a conventional system of measurement depending on what aspect of radiation requires measurement.

BOX 8.1   RADIATION MEASUREMENT TERMS

- Absorbed dose is the amount of radiation that is absorbed by the body.
- Curie (Ci) is a unit of measurement of radioactivity where $1 \text{ Ci} = 3.7 \times 10^{10}$ decays per second.
- Exposure is the amount of radiation to which the body is exposed.
- Radiation absorbed dose (RAD) is a measure of the absorbed dose of ionizing radiation.
- RAD = 100 ergs per gram = 0.01 gray (Gy).
- Radioactive half-life is the time required for isotope radioactivity to decrease by 50 percent.
- Roentgen equivalent man (Rem) is the dosage of any ionizing radiation that will cause biological injury to human tissue equal to the injury caused by one roentgen of x-ray or gamma ray dosage with one rem equivalent to 0.01 Sievert (Sv)
- Roentgen is the unit of measure for the quantity of radiation produced by gamma or x-rays.

A. OSHA Ionizing Radiation Standard (29 CFR 1910.1096)

The Occupational Safety and Health Administration (OSHA) standard covers X-ray equipment, accelerators, accelerator-produced materials, electron microscopes, and naturally occurring radioactive materials such as radium. OSHA exposure standards for whole body radiation should never exceed three rem/quarter (year). Lifetime or cumulative exposure should not exceed five (N-18) rem. OSHA regulates exposure to all ionizing radiation for sources not under Nuclear Regulatory Commission (NRC) jurisdiction. OSHA and NRC signed a Memorandum Agreement in 1989 that outlined the compliance authority of both agencies. This coordinated interagency effort helps prevent gaps in protecting workers and avoids duplication of effort.

The degree of human exposure depends on the amount of radiation, duration of exposure, distance from the source, and the type of shielding used. OSHA recommends workers wear a badge or device to support long-term exposure monitoring efforts. OSHA recommends a passive dosimeter for personnel working with x-ray equipment, radioactive patients, or radioactive materials. Depending on the work situation, body badges may be worn at the collar, chest, or waist level. Personnel working in high-dose fluoroscopy settings need to wear two badges for monitoring purposes. Personnel exposed to beta and gamma doses should wear ring devices on the hand nearest the radiation source. Lead aprons and gloves offer some protection for employees and patients exposed to a direct x-ray field. Assign a specific person to ensure proper maintenance of all portable x-ray machines. For preventive and corrective maintenance information for x-ray machines, refer to 21 CFR 1000, Radiological Health. For information about exposure limits, refer to the OSHA Ionizing Radiation standard found in 29 CFR 1910.1096. Employers must supply appropriate personnel monitoring equipment such as film badges, pocket chambers, pocket dosimeters, or film rings.

B. Restricted Areas

OSHA requires the designation of restricted areas to protect individuals from radiation exposure. The term "unrestricted area" refers to any area where access is not controlled by the employer for purposes of protection of individuals from exposure to radiation. No employer can possess, use, or transport radioactive material within a restricted area that causes individuals to experience airborne radioactive material exposure in excess of limits specified in Table 1, Appendix B to 10 CFR 20. When monitoring exposure, make no allowance for the use of protective clothing or equipment, or particle size.

C. Surveys and Monitoring

A survey is an evaluation of the radiation hazards incident to the production, use, release, disposal, or presence of radioactive materials or other sources of radiation under a specific set of conditions. When appropriate, such evaluation includes a physical survey of the location of materials and equipment, and measurements of levels of radiation or concentrations of radioactive material present. Every employer should supply appropriate personnel monitoring devices such as film badges, pocket chambers, pocket dosimeters, or film rings for use by exposed employees. Personnel monitoring equipment refers to devices designed to be worn or carried by an individual for the purpose of measuring the dose received.

D. Radiation Areas

A "radiation area" refers to any area accessible to personnel in which there exists radiation at such levels that a major portion of the body could receive in any one hour a dose in excess of five millirem, or in any five consecutive days a dose in excess of 100 millirem. "High radiation area" means any area accessible to personnel in which there exists radiation at such levels that a major portion of the body could receive in any one hour a dose in excess of 100 millirem.

E. Caution Signs

Caution signs, labels, signals, and symbols should use the conventional radiation caution colors of magenta or purple on yellow background. The radiation symbol contains a three-bladed design with the words "Radiation Area." Each area with radiation should be conspicuously posted with a sign or signs bearing the radiation caution symbol. Each high radiation area should be conspicuously posted with a sign or signs bearing the radiation caution symbol and appropriate wording.

F. Airborne Radioactivity

Airborne radiation can appear in any room, enclosure, or operating area in which radioactive materials exist in concentrations in excess of the amounts specified in Table 1 of Appendix B to 10 CFR 20. Post a sign or signs bearing the radiation caution symbol and the word CAUTION in areas where exposure could occur.

G. Storage Areas

Each area containing radioactive material exceeding 10 times the quantity of such material as specified in Appendix C to 10 CFR 20 must post conspicuous signs bearing the radiation caution symbol and the word CAUTION. Secure radioactive materials stored in nonradiation areas to prevent unauthorized removal from the place of storage. Employers must never dispose of radioactive material in an improper or illegal manner. They can transfer materials to an authorized recipient.

H. Tuberculosis Exposures

Facilities should protect radiology personnel from tuberculosis (TB) exposures during x-ray procedures. Exposures to TB can occur if radiology rooms do not contain proper ventilation. Hospitals should develop written procedures for the safe handling of TB patients in all radiology areas. TB patients should wear masks and stay in radiology suites the shortest time possible. Healthcare facilities serving populations with a high prevalence of TB may need to supplement general ventilation and use additional engineering approaches.

I. Ergonomics

Radiology staff can experience work-related musculoskeletal disorders (MSDs) from constant lifting and reaching during x-ray procedures and patient transfers. Employers should assess the radiology area for ergonomic stressors. Train employees in proper lifting techniques. Avoid awkward postures and working above shoulder height. Use lift mechanical aids and ensure sufficient staffing. Provide instructions to patients on ways to help facilitate the lift or transfer procedure.

J. Slips, Trips, and Falls

Potential exists for slips and falls in the radiology area. Ensure floors don't contain slip hazards such as water, blood, vomit, or excreta. Keep aisles and passageways clear and in good repair with no obstruction across or in the aisles. Provide floor plugs for equipment to prevent the need for placing cords across pathways. Report and clean all spills immediately. Correctly maintain floors by using nonskid waxes.

K. Bloodborne Pathogens

Use gloves, masks, and gowns if blood or fluid exposure exists. Use appropriate engineering and work practice controls to limit exposure. Wear gloves to protect hands coming into contact with blood, mucous membranes, or nonintact skin. Follow proper work practices when performing vascular access procedures or when handling contaminated items or surfaces.

L. STORAGE AND HANDLING PROCEDURES

Properly calibrate radiation measurement instruments before each use. Train personnel handling and exposed to radiation wastes. Each department that generates radioactive wastes should develop written procedures that cover handling, transportation, and disposal. Properly secure and store all waste materials and designate controlled areas. Dispense or draw materials only behind a protective barrier. Label refrigerators that contain stored materials. Notify the radiation control officer when receiving a contaminated shipment.

M. MEDICAL RADIOACTIVE MATERIALS

A radionuclide refers to any type of radioactive material including elements and isotopes of elements. Most radioactive materials used in nuclear medicine consist of isotopes since individual medical treatment may require an isotope with specific radioactive properties. Radioisotopes show how the disease process alters the normal function of an organ. A patient swallows, inhales, or receives an injection of a tiny amount of a radioisotope. Cameras then reveal where the isotope accumulates in the body. Laboratory tests use radioisotopes to measure important substances in the body including thyroid hormones. Some facilities use isotopes to sterilize hospital items such as sutures, syringes, catheters, and hospital clothing otherwise destroyed by heat sterilization. Sterilization using radioisotopes can prove valuable because the process permits the items to remain in their sealed packages. NRC rules outline minimum safety requirements for workers and patients.

N. SHIELDING

The effectiveness of shielding material relates to its cross section for scattering and absorbing radiation. The radiation that manages to get through falls exponentially with the thickness of the shield. Practical radiation protection depends on juggling the three factors to identify the most cost-effective solution. Different types of ionizing radiation can behave in different ways. This results in the use of different shielding techniques. Particle radiation consists of a stream of charged or neutral particles, charged ions, and subatomic elementary particles. This includes solar wind, cosmic radiation, and neutron flux in nuclear reactors.

O. ALARP

ALARP stands for "as low as reasonably practicable." An equivalent term, ALARA, "as low as reasonably achievable," is also commonly used. The application of radiation can aid the patient by providing doctors and other healthcare professionals with a medical diagnosis. However, keeping exposure reasonably low will reduce the probability of cancers or sarcomas and eliminate skin reddening or cataracts. Any radiation exposure, no matter how small, can increase the chance of negative biological effects such as cancer. The probability of the occurrence of negative effects of radiation exposure increases with cumulative lifetime dosage.

II. NUCLEAR MEDICINE

Nuclear medicine uses small amounts of radioactive material to diagnose or treat a variety of diseases, including cancer and heart disease. Nuclear medicine or radionuclide imaging procedures help physicians diagnose medical conditions. Depending on the type of nuclear medicine, a radiotracer is injected into a vein, swallowed, or inhaled as a gas. It eventually accumulates in the organ or area of your body being examined, where it gives off energy in the form of gamma rays. This energy can then be detected by a device called a gamma camera, a positron emission tomography (PET) scanner, and/or a probe. These devices work together with a computer to measure the amount of radiotracer absorbed by the body. The devices also work to produce special pictures detailing the structure and function of organs and tissues. Nuclear medicine images superimposed by computed tomography (CT) or magnetic resonance imaging (MRI) produce special views. This practice is known as imaging fusion or co-registration. These views allow the information from two different studies to become correlated and interpreted as one image, which results in a more precise diagnosis. Manufacturers now make single photon emission computed tomography (SPECT)/CT and PET/CT units that perform both imaging studies at the same time. Iodine therapy uses radioactive material to treat cancer and other medical conditions affecting the thyroid gland. Most nuclear medicine procedures use a gamma camera, a specialized camera encased in metal that is capable of detecting radiation and taking pictures from different angles. It may be suspended over the examination table or beneath the table. New technologies make the diagnosis, management, and treatment of illnesses more sensitive, more specific, more accurate, and ultimately safer for both the patient and the technologist.

CT, MRI, and "ultrasound scans" do not involve radioactive materials. Most patients receive radioactive iodine-131 (I-131). I-131 has a half-life of eight days and emits both *beta particles* and *gamma rays*. The beta particles are responsible for killing the tumor cells. They have such a short "range" in tissue that they do not leave the patient's body, and so present no external hazard. However, if any I-131 is unintentionally ingested, the beta particles can present an internal hazard. Because I-131 is excreted in the patient's urine, saliva, and perspiration, small amounts of radioactivity may be present on surfaces in the patient's room. This "contamination" can be ingested by "surface-to-hand-to-mouth" contact. The gamma rays have properties like x-rays. They pass out of the patient's body, and therefore present a potential external hazard to bystanders. To reduce this hazard, some patients may be housed in special lead-lined rooms. Work safely by using a few simple techniques:

1. Put on shoe covers and protective gloves before entering the patient's room.
2. Work quickly, but effectively and courteously. Minimize your time in the room.
3. Maintain the greatest distance possible from the patient consistent with effective care. Radiation exposure drops off drastically with increasing distance.
4. Observe Universal Precautions while handling blood and other body fluids, especially urine.
5. Leave all trash, linens, and food trays in the room. Upon leaving the room, remove gloves and shoe covers and place them in the trash box inside the room.
6. After leaving the room, wash your hands.
7. In the event of a medical emergency involving the patient, the patient's well-being is the primary consideration. All initial measures necessary to sustain the patient should be undertaken, regardless of radiation considerations.

It may be possible to further reduce the external hazard by using portable shields. In general, lead aprons are minimally effective and their routine use during ordinary caregiving is not recommended. However, during prolonged procedures at close proximity to the patient, they can reduce exposure by about 15 percent. A few patients are treated with pure beta emitters such as phosphorus-32 (P-32), strontium-89 (Sr-89), yttrium-90 (Y-90), and samarium-153 (Sm-153). These patients usually do not require hospitalization for radiation protection purposes, since there is no external hazard. Use Universal Precautions when caring for these patients if they are hospitalized.

A.   Brachytherapy (Implant) Patients

In cases where their tumors are close to accessible body cavities, oncology patients may undergo a procedure called brachytherapy, or "implant" therapy. Implant therapy is effective in some cases of uterine, prostate, and lung cancer. In implant therapy, a sealed source of radioactive material, usually a gamma emitter such as cesium-137 or iridium-192, is placed in a body cavity close to the tumor and left in place for a prescribed period of time. During the time the implant is in place, staff entering the room is exposed to gamma rays and must take precautions. Once the treatment is completed and the implant is removed, the patient is no longer radioactive and presents no hazard. Work safely with implant therapy patients by following these techniques:

1. Put on protective gloves before entering the patient's room. Although breakage of a sealed source is an unlikely occurrence, ingestion of radioactivity is easily prevented by wearing gloves.
2. Work quickly, but effectively and courteously. Minimize your time in the room.
3. Maintain the greatest distance possible from the patient consistent with effective care. Radiation exposure drops off drastically with increasing distance.
4. Leave all trash, linens, and food trays in the room. Upon leaving the room, remove gloves and place them in the trash receptacles inside the room. The radiation safety department surveys all materials before they leave the room.
5. After leaving the room, wash your hands.

In the event a source becomes dislodged, notify the radiation oncology resident on call immediately. Do not permit others to enter the room until the source is secured. Do not attempt to handle a dislodged implant or applicator, unless you are specially trained to do so.

B. MEDICAL, REPRODUCTIVE, AND FERTILITY CONSIDERATIONS

The potential adverse health effects of ionizing radiation have been evident since the end of the 19th century, with the discovery of x-rays and the isolation of radium. Many of the early radiologists and scientists who worked with radiation personally experienced its detrimental effects, which included leukemia, skin cancer, and bone sarcomas. Controlled studies with bacteria, plants, and animals showed exposure to high doses of radiation to be carcinogenic (capable of inducing cancer), teratogenic (capable of causing birth defects), and mutagenic (capable of producing genetic mutations). Studies of people exposed to radiation during atomic bombings demonstrated the ability of ionizing radiation to produce profound medical effects on the gastrointestinal and hematopoietic systems. Over the past half-century, thousands of additional medical, scientific, and epidemiological investigations have further increased our understanding of the biological and medical effects of ionizing radiation, and at what radiation doses these effects can be expected to occur. For this reason, we have been able to establish with some certainty the levels of exposure that can be sustained without significantly increasing the risk of harm to an individual, or his or her offspring.

III. NUCLEAR REGULATORY COMMISSION

The NRC, established by the Energy Reorganization Act of 1974, ensures civilian uses of nuclear substances meet safety, environmental, and security laws. The NRC accomplishes its mission through standards setting, rulemaking, inspections, and enforcement actions. The commission also conducts technical reviews, studies, and public hearings. The NRC issues authorizations, permits, and licenses to ensure nuclear safety. The NRC issues five-year licenses to healthcare organizations that adhere to prescribed safety standards. To apply for a license, organizations should identify authorized users, designate a radiation safety officer (RSO), and identify the address or location of radioactive materials. Any changes require the organization to file for a license amendment. Some vendors must verify that facilities receiving radioactive materials possess a license. Some radionuclide licenses apply to generators of infectious and medical wastes. A general license is issued to medical practices, clinical laboratories, and healthcare facilities. A specific use license is required for physicians in private practice. Medical use pertains to human administration of radioactive substances or radiation. A broad scope license can be issued to facilities that provide patient care and conduct research using radioactive materials.

A. NRC PERFORMANCE-BASED STANDARDS

Some NRC regulations contain detailed procedures and others leave procedural details to the licensee. Approval of the license remains contingent on NRC evaluation of the proposal submitted with the application. Report all occurrences of therapeutic or diagnostic misadministration to the NRC Regional Office within 24 hours. The NRC requires that organizations meet all radiation protection standards as outlined in 10 CFR 20. Persons working or frequenting areas with radioactive materials must understand the basic right to know information contained in 10 CFR 19, *Notices, Instructions, and Reports to Workers: Inspections and Investigations.* "Agreement states" refer to states with a formal agreement with the NRC pursuant to Section 274 of the Atomic Energy Act. Under this agreement, the NRC relinquishes regulatory control over certain byproducts, sources, and special nuclear materials used within each state. NRC periodically assesses compatibility and adequacy of the agreement state enforcement efforts.

B. RADIATION CONTROL PLANNING

Each NRC licensee should maintain a written management plan that requires the participation of all users, organizational administrators, and RSOs. Develop procedures to inform workers about the types and amounts of materials used, dosing information, safety precautions, recurring training, and continuing education. Develop guidelines to keep doses as low as reasonably achievable. The RSO should develop and implement written procedures to cover the purchase, storage, use, and disposal of radioactive byproducts. Develop procedures to investigate all incidents, mishaps, accidents, or other deviations from prescribed procedures.

C. RADIATION SAFETY COMMITTEE

The Radiation Safety Committee must approve or disapprove use of radioactive material within the licensed institution. Decisions must consider the radiological health and safety of patients and staff. Other committee duties include prescribing special conditions for the proposed use of radioactive materials including bioassay requirements, physical examinations, and the minimum level of training required. The committee should address topics such as film badge return rates, exposure guidelines, safety, quality, and regulatory compliance. The committee should review records and reports submitted by the RSO. The committee recommends actions to take for ensuring the safe use of radioisotopes. The committee must maintain a written record of all actions taken and members must demonstrate their competency or expertise in radiation safety. Committee membership should include expertise in the areas of diagnostic radiology, clinical pathology, and nuclear medicine. Committees should meet at least quarterly.

D. RADIATION SAFETY OFFICER

The RSO must qualify to serve based on education, training and experience. The RSO should advise others on safety matters pertaining to ionizing radiation and supervise radiation safety efforts. The officer implements the policies established by the Radiation Safety Committee and supervises all aspects of radiation measurement and protection activities. Duties include monitoring activities, maintenance of exposure records, survey methods, waste disposal, and radiological safety practices. The RSO should monitor the education of all users of ionizing radiation. The RSO function must also maintain an inventory of all radioisotopes and ionizing radiation producing devices, and must ensure documentation of radiation surveys and exposures of personnel. He or she should promptly report to the committee all radiation hazards, serious infractions of rules, or other items relevant to radiation safety. Organizations must maintain the radioisotope inventory, all receipt and disposition logs, radiation survey records, and any NRC documentation required by 10 CFR 20.401.

E. REGULATORY AND OCCUPATIONAL DOSE CONSIDERATIONS

Contact with radionuclide patients should never result in adverse health effects. However, healthcare organizations must comply with state and federal regulations regarding the possession and use of radioactive materials. Operations at hospitals should never pose a risk to the public. Procedures should keep exposures less than one hundred millirems in one year. While some patient treatments can meet these criteria, others require extra shielding. Similarly, the special collection of urine and trash can create environmental concerns. Hospitals should promote the concepts of ALARA. The current U.S. mandated annual occupational dose limit is 5000 millirem (5 rem). Dozens of epidemiological studies fail to demonstrate any adverse effects on health, fertility, or genetic viability at or below the occupational dose limit. However, attempt to keep exposures below 500 millirem per year, wherever feasible.

BOX 8.2 WAYS TO REDUCE HUMAN RADIATION EXPOSURE

- **Shielding:** Use proper barriers to block or reduce the penetration of ionizing radiation
- **Time:** Spend less time in radiation fields
- **Distance:** Increase the distance between radioactive sources and workers or population
- **Amount:** Reduce the quantity of radioactive material for a practice

---

F. NATIONAL COUNCIL ON RADIATION PROTECTION

This group, created by Congress, collects, analyzes, develops, and disseminates information and recommendations on radiation quantities, measurements, and units. The National Council on Radiation Protection (NCRP) publishes maximum permissible levels of external and internal radiation. The major handbook is entitled, *Maximum Permissible Body Burdens and Maximum Permissible Concentrations of Radionuclides in Air and Water for Occupational Exposure* and *Review of the Current State of Radiation Protection Philosophy*. The NCRP suggests an annual permissible whole body dose of 5 rem per year, with 3 rem permitted within a 13-week period.

IV. FOOD AND DRUG ADMINISTRATION AND RADIATION SAFETY

The Food and Drug Administration (FDA) Center for Devices and Radiological Health (CDRH) ensures that the public and professionals remain informed of the risks posed by different types of radiation emissions and radiation-emitting products. CDRH also seeks to ensure that each patient receives the appropriate radiation dose using the appropriate, medically necessary imaging exam at the appropriate time. These efforts recognize the role everyone has in promoting radiation safety. Manufacturers should design safe products. Users should know how to appropriately use radiation-emitting electronic products and understand radiation safety and protection principles. Patients and consumers should know basic radiation risk and protection concepts. Regulators must collect and disseminate appropriate information and take action on this information as necessary. The FDA continues to take efforts to reduce the risks associated with medical uses of ionizing radiation, in order to maximize their benefits. The FDA Initiative to Reduce Unnecessary Radiation Exposure from Medical Imaging describes actions that support the benefits of medical imaging while minimizing the risks. The FDA is also exploring steps to improve patient safety in radiation therapy. CDRH collaborates with the Conference of Radiation Control Program Directors (CRCPD) in a unique federal-state partnership to characterize the radiation doses patients receive and to document the state of the practice of diagnostic radiology. Each year the Nationwide Evaluation of X-ray Trends (NEXT) survey selects a particular radiological examination for study and captures radiation exposure data from a nationally representative sample of U.S. clinical facilities. CDRH staff compiles, analyzes, and publishes survey results on population exposure, radiographic and fluoroscopic technique factors, diagnostic image quality, and film processing quality. Surveys should be repeated periodically to track trends as technology and clinical practices change. Since 1973, NEXT has been conducting surveys on examinations related to the adult chest, abdomen, lumbosacral (LS) spine, upper gastrointestinal fluoroscopy, mammography, CT of the head, dental radiography, and pediatric chest radiography.

A. RADIOLOGICAL SOCIETY OF NORTH AMERICA

The Radiological Society of North America (RSNA), an association of more than 40,000 radiologists, radiation oncologists, medical physicists, and related scientists, promotes excellence in radiology through education and by fostering research, with the ultimate goal of improving patient care. The society seeks to provide radiologists and allied health scientists with educational sessions and materials of the highest quality, and to constantly improve the content and value of these educational activities. The Society also seeks to promote research in all aspects of radiology and related sciences, including basic clinical research in the promotion of quality healthcare.

B. AMERICAN COLLEGE OF RADIOLOGY

The American College of Radiology (ACR) members strive to improve quality patient care and remain committed to the advancement of the science of radiology. ACR has established guidelines, standards, and accreditations to provide the foundation for achieving quality patient care. ACR accreditation offers radiologists and other providers the opportunity for peer review of their facility's staff qualifications, equipment, quality control, and resultant image quality. The ACR actively promotes the causes and issues of radiology professionals and their patients at both the federal and state levels and offers a selection of highly respected continuing education materials as well as the opportunity for physicians to learn the latest, cutting-edge imaging techniques and image-guided procedures at the one-of-a-kind ACR Education Center.

V. RADIOACTIVE WASTE MANAGEMENT

Radioactive waste exists in solid, liquid, or gas forms. Solid waste can include rags and papers from cleanup operations, solid chemicals, contaminated equipment, experimental animal carcasses, or human or experimental animal fecal matter. Consider and evaluate properties of waste when determining the method of disposal. Specific disposal methods vary according to the material involved and licensing authority of users. Radioactive waste typically remains onsite until its half-life is spent and it is no longer considered hazardous. Segregate low-level radioactive waste and properly label it, considering the isotope, form, volume, laboratory origin, activity, and chemical composition. Proper labeling and handling are legally required and make waste management decisions easier. Never mix radioactive and other hazardous waste. Retain suppliers that accept return of isotope containers.

Develop a plan to ensure disposal of radioactive waste meets government guidelines and regulations. The plan should contain procedures for waste containing radioactive materials as defined by the NRC. Develop an emergency plan to use in response to a radiation accident or incident. Keep radioactive waste segregated, centrally processed, and properly labeled. Secure radioactive waste storage areas and identify with a radioactive hazard symbol. Consider using sensors to detect radioactive levels in trash and medical waste prior to its leaving the facility. Return all isotope containers to the appropriate distributor.

VI. NONIONIZING RADIATION (29 CFR 1910.97)

This form of radiation lacks the energy of ionizing radiation and causes damage by vibrating the atoms or molecules causing heating by friction. Examples of nonionizing radiation include lasers, microwave (MW)/radio-frequency (RF)-generating devices, ultraviolet lamps, MRI machines, cell phones, and other electrical devices that produce electromagnetic fields. Health effects include retinal and skin damage. Electromagnetic radiation has different effects on humans depending on the wavelength and type of radiation involved. Low-frequency radiation such as that generated by broadcast radio and shortwave radio has generally been considered as not dangerous. Some new information suggests that exposure to electric power frequencies could pose an adverse impact on human health. Exposure to MW radiation can occur in healthcare facilities. Some exposure risks include microwave ovens, cancer therapy procedures, thawing organs for transplantation, sterilizing ampoules, and enzyme activation in research animals. The greatest hazard associated with exposure to MW radiation is thermal heating. The exposure limit for MW energy is expressed in voluntary language and has been ruled unenforceable by OSHA. The OSHA Construction Standard does specify the design of an RF warning sign and 29 CFR 1926.54 limits worker exposure.

A. ULTRAVIOLET RADIATION

Ultraviolet (UV) radiation can emit from germicidal lamps, during some dermatology treatments, from nursery incubators, and even from some hospital air filters. Overexposure can result in skin burns and serious eye damage. Long-term exposure can contribute to accelerated skin aging and increased risk of skin cancer. The National Institute for Occupational Safety and Health (NIOSH) recommendations for UV exposure range from 200 to 400 nanometers, depending on length of exposure. All UV sources capable of causing eye or skin burns should be interlocked so that direct viewing or bodily exposure is not possible. The total intensity of UV light from lamps and reflecting surfaces should not exceed the levels specified in the latest edition of the American Conference of Governmental Industrial Hygienists (ACGIH) reference "Threshold Limit Values for Chemical Substances and Physical Agents and Biological Exposure Indices."

B. RADIO FREQUENCY AND MICROWAVE RADIATION

Consider RF and MW as nonionizing radiation sources with insufficient energy to ionize atoms. The primary health effect of RF/MW energy comes as a result of heating. The absorption of RF/MW energy varies with frequency. MW radiation is absorbed near the skin, whereas RF radiation may absorb in deep body organs. Exposure standards of Western countries are based on preventing thermal problems; however, research continues on possible nonthermal effects. Use of RF/MW radiation includes radios, cellular phones, heat sealers, vinyl welders, high-frequency welders, induction heaters, communications transmitters, radar transmitters, ion implant equipment, MW drying equipment, sputtering equipment, and glue curing. The warning symbol for RF hazards should consist of a red isosceles triangle above an inverted black isosceles triangle, separated and outlined by an aluminum colored border. The upper triangle should contain the following wording: WARNING RADIO-FREQUENCY RADIATION HAZARD.

C. WIRELESS MEDICAL TELEMETRY

Wireless medical telemetry helps monitor patient physiological parameters over a distance via RF communications between a transmitter worn by the patient and a central monitoring station. These devices give the advantage of allowing patient movement without tethering the patient to a bedside hard-wired connection. The Wireless Medical Telemetry Service (WMTS) report and order sets aside the frequencies of 608 to 614 MHz, 1395 to 1400 MHz, and 1429 to 1432 MHz for primary or co-primary use by wireless medical telemetry users. A key feature of the WMTS is the provision for establishment of a Frequency Coordinator to maintain a database of user and equipment information to facilitate sharing of the spectrum and to help prevent interference among users of the WMTS. The Federal Communications Commission (FCC) order also provides a definition for wireless medical telemetry that is consistent with the recommendations made in 1999 by the American Hospital Association (AHA) Task Group on Wireless Medical Telemetry. The FCC defines wireless medical telemetry as "the measurement and recording of physiological parameters and other patient-related information via radiated bi- or unidirectional electromagnetic signals." The FCC order also describes the requirements for users of the new WMTS. Eligible WMTS users are limited to authorized healthcare providers, which includes licensed physicians, healthcare facilities, and certain trained and supervised technicians. The healthcare facilities eligible for the WMTS include those that offer services for use beyond 24 hours, including hospitals and other medical providers. Do not include ambulances and other moving vehicles within this definition.

D.   FDA/CDRH RECOMMENDATIONS FOR EMC/EMI IN HEALTHCARE FACILITIES

CDRH receives many inquiries from healthcare organizations, medical device manufacturers, clinicians, and the public seeking information about experiences with and prevention of electromagnetic interference (EMI) with medical devices. The following information is intended to help minimize the risks associated with medical device EMI and promote electromagnetic compatibility (EMC) in healthcare facilities. Some recommendations for healthcare facilities to follow include

- Making use of available resources such as EMC professionals and publications and Internet web pages on the subject of medical device EMC
- Assessing the electromagnetic environment of the facility and identifying critical medical device use areas
- Managing the electromagnetic environment, RF transmitters, and all electrical and electronic equipment, including medical devices, to reduce the risk of medical device EMI and achieve EMC
- Coordinating the purchase, installation, service, and management of all electrical and electronic equipment used in the facility to achieve EMC
- Educating healthcare facility staff, contractors, visitors, and patients about EMC and EMI and how they can recognize medical device EMI and help minimize EMI risks
- Establishing and implementing written policies and procedures that document the intentions and methods of the healthcare institution for reducing the risk of medical device EMI and achieving EMC

E.   CONSENSUS STANDARDS

The American National Standards Institute (ANSI) publishes consensus standards on RF exposures and measurements. The Institute of Electrical and Electronics Engineers (IEEE) Standards Coordinating Committee 28 is the secretariat for ANSI for developing RF standards. It is also the parent organization for the IEEE Committee on Man and Radiation (COMAR), which publishes position papers on human exposure to electromagnetic fields. ANSI C95.1-1992 (Safety Levels with Respect to Human Exposure to Radio Frequency Electromagnetic Fields [200 kHz – 100 GHz]) includes different exposure limits for controlled and uncontrolled sites. FCC OET Bulletin #65 (August 1997) Appendix A provides a table and figure of RF exposure limits adopted by the FCC. The FCC received concurrence for these limits from other government agencies, including OSHA and NIOSH, with the reservation that induced current limits be added to the FCC standard.

F.   OTHER RECOGNIZED STANDARDS

*   IEC 60601-1-21, Medical Electrical Equipment - Part 1-2: General Requirements for Safety - Collateral standard: Electromagnetic Compatibility - Requirements and Tests (Edition 2:2001 with Amendment 1:2004; Edition 2.1 [Edition 2:2001 consolidated with Amendment 1:2004])
*   AAMI/ANSI/IEC 60601-1-22, Medical Electrical Equipment - Part 1-2: General Requirements for Safety - Collateral Standard: Electromagnetic Compatibility - Requirements and Tests
*   ANSI/RESNA WC/Vol. 2-19983, Section 21, Requirements and test methods for electromagnetic compatibility of powered wheelchairs and motorized scooters
*   ANSI C63:19:20014, Methods of Measurement of Compatibility between Wireless Communication Devices and Hearing Aids

VII.   LASERS AND ELECTROSURGERY

Light amplification by stimulated emissions of radiation or laser can pose a risk to healthcare workers and patients. A laser emits electromagnetic radiation in the visible spectrum. Laser use is increasing at a very fast pace in the healthcare environment. Lasers used in radiology departments help align patients for treatment. Eye safety is the number one concern for anyone working with or near a laser. Though rare, laser eye injuries become permanent. The nature of eye damage depends on the type, power, and duration of the laser exposure. The most common type of injury results when a light beam heats the retina and causes loss of vision in a point of a person's field of vision. A beam from a pulse laser can cause an explosion in the retina and result in severe damage. A laser with enough energy can cause retina cell death. Any damage is permanent but not as severe as thermal or acoustic damage. Lasers striking the skin can result in erythematic, blistering, and charring. The extent of the damage depends on wavelength, power, and length of exposure. Lasers also use high voltage and pose electrical risks. The FDA Bureau of Radiological Health under 21 CFR 1040 regulates laser performance. Lasers receive calibration by the manufacturer, but always check the laser system before each procedure and during extended procedures. Classifications of lasers should coincide with actual measurement of output. Only personnel trained in laser technology should make measurements. Engineering controls, such as protective housings, remote controls, or enclosed laser beam paths, ensure protection for laser operators except when the operator needs to set up, adjust, or maintain the beam. These technicians experience the highest risk for serious injury. The Laser Safety Officer (LSO) must take actions to ensure monitoring and enforcing the control of laser hazards. These controls apply to operation, maintenance, and service. Lasers and laser systems are classified on the basis of the level of the laser radiation accessible during intended use.

A.   LASER SAFETY

The primary responsibility of a perioperative nurse during a laser procedure is keeping the patient safe. Safety hazards are inherent with laser usage, but adherence to proper procedures lowers injury risks. When perioperative nurses receive education in laser science and safety, they can recognize potential hazards and help ensure adherence to safety parameters. Class 3b and 4 laser exposures usually occur from unintentional operation or when users fail to follow proper controls. The high electrical energy used to generate the beam is a potential shock hazard. Direct beam exposure can cause burns to skin and eyes possibly resulting in blindness. Electric shock and fire also pose potential hazards when using lasers. Primary worker protection measures include using effective eye protection and properly shielding high-energy beams. Ensure availability of an approved fire extinguisher. Identify laser use areas and post warning signs. Personnel should prevent laser beams from coming into contact with combustible, flammable, and reflective materials. Ensure personnel using or exposed to lasers become part of the eye health medical surveillance system.

B. LASER SAFETY OFFICER

Appoint an LSO when personnel use Class 3 or 4 lasers. Ensure the LSO possesses the authority to monitor and enforce the laser safety requirements. The LSO administers the overall laser safety plan, including confirming the classification of lasers and assuring the use of proper control measures. The LSO also approves substitute controls and standard operating procedures (SOPs). The LSO recommends and/or approves eyewear and other protective equipment, specifies appropriate signs and labels, provides proper laser safety training, and conducts medical surveillance. The LSO should receive detailed training on laser fundamentals, laser bio-effects, exposure limits, and classifications, as well as control measures including area controls, eyewear, barriers, and medical surveillance requirements.

C. LASER STANDARDS

No federal requirements exist for safety during laser procedures. Hospital operating rooms, surgery centers, and physician practice-based surgery suites should comply with the recommended consensus safety standards. The General Duty Clause permits OSHA to cite employers for not providing a place of employment free from recognized hazards. NIOSH believes that potential hazards exist from smoke generated by lasers. Studies indicate formaldehyde, hydrogen cyanide, and benzene can occur in surgical smoke emitted from lasers. NIOSH has issued suggestions for the use of smoke evacuation units, preferably vented to the outside, as well as protective equipment to be worn by personnel servicing or changing filters on smoke evacuation devices. The FDA's CDRH regulates lasers approved for the market. These agencies also regulate which procedures lasers can perform and the ancillary supplies, including fibers and hand pieces, authorized for sale. Laser injury incidents fall under the Safe Medical Device Act reporting requirements. The key safety standards for laser use are ANSI Z136.1-1993: *American National Standard for the Safe Use of Lasers* and ANSI Z136.3-1996: *American National Standard for the Safe Use of Lasers in the Healthcare Environment.* OSHA regulates work exposures by using the "general duty clause" and CFR 1910.132, which addresses face and eye protection. ACGIH publishes recommendations about how to reduce occupational exposures. NIOSH recommends that organizations appoint an LSO in facilities where laser use warrants extra precautions. The Laser Institute of America also publishes information on safely using lasers and is the source organization for the ANSI standards. The Association of periOperative Registered Nurses (AORN) addresses laser safety in its *Recommended Practices for Laser Safety in Practice Settings.* ANSI Standard Z136.3 remains the recognized national standard for laser safety in healthcare organizations. However, AORN standards function as the optimal standards of perioperative nursing practice. For the most part, both sets of standards communicate the same information regarding laser safety and strongly promote laser safety education and training for all individuals present during a laser procedure. Individuals must complete this training before being assigned to work with lasers. Yearly reinforcement of the information and recredentialing should also take place.

D. LASER CLASSIFICATIONS

New classes—1m, 2m, and 3r—will further delineate the danger potential of medical and industrial lasers. Laser energy is light energy. Consider all Class 3 and Class 4 lasers as nonionizing, which indicates that laser energy does not cause molecular changes to tissue of the operator or others in close proximity. A pregnant staff member or physician need not fear that laser energy will cause harm to her fetus. Beam-related safety hazards include eye injuries, fire and thermal injuries, and smoke plume. Consider any electrical hazards as nonbeam related. Refer to the IEC 60825-1 standard for requirements of the classification system. Classes 2 and higher must contain the triangular warning label. Lasers may require other labels in specific cases indicating laser emission, laser apertures, skin hazards, and invisible wavelengths.

1. Class 1 Laser

A Class 1 laser is safe under all conditions of normal use. Never exceed the maximum permissible exposure (MPE) limit. This class includes high-power lasers within an enclosure that prevents exposure to the radiation and that cannot be opened without shutting down the laser. For example, a continuous laser at 600 nm can emit up to 0.39 mW, but for shorter wavelengths, the maximum emission is lower because of the potential of those wavelengths to generate photochemical damage. The maximum emission is also related to the pulse duration in the case of pulsed lasers and the degree of spatial coherence.

2. Class 1M Laser

A Class 1M laser is safe for all conditions of use except when passed through magnifying optics such as microscopes and telescopes. Class 1M lasers produce large-diameter beams, or beams possess divergent capabilities. Never exceed the MPE for a Class 1M laser unless using focusing or imaging optics to narrow the beam. Classify a laser as Class 1M if the total output power remains below Class 3B. Lasers with the power that can pass through the pupil of the eye stays within Class 1.

3. Class 2 Laser

A Class 2 laser is safe because the blink reflex will limit the exposure to no more than 0.25 seconds. It only applies to visible-light lasers (400–700 nm). Limit Class 2 lasers to 1 mW continuous wave or more if the emission time remains less than 0.25 seconds or if the light does not perform as spatially coherent. Intentional suppression of the blink reflex could lead to eye injury.

4. Class 2M Laser

A Class 2M laser is safe because of the blink reflex if not viewed through optical instruments. As with Class 1M, this applies to laser beams with a large diameter or large divergence, for which the amount of light passing through the pupil cannot exceed the limits for Class 2.

5. Class 3R Laser

A Class 3R laser is considered safe if handled carefully, with restricted beam viewing. With a Class 3R laser, the MPE can be exceeded but with a low risk of injury. Visible continuous lasers in Class 3R must be limited to 5 mW. For other wavelengths and for pulsed lasers, other limits apply.

6. Class 3B Laser

A Class 3B laser poses hazards when eyes receive direct exposures. Diffuse reflections such as from paper or other matte surfaces are not harmful. Continuous lasers in the wavelength range from 315 nm to far infrared must be limited to 0.5 W. For pulsed lasers between 400 and 700 nm, the limit is 30 mJ. Other limits apply to other wavelengths and to ultra-short pulsed lasers. Protective eyewear is typically required where direct viewing of a class 3B laser beam may occur. Class 3B lasers must be equipped with a key switch and a safety interlock.

7. Class 4 Laser

Class 4 lasers include all lasers with beam power greater than Class 3B. By definition, a Class 4 laser can burn the skin, in addition to potentially devastating and permanent eye damage as a result of direct or diffuse beam viewing. These lasers may ignite combustible materials, and thus may represent a fire risk. Class 4 lasers must be equipped with a key switch and a safety interlock. Most entertainment, industrial, scientific, military, and medical lasers are in this category.

E. Laser Plumes

During surgical procedures that use a laser or electrosurgical unit, the thermal destruction of tissue creates a smoke byproduct. An estimated 500,000 workers (surgeons, nurses, anesthesiologists, and surgical technologists) are exposed to laser or electrosurgical smoke each year. NIOSH research studies have confirmed that this smoke plume can contain toxic gases and vapors such as benzene, hydrogen cyanide, formaldehyde, bio-aerosols, cellular material including blood fragments, and viruses. At high concentrations, the smoke causes ocular and upper respiratory tract irritation in healthcare personnel, and creates visual problems for the surgeon. The smoke produces an unpleasant odor and may possess mutagenic potential. Surgical smoke may possess potential for generating infectious viral fragments. Use portable smoke evacuation devices and room suction systems. Keep the smoke evacuation devices or room suction hose nozzle inlet within two inches of the surgical site to effectively capture airborne contaminants. Keep smoke evacuation devices activated at all times when airborne particles are produced during all surgical or other procedures. Consider all tubing, filter, and absorbers as infectious waste and dispose of them appropriately. Install new filters and tubing before each procedure. Inspect smoke evacuation systems regularly to prevent possible leaks. Use Universal Precautions as required by the OSHA Bloodborne Pathogens Standard. For additional information refer to *Control of Smoke from Laser/Electric Surgical Procedures*: DHHS (NIOSH) Publication No. 96-128.

F. Laser Skin Protection

If a potential exists for skin exposure to ultraviolet lasers (200–400 nm), use skin covers and/or sunscreen. Most gloves will provide some protection against laser radiation. Tightly woven fabrics and opaque gloves provide the best protection. A laboratory jacket or coat can provide protection for the arms. For Class 4 lasers, give consideration to using flame-resistant materials. Use protective clothing when exposed to levels of radiation that exceed exposure limits for the skin.

G. Fire Prevention Tips during Laser Surgery

Place the laser in standby mode whenever it is not in active use. Activate the laser only when the tip is under the surgeon's direct vision. Allow only the person using the laser to activate it. Deactivate the laser and place it in standby mode before removing it from the site. When performing laser surgery through an endoscope, pass the laser fiber through the endoscope before introducing the scope into the patient. This will minimize the risk of damaging the fiber. Before inserting the scope in the patient, verify the fiber's functionality. During lower airway surgery, keep the laser fiber tip in view and make sure it is clear of the end of the bronchoscope or tracheal tube before laser emission. Use appropriate laser resistant tracheal tubes during upper airway surgery. Follow the directions in the product literature and on the labels, which typically include information regarding the tube's laser resistance, use of dyes in the cuff to indicate a puncture, use of a saline fill to prevent cuff ignition, and immediate replacement of the tube if the cuff becomes punctured.

VIII. MAGNETIC RESONANCE IMAGING

MRIs provide detailed images of organs and tissues throughout the body without the need for x-rays. Instead, an MRI uses a powerful magnetic field, radio waves, a rapidly changing magnetic field, and a computer to create images that show whether or not there is an injury or some disease process present. For this procedure, the patient is placed within the MRI scanner. The magnetic field aligns atomic particles called protons that exist in most of the body's tissues. Radio waves then cause these particles to produce signals received within the MR scanner. The signals become specially characterized by using a changing magnetic field, and a computer processor creates very sharp images of tissues as "slices" that can be viewed in any orientation. An MRI exam causes no pain, and the magnetic fields produce no known tissue damage of any kind. The MR scanner may make loud tapping or knocking noises at times during the exam; using earplugs prevents problems that may occur with this noise. Key safety concerns involve the use of strong magnetic fields, RF energy, time-varying magnetic fields, cryogenic liquids, and magnetic field gradients. Magnetic fields from large bore magnets can literally pick up and pull large ferromagnetic items into the bore of the magnet. Take caution to keep all ferromagnetic items away from the magnet. The kinetic energy of such an object being sucked into a magnet can smash an RF imaging coil. Similar forces work on ferromagnetic metal implants or foreign matter in those being imaged. These forces can pull on these objects, cutting and compressing healthy tissue. For these reasons individuals with foreign metal objects such as shrapnel or older ferromagnetic implants are not imaged. There exist additional concerns regarding the effect of magnetic fields on electronic circuitry, specifically pacemakers. An individual with a pacemaker walking through a strong magnetic field can induce currents in the pacemaker circuitry, which will cause it to fail and possibly cause death. Magnetic fields will also erase credit cards and magnetic storage media.

A.   FDA MRI Safety Guidelines

The guidelines state that field strengths not exceeding 2.0 Tesla may be routinely used. People with pacemakers should never enter magnetic fields greater than 5 gauss. A 50 gauss magnetic field will erase magnetic storage media. The RF energy from an imaging sequence can cause heating of the tissues of the body. The FDA recommends limited exposure to RF energy. The specific absorption rate (SAR) serves as the limiting measure. The formula for this is SAR = joules of RF/second/kg of body weight = watts/kg. The recommended SAR limitations depend on the anatomy being imaged. The SAR for the whole body should be less than 0.4 W/kg. It should be less than 3.2 W/kg averaged over the head. Any pulse sequence should not raise the temperature by more than 1 degree Celsius (C) and no greater than 38 degrees C in the head, 39 degrees C in the trunk, and 40 degrees C in the extremities. Some RF coils, such as surface coils, contain failure modes that can cause burns to the patient. Take care to keep these coils in proper operating order. The FDA recommendations for the rate of change of a magnetic field state that the dB/dt for the system should be less than that required to produce peripheral nerve stimulation. Imaging gradients do produce high acoustic noise levels.

BOX 8.3   MRI SAFETY AND COMPATIBILITY STANDARDS

- ASTM F2052-00 Standard Test Method for Measurement of Magnetically Induced Displacement Force on Passive Implants in the Magnetic Resonance Environment
- ASTM F2119-01 Standard Test Method for Evaluation of MRI Artifacts From Passive Implants
- IEC 601-2-33 - Medical Electrical Equipment - Part 2: Particular requirements for the safety of magnetic resonance equipment for medical diagnosis

B.   Other MRI Safety Recommendations

The ACR has made safety recommendations for the use of MRI machines. The new recommendations include restricting access to MRI rooms, appointing a special director of hospital MRI facilities, and educating those working near or in an MRI department about safety. Consider patients with certain implanted devices, such as many types of intra-cranial aneurysm clips, as contraindicated from MRI imaging since the torque and displacement forces produced on the device can result in the tearing of soft tissues. Other implants, such as certain cardiac pacemakers, can function erratically even in relatively weak magnetic fields. In device labeling for pacemakers, MRI is listed as a contraindication. Prohibit individuals with implanted pacemakers from entering the MRI procedures room or coming within the five gauss line around the scanner. With regard to medical devices, electrical currents may occur due to conductive metal implants, such as skull plates, and hip prostheses.

C. MRI BURNS

When using conductive patient leads during MRI scanning, ensure that the leads do not form loops. Looped patient leads or devices such as the halo device used for spinal immobilization can pick up RF energy resulting in induced currents, heating of the material, and as a result, potentially severe patient burns. To reduce the possibility of burns, thermally insulate electrically conductive material in the bore of the magnet from the patient using blankets or sheets. Other steps to prevent burns include the following:

- Ensure conductive cables are not looped and that cables do not cross each other
- Place sensors—such as those for pulse oximetry—as far as possible from the RF coils
- Use manufacturer-supplied padding, rather than blankets or sheets, to prevent patients from contacting the magnet bore
- Regularly check all sensors, cables, and MR accessories such as RF coils and cables for any breaks in insulation
- Regularly inspect items provided for patient comfort, such as headphones and video goggles, for signs of damage

Even when steps such as these are taken, heating can occur during an MRI scan, so MRI technicians must ask patients to signal if they feel undue heat during the scan.

IX. OTHER CLINICAL RISKS

A. CT RADIATION DOSES

CT is fast, reliable, and convenient, but its comparatively high x-ray dose has only just begun to attract attention. In the United States, it has been estimated that CT could be responsible for about 6,000 additional cancers a year, roughly half of them fatal (ECRI Institute, 2008).

To ensure patients are not unnecessarily exposed to high dose levels, consider the following:

- The expected benefits of a CT need to outweigh the radiation risks and CT referral guidelines should be regularly reviewed, particularly where children are concerned.
- Scanning protocols should be optimized to minimize doses.
- Those performing CT exams should be specifically trained for CT and maintain their training.
- Monitoring CT use and dose should be part of normal quality control and equipment maintenance efforts.
- Referring clinicians should have easy access to information regarding the dose and the cancer risk associated with CT.

B.   Fiber-Optic Light Burns

Fiber-optic light sources—used in devices such as endoscopes, retractors, and headlamps—are often referred to as "cold" light sources. This can be misleading. There are two main sources of burns from fiber-optic lights:

- The light itself: This is usually caused when a clinician places the endoscope or the distal end of the fiber-optic cable on the patient without shutting off the light source.
- Heated cable connections: This is typically caused when the diameter of the light cable is too large for the light post on the connected device. The light then contacts the metal portion of the light post, rather than the fibers within, heating the connection. If the connection contacts skin, a burn may result.

To reduce the risk of burns, consider the following:

- Never place the endoscope or the end of the light cable on a patient or on flammable objects.
- Turn off the light source or place it in standby mode before removing the cable from the light source or the instrument from the cable.
- Use only the minimum output needed to perform the procedure.
- Only use light sources that incorporate safety features, such as those that power up in standby mode or at very low output settings.

C.   Sonography

As sonographers work with ultrasound equipment they may incur a risk for developing work-related MSDs. Sonographers with heavy workloads and those with experience in the profession can incur risks. According to the Society of Diagnostic Medical Sonography (SDMS), sonographers on average experience pain or other disorders within five years of entering the profession. Sonographers can experience a variety of ergonomics-related risks when they performing specific tasks such as

- Transporting patients and equipment
- Positioning patients and equipment
- Using and orientating ultrasound equipment

Engineering, administrative, and work practice controls such as room layout and equipment placement, scheduling, staffing, patient assessment, training, and work practices may also need consideration to reduce the risk of developing an injury.

X.   LABORATORY SAFETY

Hospitals operate in a variety of clinical laboratory functions. The size and scope of these lab functions will vary depending on the size and type of organization or facility. Many outside clinical laboratories also serve a variety of medical organizations and practices.

A.   Clinical Laboratory Improvement Amendments

The Centers for Medicare & Medicaid Services (CMS) regulates all laboratory testing (except research) performed on humans in the United States through the Clinical Laboratory Improvement Amendments (CLIA). In total, CLIA covers approximately 200,000 laboratory entities. The

> **BOX 8.4   CLINICAL LAB FUNCTIONS IN SOME HOSPITAL SETTINGS**
>
> - **Pathology:** Processes and tests tissue removed during surgical procedures
> - **Cytology:** Processes specimens to determine abnormalities in cell structure
> - **Chemistry:** Analyzes body fluids to determine glucose, protein, enzyme, and hormone levels
> - **Serology:** Analyzes body fluids for antigens and antibodies
> - **Hematology:** Analyzes blood to determine info relating to red cells, white cells, and platelets
> - **Microscopy:** Analyzes urine and body fluids
> - **Microbiology:** Analyzes specimens to determine causes of infection

Division of Laboratory Services, within the Survey and Certification Group, under the Center for Medicaid and State Operations (CMSO), has the responsibility for implementing the CLIA Program. The objective of the CLIA is to ensure quality laboratory testing. Although all clinical laboratories should be properly certified to receive Medicare or Medicaid payments, CLIA has no direct Medicare or Medicaid responsibilities.

B. JOINT COMMISSION LABORATORY ACCREDITATION

Since 1979, the Joint Commission has been accrediting hospital laboratory services. It has been accrediting independent laboratories since 1995. The Joint Commission accredits approximately 3,000 clinical laboratories. CMS officially recognizes the Joint Commission Laboratory Accreditation Program as meeting the requirements of CLIA '88. CLIA regulations require that all laboratories be surveyed on a two-year cycle. Joint Commission standards and CLIA regulations require that laboratories use CMS-approved proficiency testing for all regulated tests conducted by the lab. CLIA requires that a laboratory's proficiency testing results be monitored on an ongoing basis. Achieving accreditation sends a strong statement about a lab's commitment to provide services of the highest quality. The Joint Commission uses a tracer methodology that reviews the entire scope of the laboratory testing process, including pre- and post-analytical processes that occur outside the laboratory. The tracer system follows results to the bedside. Joint Commission lab surveys also evaluate other areas such as environment of care management, emergency management, infection control, and adherence to National Patient Safety Goals.

C. COLLEGE OF AMERICAN PATHOLOGISTS ACCREDITATION PROGRAM

The College of American Pathologists (CAP) Laboratory Accreditation is an internationally recognized program and the only one of its kind that utilizes teams of practicing laboratory professionals to serve as inspectors. Designed to go well beyond regulatory compliance, accreditation helps laboratories achieve the highest standards of excellence to positively impact patient care. Accreditation standards focus on detailed checklist requirements. The checklists provide a quality blueprint for laboratories to follow. CAP Laboratory Accreditation meets the needs of a variety of laboratory settings from complex university medical centers to physician office-based laboratories. Because of its comprehensive nature, CAP accreditation can help achieve a consistently high level of service throughout an institution or healthcare system. CMS grants CAP Laboratory Accreditation deeming authority. The Joint Commission recognizes CAP and accreditation helps organizations meet state certification requirements. CAP also provides laboratory accreditation to forensic urine drug testing and reproductive laboratories, cosponsored with the American Society for Reproductive

Medicine (ASRM). The goal of CAP Laboratory Accreditation focuses on improving patient safety by advancing the quality of pathology and laboratory services. Upon successful completion of the inspection process, the laboratory receives CAP accreditation to become part of an exclusive group meeting the highest standards of excellence.

This accreditation program seeks to improve patient safety and reduce laboratory-related risks. The accreditation conforms to ISO 15189:2007 requirements. Laboratories accredited to the ISO standard are well positioned to rapidly respond to the changing healthcare environment and demonstrate measurable quality to their customers. CAP serves as a medical society serving more than 17,000 physician members and the laboratory community throughout the world. It is the world's largest association composed exclusively of board-certified pathologists and pathologists-in-training and is widely considered the leader in laboratory quality assurance. CAP is an advocate for high quality and cost-effective medical care. ISO promotes standardization facilitating the international exchange of goods and services, and developing cooperation in the spheres of intellectual, scientific, technological and economic activity.

D. COLA Accreditation

The Commission on Office Laboratory Accreditation (COLA) was founded in 1988 as a private alternative to help laboratories stay in compliance with the new CLIA regulations. In 1993, CMS granted COLA deeming authority under CLIA, and in 1997 the Joint Commission recognized COLA's laboratory accreditation program. After 35,000 surveys in which COLA's practical, educational accreditation methods helped physician office laboratories stay in compliance with CLIA, COLA has expanded its program offerings to include hospital and independent laboratories. COLA is approved by CMS to accredit laboratories in the following specialties: (1) chemistry, (2) hematology, (3) microbiology, (4) immunology, (5) immune-hematology and transfusion services, and (6) pathology (cytology, histopathology, and oral pathology).

E. OSHA Laboratory Standard (29 CFR 1910.1450)

The OSHA standard requires labs to produce a Chemical Hygiene Plan that addresses the specific hazards found in its location, and its approach to them. The standard emphasizes the use of safe work practices and appropriate worker protection as required by the laboratory environment. The standard covers all chemicals that meet the definition of a health hazard as defined in the Hazard Communication Standard published in 29 CFR 1910.1200. The standard does not specify work practices necessary to protect employees from potential hazards associated with chemical use but does require that physical hazards be addressed in the employer's training program. The OSHA Lab Standard requires continued compliance with all published permissible exposure limits (PELs) and with the employer's written chemical hygiene plan. The standard requires that special consideration be given for particularly hazardous substances including some selected carcinogens, reproductive toxins, and substances containing a high degree of acute toxicity. The laboratory standard 29 CFR 1910.1450 can apply to clinical hospital laboratories because of the use of hazardous chemicals. Laboratory use of hazardous chemicals means handling or use of such chemicals in which all of the following conditions are met: (1) chemical manipulations carried out on a laboratory scale, (2) multiple chemical procedures or chemicals used, (3) procedures involved not part of a production process, and (4) protective laboratory practices and equipment used to minimize employee exposures. Please note that according to an OSHA interpretation, the standard does not apply to a pharmacy operation mixing cytotoxic drugs.

BOX 8.5   KEY REQUIREMENTS OF THE OSHA LAB STANDARD

- Conduct employee exposure monitoring (under certain conditions)
- Develop SOPs
- Emphasize requirements of the HAZCOM Standard for employee training
- Arrange for medical consultations and examinations
- Develop a Chemical Hygiene Plan and appoint a Chemical Hygiene Officer
- Provide hazard identification information such as Safety Data Sheets (SDSs) and labeling requirements
- Ensure chemical fume hood performance certifications

BOX 8.6   RELEVANT OSHA STANDARDS FOR LABORATORIES

- Personal Protective Equipment (29 CFR 1910 Subpart I)
- General PPE Requirements (29 CFR 1910.132)
- Respiratory Protection (29 CFR 1910.134)
- Toxic and Hazardous Substances (29 CFR 1910 Subpart Z)
- Hazard Communication (29 CFR 1910.1200)
- Bloodborne Pathogens (29 CFR 1910.1030)
- Occupational Exposure to Hazardous Chemicals in Laboratories (29 CFR 1910.1450)

BOX 8.7   OTHER RELEVANT LAB STANDARDS & RESOURCES

- ANSI Z358.1-2004, Emergency Eyewash and Shower Equipment, contains provisions regarding the design, performance, installation, use, and maintenance of various types of emergency equipment. In addition to these provisions, some general considerations apply to all emergency equipment.
- ANSI Z9.5-2003, Laboratory Ventilation, is intended for use by employers, architects, occupational and environmental health and safety professionals, and others concerned with the control of exposure to airborne contaminants. The book includes new chapters on performance tests, air cleaning, preventative maintenance, and work practices. It also highlights the standard's requirements and offers good practices for laboratories to follow. The book also offers referenced standards and publications, guidance on selecting laboratory stack designs, an audit form for ANSI Z9.5, and a sample table of contents for a laboratory ventilation management plan.
- ANSI/ASHRAE 110-1995, Method of Testing the Performance of Laboratory Hoods, specifies a quantitative test procedure for evaluation of a laboratory fume hood. A tracer gas is released at prescribed rates and positions in the hood and monitored in the breathing zone of a mannequin at the face of the hood. Based on the release rate of the tracer gas and average exposure to the mannequin, a performance rating is achieved.
- NFPA 45, Standard on Fire Protection for Laboratories Using Chemicals, 2004 Edition, applies to laboratories that handle hazardous chemicals.
- NIOSH Pocket Guide to Chemical Hazards (NPG) provides a source of general industrial hygiene information on several hundred chemicals and classes for workers, employers, and occupational health professionals.
- NIOSH Master Index of Occupational Health Guidelines for Chemical Hazards summarizes pertinent information about the properties and hazards of many chemicals found in laboratories.

(Continued)

BOX 8.7   OTHER RELEVANT LAB STANDARDS & RESOURCES (*Continued*)

- The AIHA Laboratory Health & Safety Committee provides resources to aid in prevention, identification, and control of potential exposures to chemical, biological, ergonomic, ionizing and nonionizing radiation, and physical hazards in laboratories.
- The Hazardous Substances Data Bank (HSDB), National Library of Medicine (NLM), provides a toxicology data file that focuses on the toxicology of potentially hazardous chemicals. It is enhanced with information on human exposure, industrial hygiene, emergency handling procedures, environmental fate, regulatory requirements, and related areas.
- Biosafety in Microbiological and Biomedical Laboratories, 5th Edition, provides guidance for implementing safety and hazard control practices in research labs.

F. SUMMARY OF LAB SAFE WORK PRACTICES

Supervisors must prohibit mouth pipetting or suctioning of blood related materials. Restrict eating, drinking, smoking, applying cosmetics or lip balm, or handling contact lenses in areas with a reasonable likelihood of occupational exposure to bloodborne pathogens. Never store food or drink in refrigerators, freezers, shelves, cabinets, or on countertops or bench tops where blood or other potentially infectious materials exist. Use splatter guards to prevent splashing from reaching employees. Use hands-free sensor-controlled automatic sinks with foot, knee, or elbow controls. Other safety suggestions include using centrifuge tubes with caps, working in appropriate biological safety cabinets (BSCs), checking daily for proper air exchange and air flow, and maintaining records of ventilation systems and other equipment. Workers should never stand on chairs, lab stools, boxes, or drums to reach high shelves or the ceiling area. Use stepladders or step stools specially designed for such purposes. Wash hands and arms several times during the course of the day to remove bits of irritating chemicals, animal dander, or biohazards. Maintain adequate ventilation at all times. Check hood drafts regularly, and direct questions about the proper functioning of the hood to the maintenance department or the chemical hygiene officer. Stay aware that static electricity can develop when transferring poor conductor liquids from one container to another. Watch for electrical charges that could develop when compressed gases release rapidly from a cylinder. These charges can jump air gaps and form sparks, which may ignite flammable vapors or gases. Ensure the proper grounding of cylinders by connecting the container and receiver by a ground wire. Electrical charges may also build up on personnel wearing shoes with rubber or plastic soles. Report sluggish drains to the maintenance department immediately. Never pour chemicals down a drain that could interact with the pipe material or cause a chemical reaction. Never pour any flammable materials down a drain. Refer to NFPA 45, Fire Protection Standard for Laboratories Using Chemicals, for information on construction, ventilation, and fire protection requirements.

G. CENTRIFUGES

Cover all centrifuges during operation. Centrifuge tubes should fit the metal buckets and should never contain defects or cracks. Cushions at the bottom of the cups should be in good condition. Implement an inspection and maintenance schedule for centrifuges and associated equipment installed in the laboratory. Equip all electrical heating equipment with over-temperature shutoff controls. Make thermal gloves, beaker and crucible tongs, and test tube holders available for handling hot items.

H. LABORATORIES AND THE OSHA BLOODBORNE PATHOGENS STANDARD

Take appropriate actions to prevent exposure of laboratory employees to bloodborne pathogens while handling contaminated lab samples such as blood or other body fluids. OSHA sets additional requirements for biomedical research labs and facilities. Exposure of laboratory employees to bloodborne pathogens can occur while handling contaminated lab samples such as blood or other body fluids. Require workers to wear appropriate personal protective equipment (PPE) as required by the OSHA standard. The type and amount of PPE depends on the anticipated exposure. Post a hazard warning sign incorporating the universal biohazard symbol on all access doors when potentially infectious materials, including infected animals, remain present in the work area. The hazard warning sign must meet OSHA 29 CFR 1910.1030 requirements. Conduct all activities involving other potentially infectious materials using a BSC or other physical containment device within the containment module. Prohibit work with other potentially infectious materials on open benches. Use only certified BSCs or other appropriate combinations of personal protection or physical containment devices. Detail any requirements for special protective clothing, respirators, centrifuge safety cups, sealed centrifuge rotors, and containment caging for animals to prevent exposures to droplets, splashes, spills, or aerosols. Recommend that sinks permit foot, elbow, or automatic operation and locate sinks near an exit door in the work area. Each laboratory must provide a hand-washing and eyewash facility that is readily available within the work area. Establish controls to prevent employee exposure from needle stick injuries or cuts from sharp objects when working with specimens, centrifuge tubes, or overfilled sharps containers. Use engineering controls such as safer needle devices and work practice controls include altering the way a task is performed to reduce the chance of injury. OSHA, FDA and NIOSH now recommend not using glass capillary tubes due to exposure risks during breakage. Human immunodeficiency virus (HIV) and hepatitis B virus (HBV) research laboratories must only use hypodermic needles and syringes for parenteral injection and aspiration of fluids from laboratory animals and/or diaphragm bottles. Research facilities must only use needle locking syringes or disposable syringe needle units for the injection or aspiration of other potentially infectious materials.

I. TUBERCULOSIS

Require the use of appropriate controls such as limiting access, sealing windows, providing directional airflow, preventing recirculation of laboratory exhaust air, and filtering exhaust air. Use BCSs whenever working with infectious materials that could aerosolize. Processes that can expose employees to aerosolized materials include (1) pouring liquid cultures, (2) using fixed volume automatic pipetting devices, (3) mixing liquid cultures with a pipette, (4) preparing specimens and culture smears, and (5) dropping and spilling tubes containing suspensions of bacilli.

J. MORGUE

Employee exposures include biological risks from infectious diseases and agents such as staph, strep, and HBV. Formaldehyde exposures from contact with cadavers pose a real hazard. Use appropriate PPE. Additional protections may apply during autopsies. Ensure the operation of appropriately designed ventilation systems. Locate local vacuum systems for power saws in the morgue. Provide appropriate ventilation systems such as downdraft tables that capture the air around the cadaver. Use splatter guards to prevent splashes from reaching employees. Require the use of surgical caps or hoods and shoe covers or boots when anticipating high contamination.

K. CHEMICAL AND FIRE HAZARDS

Laboratory workers can experience routine exposure to hazardous chemicals such as acetone, carbon monoxide, formaldehyde, hydrogen sulfide, mercury, nitric acid, and xylene. Many exposures occur annually in laboratories, resulting in chemical-related illnesses such as dermatitis, eye irritation, and even fatal pulmonary edema. Employers must include information on additional protective measures for work that involves carcinogens, reproductive toxins, and acutely toxic substances.

Establish a designated area with appropriate warning signs of the hazards present. Provide information on safe and proper use of a fume hood or equivalent containment device. Develop procedures for decontaminating the designated area, including the safe removal of contaminated waste and biohazards. Employers must ensure that hazardous chemical container labels are not removed or defaced. SDSs must accompany incoming shipments of chemicals and employers must make them made available to exposed employees. Develop plans and conduct drills to prepare for emergencies such as fire, explosion, accidental poisoning, chemical spill, vapor release, electric shock, bleeding, and personal contamination. Employers must provide appropriate safety equipment and first aid kits. Safety equipment may include fire extinguishers, fire blankets, automated external defibrillators (AEDs), safety showers, eyewash fountains, spill control materials, and fume exhaust hoods. Test or check safety equipment on a monthly basis.

L. CHEMICAL EXPOSURE RESPONSE

Determine exposure outcomes by considering the following issues: (1) route of exposure, (2) physical properties of the chemical, and (3) individual susceptibility to the chemical. Report all exposure incidents to the laboratory manager or supervisor, or principal investigator, regardless of severity. When decontaminating skin, immediately flush with water for at least 15 minutes. Use shower and eyewash stations as appropriate. Check the SDS to determine if any delayed effects should be expected. Discard contaminated clothing or launder them separately from other clothing. Consider formaldehyde as a suspected carcinogen. When any possibility exists that employee's eyes may be splashed with solutions containing 0.1 percent or greater formaldehyde, the employer must provide acceptable eyewash facilities within the immediate work area for emergency use.

M. STANDARD OPERATING PROCEDURES

Develop SOPs relevant to safety and health considerations to be followed when laboratory work involves the use of hazardous chemicals. This is especially the case if your lab operations include the routine use of select carcinogens, reproductive toxins, and substances of acute toxicity. SOPs can function as standalone documents or supplemental information included as part of research notebooks, experiment documentation, or research proposals. The key idea with laboratories having SOPs is to ensure a process is in place so that an experiment is well thought out and includes and addresses relevant health and safety issues.

N. LABORATORY EQUIPMENT

Laboratory equipment may include refrigerators, centrifuges, microscopes, glassware, vacuum systems, stirring and mixing devices, heating devices, and autoclaves. Laboratory workers should understand all potential equipment hazards, and should follow cleaning, maintenance, and calibration schedules. Electrically powered equipment such as hot plates, stirrers, vacuum pumps, electrophoresis apparatus, lasers, heating mantles, ultrasonic devices, power supplies, and microwave ovens can pose significant hazards when mishandled or not maintained. Require grounded plugs on all electrical equipment and ground fault interrupters (GFIs) at needed locations. Compressed gases can pose toxic, flammable, oxidizing, corrosive, inert, or a combination of hazards. Use appropriate care when handling and storing compressed gas cylinders.

O. MICROTOME SAFETY

Modern microtomes are precision instruments designed to cut uniformly thin sections of a variety of materials for detailed microscopic examination. For light microscopy, the thickness of a section can vary between 1 and 10 microns. All microtomes consist of a base or the microtome body, knife attachment and knife, and the material or tissue holder. With most microtomes a section is cut by advancing the material holder towards the knife while the knife is held rigidly in place. The cutting action can occur in either in a vertical or horizontal plane. Microtome knives constitute one of the ever-present and continuing hazards faced by medical laboratory personnel in the production of quality sections for diagnosis. Sharp microtome knives pose cutting hazards to users. However, injuries need not occur if workers take precautions and handle microtome knives with care and respect at all times. Always use knife guards on microtomes and carry a solid microtome knife in its box. Attach a handle before removing the knife from the box and never attempt to catch a dropped knife. Take extra care when tightening the screws used for holding disposable blades firmly in blade holders. Recommend workers take tetanus boosters every 5 to 10 years. Glass knives, prepared on knife-making machines by breaking hardened glass under pressure, pose an extra hazard. Splintering may occur when making new glass knives and safety glasses should be worn to prevent eye damage.

P.   PRESSURE AND VACUUM SYSTEMS

Working with hazardous chemicals at high or low pressures requires planning and special precautions. Implement procedures to protect against explosion or implosion through appropriate equipment selection and the use of safety shields. Take care when selecting glass apparatus and ensure that selected models can safely withstand designated pressure extremes. Always provide guards on all vacuum pumps. Vacuum work can result in an implosion and the possible hazards of flying glass, splattering chemicals, and fire. Correctly install all vacuum operations and know the potential risks.

Q.   FUME HOODS AND LABORATORY VENTILATION

A well-designed chemical fume hood, when properly installed and maintained, offers a large degree of protection to the user. A fume hood functions as a ventilated enclosure that contains gases, vapors, and fumes to prevent their release in the laboratory. Hoods can also limit the effects of a spill by partially enclosing the work area and drawing air into the enclosure by means of an exhaust fan. An exhaust fan situated on the top of the laboratory building pulls air and airborne contaminants into the hood through ductwork and exhausts them to the atmosphere. In a well-designed, properly functioning fume hood, only about 0.0001 to 0.001 percent of the material released within the hood actually escapes from the hood and enters the laboratory. Base the necessity of a fume hood by conducting a hazard analysis. The analysis should include a review of the quantity and toxicity of the materials used and the experiment conducted. Also consider volatility of the materials present, potential for their release, the number and sophistication of manipulations, and the skill of the lab person performing the work. Many laboratories use equipment and apparatus that can generate airborne contaminants that cannot be controlled by a fume hood. Examples include gas chromatographs, ovens, and vacuum pumps.

R.   EMPLOYEE TRAINING

Employers must provide training on safe lab work practices. Topics should include issues such as hazard awareness, handling of chemicals, procedures, and PELs. Provide workers with information on health and hygiene, physical hazards, electrical safety, emergency procedures, PPE, working alone in the laboratory, security, and handling visitors. Conduct worker training prior to initial assignment and whenever new exposure situations occur. Training must include location of the facility hygiene plan and the requirements of the OSHA Laboratory Standard. Provide information about OSHA PELs and recommended exposure limits (RELs) if no regulatory standard applies. Educate workers on the procedures for handling, storing, and disposing of hazardous chemicals. Review the elements of the chemical hygiene plan. Review specific employer procedures, engineering controls, work practices, and PPE. Ensure employees understand detection methods, observation guidelines, and monitoring procedures. Training should emphasize visual appearances and presence of odors that can help detect the presence of hazardous chemicals. Consider conducting regular departmental safety meetings to discuss the results of inspections and aspects of laboratory safety.

S.   MEDICAL EXAMINATIONS AND CONSULTATIONS

The OSHA Laboratory Standard does not mandate medical surveillance for all laboratory workers. The employer must provide workers an opportunity for medical attention. This includes follow-up examinations and treatment recommended by an examining physician when an employee exhibits signs or symptoms associated with exposure to a hazardous chemical or the worker is routinely exposed above the action level or PEL for a regulated substance. Offer medical consultation to any employee potentially exposed through a spill, leak, or explosion of a hazardous chemical or substance. Employers must provide information about the hazardous chemical, conditions under which the exposure occurred, and a description of symptoms experienced by the worker. They must also obtain from a treating physician any written opinion requiring follow-up examinations or medical tests.

T.   SUPERVISOR RESPONSIBILITIES

Supervisors must ensure that employees know, understand, and follow the chemical hygiene plan and related SOPs. Supervisors must ensure availability of proper PPE and train employees in its proper use. Perform quarterly chemical hygiene and housekeeping inspections. Perform semiannual chemical inventories of all laboratories and storage areas. Determine PPE for the procedures and chemicals in use in the area. Supervisors must conduct self-inspections to assure that healthful working conditions are regulatory compliant.

U.   SAFETY PERSONNEL RESPONSIBILITIES

Safety personnel can help lab safety by providing training, resources, and consultation for a variety of laboratory safety issues, including chemical safety, biological safety, electrical safety, laser safety, radiation safety, and other topics. They can also review the chemical hygiene plan, help develop and maintain laboratory safety manuals, conduct exposure monitoring, inspect fume hoods, and perform safety audits.

V.   LABORATORY PERSONNEL SAFETY RESPONSIBILITIES

Lab personnel must plan and conduct each laboratory operation in accordance with the chemical hygiene plan. Employees must also keep lab work areas in good order. Employers must educate personnel on how to correctly select and use required PPE. Provide employees with a system to report exposures, injuries, or problems to supervisors or the chemical hygiene officer.

W.   ANIMAL RESEARCH FACILITIES

Use care when handling animals to avoid being bitten or scratched. Use proper restraining or protective devices whenever possible. Wear protective gloves when dissecting or conducting necropsy. Use first-aid procedures to treat animal bites and scratches and report all incidents immediately. Immediately report allergic reactions to animals or to the drugs used in treating animals. When using animals to study the progress of disease, it is the responsibility of the supervisor to explain methods of protection to all workers. Thoroughly disinfect the living area of infected animals. Render all animal carcasses noninfectious by autoclaving or incineration. Animal Biosafety Levels 1, 2, 3, and 4 provide increasing levels of protection to personnel and the environment.

One additional biosafety level designated BSL-3-Agriculture addresses activities involving large or loose-housed animals and/or studies involving agents designated as High Consequence Pathogens by the United States Department of Agriculture (USDA).

X.   LABORATORY HAZARDOUS WASTE DISPOSAL

Lab workers must know waste characteristics, proper packaging standards, labeling requirements, and waste collection or containment policies. Labs should maintain chemical inventory to avoid purchasing unnecessary quantities of chemicals and develop a program for dating stored chemicals. Other helpful information needed to ensure proper disposal includes the following: (1) procedures for drain disposal of chemical waste, (2) policies on disposing of empty chemical containers, and (3) knowledge of federal, state, and local regulations for proper disposal. Determine procedures for special types of waste including batteries, mercury-containing items, used oil, and recycling chemicals.

Y.   AUTOCLAVES

Autoclaves provide steam sterilization using a combination of temperature and pressure for a set time. A common autoclave cycle is 121 degrees Celsius at 15 psi for 20 minutes; however, settings on individual autoclave machines can vary slightly. While autoclaving does sterilize products, it does not clean the product, so take measures to clean labware to remove any soil or debris. Use a manual cleaner, ultrasonic cleaner, or laboratory washer. Because of the high temperature required for autoclaving, some plastics cannot survive autoclaving due to heat requirements. These would include polyethylene, polystyrene, or polyurethane. It is important for anyone using high-temperature sterilization processes to determine if their lab ware can tolerate high heat applications.

Z.   GOOD CLINICAL LABORATORY PRACTICES

To maintain a good clinical laboratory practices (GCLP) environment for clinical trials, labs must implement all key GCLP practices. These elements include (1) developing effective organization and personnel management, (2) designing testing facilities, (3) appropriately validating assays, (4) using relevant positive and negative controls for the assays, and (5) creating a system for recording, reporting, and archiving data. Conducting audits helps ensure compliance with GCLP guidance. GCLP compliance will help laboratories ensure production of accurate, precise, and reproducible data to support sponsor confidence and withstand regulatory agency review. GCLP embraces both research and clinical labs. GCLP standards encompass applicable portions of 21 CFR Parts 58 (GLP) and 42 CFR part 493 (Clinical Laboratory Improvement Amendments -CLIA). Due to the ambiguity of some parts of the CFR regulations, the GCLP standards are described by merging guidance from regulatory authorities as well as other organizations and accrediting bodies, such as the CAP and the ISO. The GCLP standards provide a single, unified document to guide the conduct of laboratory testing for human clinical trials. The intent of GCLP guidance is to help laboratories ensure the quality and integrity of data. The guidelines also encourage accurate reconstruction of experiments, monitoring of data quality, and comparing of test results.

AA.   LABORATORY PHYSICAL ENVIRONMENTS

Provide an environment in which laboratory testing does not compromise the safety of the staff, or the quality of the pre-analytical, analytical, and post-analytical processes. Design the laboratory to assure proper equipment placement, adequate ventilation, and sufficient storage areas. Ensure lab design supports archiving of data in a secure fireproof, fire-resistant, or fire-protected environment. Provide access to authorized personnel only. Laboratory designs must provide sufficient work areas to support worker effectiveness and safety. Maintain the laboratory's ambient temperature and humidity to ensure equipment and testing remains in tolerance limits established by the manufacturer. Use ambient temperature logs to document the acceptable temperature range, record actual temperatures, and provide documentation of corrective action taken to maintain acceptable temperature ranges. Clean and maintain all floors, walls, ceilings, and bench tops in the laboratory.

XI. DRUG HAZARDS

Because of the complexity of medications, including specific indications, effectiveness of treatment regimens, medication safety, and patient compliance issues, many pharmacists practicing in hospitals gain more education and training after pharmacy school through a pharmacy practice residency, which is sometimes followed by another residency in a specific area. Those pharmacists, often referred to as clinical pharmacists, specialize in various disciplines of pharmacy. For example, pharmacists can specialize in hematology/oncology, HIV/AIDS, infectious disease, critical care, emergency medicine, toxicology, nuclear pharmacy, pain management, psychiatry, anti-coagulation clinics, herbal medicine, neurology/epilepsy management, pediatrics, neonatal pharmacists, and more.

Hospital pharmacies usually stock a larger range of medications, including more specialized medications, than would be feasible in the community setting. Most hospitals dispense medications as a unit dose, or a single dose of medicine. Hospital pharmacists and trained pharmacy technicians compound sterile products for patients including total parenteral nutrition (TPN), and other medications given intravenously. This is a complex process that requires adequate training of personnel, quality assurance of products, and adequate facilities. Some hospital pharmacies outsource high-risk preparations and some other compounding functions to companies who specialize in compounding. The high cost of medications and drug-related technology, combined with the potential impact of medications and pharmacy services on patient safety make it imperative that hospital pharmacies perform at the highest level possible.

A. CLINICAL PHARMACY

Clinical pharmacists provide direct patient care services that optimize the use of medication and promote health, wellness, and disease prevention. Clinical pharmacists care for patients in all healthcare settings, but the clinical pharmacy movement initially began inside hospitals and clinics. They often collaborate with physicians and other healthcare professionals to improve pharmaceutical care, and now serve as an integral part of the interdisciplinary approach to patient care. They work collaboratively with physicians, nurses, and other healthcare personnel in various medical and surgical areas, and often participate in patient care rounds and drug product selection. In most hospitals in the United States, potentially dangerous drugs that require close monitoring are dosed and managed by clinical pharmacists. Pharmacies must comply with Joint Commission, American Osteopathic Association, or other Medicare accreditation/certification standards. OSHA regulates worker safety in pharmacies, including exposure to hazardous materials and drugs. In hospital settings, the pharmacy plays a key role in providing quality care. The pharmacy serves other functions including directing special drug programs, providing services to satellite locations, and managing drug information systems. The pharmacy department also plays a key role in preventing medication errors. The onsite licensed pharmacists prepare pharmacy compounds and mix all sterile medications, intravenous admixtures, or other drugs, except in emergency situations.

B. GENERAL SAFETY CONSIDERATIONS

Use safety materials and equipment while preparing hazardous medications. Avoid contamination by using clean or sterile techniques as appropriate. Maintain clean, uncluttered, and functionally separate areas for product preparation. Visually inspect the integrity of the medications. Workers not aware of proper work practices and controls may be exposed to hazardous drugs through the skin, mouth, or by inhalation. The OSHA Technical Manual and new NIOSH Guidelines provide guidance regarding the safety, use, administration, storage, and disposal of hazardous drugs.

C. Pharmacy Safety

The American Society of Healthcare System Pharmacists recommends that pharmacies develop and implement a safety management program. Pharmacy representation on the safety committee is important not only to the department, but also to the effectiveness of the safety committee. Concern over safe handling of medication, errors, and hazardous drug safety has increased the pharmacy's presence on the committee. The pharmacy director must ensure that the department conducts an effective orientation and on-the-job training program to address

- The importance of practicing safety on the job
- The department's disaster planning and emergency response roles
- Hazards found in specific jobs or processes
- Organizational/departmental safety policies and procedures

Pharmacy personnel must be familiar with the organization's emergency management plan, and must be trained in the department's responsibilities in supporting the plan. The department should develop a plan for obtaining and distributing drugs during emergency situations. Pharmacy staff members should know

- Fire identification and reporting procedures
- The classes and hazards of fire
- How to activate the fire alarm and notify others
- How to select and use the proper fire extinguisher
- Techniques for controlling smoke and fire
- Evacuation routes and egress responsibilities

D. OSHA Hazard Communication Standard

OSHA requires a written plan that contains information related to worker training, warning labels, and access to SDSs. Employees must understand the requirements of the Hazard Communication Standard, including operations or procedures with hazard exposures. The Hazard Communication Standard applies to drugs and pharmaceuticals that the manufacturer has determined to be hazardous. It could also apply to hazardous substances known to be present in the workplace that employees may be exposed to under normal conditions or in a foreseeable emergency. The exemptions to the standard include

- Drugs that are in solid, final form for direct administration to the patient (final form exemption would also apply to tablets or pills that are occasionally crushed, if the pill or tablet is not designed to be dissolved or crushed prior to administration)
- Consumer products subjected to the labeling requirements of the terms as defined in the Consumer Product Safety Act and the Federal Hazardous Substances Act

E. Pharmacy Ergonomics

Pharmacy personnel may develop MSDs such as carpel tunnel syndrome or tendonitis from activities that involve repetitive tasks, forceful exertions, awkward postures, or contact stress. Use assistive devices if possible. Modify pharmacy tasks to decrease incidence of work-related MSDs. Redesign the process to incorporate variation into the task. Ergonomically comfortable workstations should include wrist pads, adjustable padded chairs, keyboard trays, and monitors that are at a comfortable height.

F. WORKPLACE VIOLENCE

Pharmacists may be exposed to workplace violence due to the availability of drugs and money in the pharmacy area. OSHA recommends that employers establish and maintain a violence prevention program. Install Plexiglas® in the payment window in the pharmacy area. Provide better visibility and lighting in the pharmacy area. Conduct training for staff in recognizing and managing hostile and assaultive behavior. Implement security devices such as panic buttons, beepers, surveillance cameras, alarm systems, two-way mirrors, card-key access systems, and security guards.

G. GENERAL MEDICATION LABELING

Standardize labeling to meet organization policy, applicable law, or practice standards. Properly label all medication when prepared if not administered immediately. Appropriately label any container, including plastic bags, syringes, bottles, or boxes, that can be labeled and secured. Label with the drug name, strength, and amount, if not apparent from the container. Include expiration date when not used within 24 hours. Label compounded IV admixtures and nutrition solutions with the date prepared and the diluents. When preparing medications for multiple patients or if the preparing person will not administer the medication, include the patient name and location on the label.

H. CLOSED PHARMACY PROCEDURES

Develop a process for providing medications to meet patient needs when the pharmacy is closed. Store the medications in a night cabinet, automated storage and distribution device, or a selected section of the pharmacy. Only permit trained designated prescribing professionals or nurses to access medications. Establish quality control procedures such as a second check by another individual or a secondary verification such as bar coding to prevent medication retrieval errors. Arrange for a qualified pharmacist to stay available on call or at another location to answer questions or provide for medications beyond those accessible to nonpharmacy staff. Implement changes as needed to reduce the amount of times nonpharmacist healthcare professionals obtain medications after the pharmacy is closed.

I. DRUG RECALLS OR SAFETY ALERTS

When the organization has been informed of a medication recall or discontinuation by the manufacturer or the FDA for safety reasons, retrieve the medications within the organization and handle per organization policy, law, or regulation. Recalls generally occur by lot number. An organization may retrieve all lots of a recalled medication instead of recording and identifying medications by their lot number. When the organization has been informed of a medication recall or discontinuation by the manufacturer or the FDA for safety reasons, notify everyone that orders, dispenses, and administers of the recalled or discontinued medications.

J. HIGH-RISK MEDICATIONS

Refer to lists of high-risk or high-alert drugs available from the Institute for Safe Medication Practices (ISMP) or the United States Pharmacopeia (USP). The organization should develop a list of high-risk or high-alert drugs based on its utilization patterns, administered drugs, and internal data about medication errors. High-risk drugs may include investigational drugs, controlled medications, drugs not approved by the FDA, medications with a narrow therapeutic range, psychotherapeutic medications, and look- or soundalike medications. As appropriate to the services provided, the organization should develop processes for procuring, storing, ordering, transcribing, preparing, dispensing, administering, and monitoring high-risk or high-alert medications.

K. INVESTIGATIONAL MEDICATION SAFETY

The organization protects the safety of patients participating in investigational or medication studies by ensuring adequate control and support. The organization should show sensitivity to the use of particular populations for experimentation and research, and review all investigational medications to evaluate safety. Develop a written process for reviewing, approving, supervising, and monitoring investigational medication use. Review and accommodate, as appropriate, the patient's continued participation in the protocol. Specify that the pharmacy controls the storage, dispensing, labeling, and distribution of the investigational medications.

L. EVALUATION OF MEDICATION MANAGEMENT

Evaluate the medication management system for risk points and identify areas to improve safety. Routinely evaluate literature for new technologies or successful practices demonstrated to enhance safety for improving the medication management system. Review internally generated reports to identify trends or issues within the system. When the organization receives a medication recall or discontinuation notice for safety reasons, identify all patients who received the medication. Return any medications when allowed under law, by regulation, and organization policy. Control and account for previously dispensed but unused, expired, or returned medications. The pharmacy remains responsible for controlling and accounting for all unused medications returned to the pharmacy.

M. DRUG QUALITY AND STORAGE

The American Society of Healthcare System Pharmacists (ASHP) sets guidelines on drug quality and specifications. Pharmacy procedures should require that all drugs and medications meet the standards of the U.S. Pharmacopeia/National Formulary (USP/NF). Drugs not included in the USP/NF should be approved by the FDA. Obtain drugs from known sources that meet identity, purity, and potency requirements. Drugs should comply with FDA current manufacturing practices. Store drugs for external use separately from medications taken internally. Never keep respiratory care drugs and those used to prepare irrigation solutions with other injectable drugs. Never store large quantities of acids or other hazardous materials close to floor level. Never store large or heavy drug containers in lower shelves. Identify hazardous storage areas and post appropriate caution or warning signs. Never store drugs in a refrigerator that contains food or drink. Consider the following factors when assessing sources of drugs and medications:

- Data on sterility and analytical controls
- Bioavailability and bioequivalence information
- Information about raw materials and finished products
- Miscellaneous information on the quality of the drug or medication

N. HAZARDOUS DRUG SAFETY

NIOSH released new hazardous drug guidelines in 2004 for healthcare organizations. OSHA currently enforces safety issues using the general duty clause and existing standards dealing with hazard communication and PPE. Consider all drugs with toxic, irritating, sensitizing, or organ-targeting characteristics as hazardous. Both clinical and nonclinical workers may experience exposures to hazardous drugs when they create aerosols, generate dust, clean up spills, or touch contaminated surfaces during the preparation, administration, or disposal of hazardous drugs. Risks and hazard exist when reconstituting powdered or lyophilized drugs and further diluting either the reconstituted powder or concentrated liquid forms of hazardous drugs. Exposures to hazardous drugs may occur through inhalation, skin contact, skin absorption, ingestion, or injection. Inhalation and skin contact/absorption remain the most likely routes of exposure, but unintentional ingestion from hand-to-mouth contact and unintentional injection through a needle stick or sharps injury can also occur. In most cases, the percentage of air samples containing measurable airborne concentrations of hazardous drugs has been low. Recently, several studies examined environmental contamination of hazardous drug preparation and administration areas. Factors that affect worker exposures include the following:

- Drug handling circumstances (preparation, administration, or disposal)
- Amount of drug prepared
- Frequency and duration of drug handling
- Potential for absorption
- Use of ventilated cabinets

O. CURRENT STANDARDS

No NIOSH RELs, OSHA PELs, or ACGIH threshold limit values (TLVs) exist for hazardous drugs. PELs, RELs, and TLVs exist for inorganic arsenic compounds, which include the antineoplastic drug arsenic trioxide. Some pharmaceutical manufacturers develop risk-based occupational exposure limits (OELs) for use in their own manufacturing setting. Look for this information on available SDSs or request it from the manufacturer. Resource Conservation and Recovery Act (RCRA) regulations require that hazardous waste be managed by following a strict set of regulatory requirements. The RCRA list of hazardous wastes includes only about 30 pharmaceuticals, nine of which classify as antineoplastic drugs. Recent evidence indicates that a number of drug formulations exhibit hazardous waste characteristics. Dispose of hazardous drug waste in a manner similar to that required for RCRA-listed hazardous waste. Hazardous drug waste includes partially filled vials, undispensed products, unused IVs, needles and syringes, gloves, gowns, underpads, contaminated materials from spill cleanups, and containers such as IV bags or drug vials that contain more than trace amounts of hazardous drugs and are not contaminated by blood or other potentially infectious waste.

P. NIOSH REVISION OF ASHP DEFINITION

The 1990 ASHP definition of hazardous drugs was revised by the NIOSH Working Group on Hazardous Drugs in 2004. Drugs considered hazardous include those that exhibit one or more of the following characteristics in humans or animals:

- Carcinogenicity
- Teratogenicity or other developmental toxicity
- Reproductive toxicity
- Organ toxicity at low doses
- Genotoxicity

Q. DEVELOPING A HAZARDOUS DRUG LIST

Compliance with the OSHA Hazard Communication Standard (HCS) entails (1) evaluating whether these drugs meet one or more of the criteria for defining hazardous drugs and (2) posting a list of the hazardous drugs to ensure worker safety. Organizations may access the NIOSH website to refer to the NIOSH listing of hazardous drugs. Hazardous drug evaluation remains a continual process. Local hazard communication plans should provide for assessment of new drugs as they enter the marketplace. Toxicological data is often nonexistent for investigational drugs. However, if the mechanism of action suggests that there may be a concern, it is prudent to handle them as hazardous drugs until adequate information becomes available to exclude them. Some drugs defined as hazardous may not pose a significant risk of direct occupational exposure because of their dosage formulation. However, they may pose a risk if solid drug formulations become altered, such as by crushing tablets or making solutions from them outside a ventilated cabinet.

R. Where to Find Information Related to Drug Toxicity

Pharmacy or nursing departments often develop lists of hazardous drugs. These comprehensive lists should include all hazardous medications routinely used or likely used by a local practice. Some of the resources that employers can use to evaluate the hazard potential of a drug include the following:

- Material SDSs
- Product labeling approved by the FDA and packaging inserts
- Special health warnings from drug manufacturers, the FDA, and other professional groups
- Reports and case studies published in medical and other healthcare profession journals
- Evidence-based recommendations from other facilities that meet the criteria for defining hazardous drugs

The NIOSH website Hazardous Drug Listing was compiled from information provided by (1) four institutions that generated lists of hazardous drugs for their respective facilities, (2) the American Hospital Formulary Service Drug Information (AHFS DI) monographs, and (3) several other sources. Institutions may want to adopt this list or compare theirs with the list on the NIOSH website.

S. 2010 NIOSH Update of Hazardous Drug Alert for Healthcare Settings

NIOSH has published an update to their 2004 Alert: *Preventing Occupational Exposures to Antineoplastic and Other Hazardous Drugs in Healthcare Settings*. This latest update adds 21 drugs to Appendix A, the list of drugs considered hazardous. The update added the following nine chemotherapy drugs to the previous list published in the 2004 alert. These chemotherapy drugs include

- Bortizomib
- Clofarabine
- Dasatinib
- Decitibine
- Nelarabin
- Pemetrexed
- Sorafenib
- Sunitinib malate
- Vorinostat

T. Hazardous Drug Safety Plan

Accomplish a workplace analysis of all hazardous drug areas. Develop, implement, maintain, and review annually the written hazardous drug safety plan designed to protect those who handle or are exposed to hazardous medications. NIOSH and OSHA provide guidance in the development of a drug safety and health plan. Nursing stations on floors where hazardous drugs will be administered

---

BOX 8.8   OTHER HAZARDOUS DRUGS LISTED IN THE 2010 UPDATE

- Alefacept
- Bosentan
- Entecavir
- Lenalidomide
- Medroxyprogesterone acetate
- Palifermin
- Paroxetine HCl
- Pentetate calcium trisodium
- Rasagiline mesylate
- Risperidone
- Sirolimus
- Zonisamide

---

should provide spill and emergency skin and eye decontamination kits. Maintain copies of relevant SDSs for guidance. Plan contents should include the following:

- Labeling, storage, spill control, and response actions
- Detailed procedures for preparation and administration
- Use and maintenance of equipment used to reduce exposures such as ventilated cabinets, closed system drug transfer devices, needleless systems, and PPE
- Work practices covering manipulation techniques
- General hygiene practices, including no eating or drinking in drug handling areas such as the pharmacy or clinical areas
- Provide both general and specific safety training in handling hazardous drugs
- Training about location and proper use of spill kits
- Ensure training meets all relevant OSHA requirements
- Procedures for cleaning and decontamination of the work areas and for proper waste handling and disposal of all contaminated materials including patient wastes

During the preparation of hazardous drugs, use a ventilated cabinet to reduce the potential for occupational exposure. Performance test methods and criteria for BSCs may be found in *Primary Containment for Biohazards: Selection, Installation and Use of Biological Safety Cabinets*, 2nd Edition, CDC/NIH, 2000. A current field certification label should be prominently displayed on the ventilated cabinet per NSF/ANSI 49.

Protocols should specify that unventilated areas such as storage closets not be used for drug storage or any tasks involving hazardous drugs. Hazardous drugs should also be stored and transported in closed containers that minimize the risk of breakage. The storage area should contain sufficient general exhaust ventilation to dilute and remove any airborne contaminants. Depending upon the physical nature and quantity of the stored drugs, consideration should be given to installing a dedicated emergency exhaust fan sufficient in size to quickly purge to the outdoors any airborne contaminants within the storage room and to prevent airborne contamination in adjacent areas in the event of a spill. Limit access to areas where personnel prepare, receive, or store hazardous drugs. Place signs to restrict entry. Design bins or shelves for storing hazardous drugs to prevent breakage and to limit risk of falling. Apply warning labels to all hazardous drug containers, shelves, and bins. Recommend hazardous drugs requiring refrigeration be stored separately from nonhazardous drugs. The most significant risk for exposure during distribution and transport is from spills, resulting from damaged containers. PPE is generally not required when packaging is intact during routine activities. Any person opening a container to unpack the drugs should wear chemotherapy gloves, protective clothing, and eye protection. Wear chemotherapy gloves when transporting the vial of syringe to the work area due to possible contamination.

U.   TRANSFER PROCEDURES

Transfers from primary packaging such as vials to dosing equipment such as infusion bags, bottles, or pumps should be carried out using closed systems whenever possible. Devices that contain the product within a closed system during drug transfers limit the potential for aerosol generation, as well as exposure to sharps. Evidence has documented a decrease in drug contaminants present within a Class II BSC when a closed system transfer device was used. However, a closed system transfer device is not an acceptable substitute for a ventilated cabinet and should only be used in conjunction with a ventilated cabinet.

V.   CAREGIVER EXPOSURE PRECAUTIONS

Exposure can occur in personnel handling patient linens and excreta from patients receiving hazardous drugs within the last 48 hours. In some cases, take precautions for up to seven days. Wear two pairs of appropriate gloves and a disposable gown. Wear face shields if a potential exists for splashing. Wash hands with soap and water after removal of gloves.

W.   ADMINISTERING AEROSOLIZED DRUGS

Ribavirin, a synthetic nucleoside with antiviral activity, can be effective against respiratory syncytial virus. It is reconstituted from a lyophilized powder for aerosol administration. Ribavirin is usually administered in the aerosolized form via mask or oxygen tent for 12–18 hours per day for three to seven days. A small particle aerosol generator creates respirable particles of 1.3-micrometer median diameter. Under current practice, excess drug is exhausted into the patient's room, causing additional exposures. Studies show that Ribavirin is a reproductive risk in rodents and rabbits. Human studies on nurses who administer the drug by oxygen tent absorbed a dose that exceeded the safety factor of the short-term daily dose level. Minor pulmonary function abnormalities did occur in healthy adult volunteers in clinical studies. When administering aerosolized drugs, additional precautions should be observed:

- Use of NIOSH-approved respirators
- If possible, administer in booths with local exhaust or isolation rooms with high efficiency particulate air (HEPA)-filtered systems
- Permit only trained personnel to administer hazardous drugs
- Require caregivers to wear disposable gloves and gowns
- Work at waist level if possible and avoid working above the head
- Warn pregnant staff or women breast-feeding to avoid contact with these drugs

Aerosolized pentamidine is FDA approved for the treatment and prophylaxis of some types of pneumonia. Pentamidine administered as an aerosol must be reconstituted from a lyophilized powder. No studies exist to evaluate the potential carcinogenic, mutagenic, or reproductive effects of pentamidine. Studies among healthcare workers reveal uptake by those personnel who administer the drug. Side effects include coughing, sneezing, mucous membrane irritation, headache, and bronchospasms.

X.   Hazardous Drug Waste

OSHA covers bags containing materials contaminated with hazardous drugs under the Hazard Communication Standard. Recommend the use of thick, leak-proof plastic bags, colored differently from other hospital bags. Use these bags for routine collection of discarded gloves, gowns, and other disposable materials. Label these bags as hazardous drug–related wastes. The OSHA Technical Manual suggests keeping waste bags inside a covered waste container clearly labeled "Hazardous Drug WASTE ONLY." At least one such receptacle should be located in every area preparing or administering hazardous drugs. Never move waste from one area to another. Seal bags when filled and tape covered waste containers. Label needle containers and breakable items of hazardous waste as "Hazardous Drug WASTE ONLY." The Bloodborne Pathogens Standard requires the use of properly labeled, sealed, and covered disposal containers for waste containing blood or other potentially infectious materials. Hazardous drug–related wastes should be disposed of according to the Environmental Protection Agency (EPA), state, and local regulations for hazardous waste. This disposal can occur at either an EPA-compliant incinerator or a licensed sanitary landfill for toxic wastes. Commercial waste disposal must be performed by a licensed company. While awaiting removal, maintain waste in a secure area in covered, labeled drums with plastic liners.

Y.   Spill Control

Spills should be managed according to workplace hazardous drug spill policies and procedures. The size of the spill might determine both who can conduct the cleanup and the decontamination. Make spill kits and other cleanup materials available in areas handling hazardous drugs. However, OSHA requires that persons who wear respirators such as those contained in some spill kits follow a complete respiratory protection program including fit testing. The written program should address the protective equipment required for differing amounts spilled, the possible spreading of material, restricted access to hazardous drug spills, and signs to be posted. Dispose of all cleanup materials in a hazardous chemical waste container in accordance with RCRA regulations.

Z.   Medical Surveillance

In addition to preventing exposure to hazardous drugs and careful monitoring of the environment, medical surveillance is an important part of a safe handling program. NIOSH recommends employees handling hazardous drugs participate in medical surveillance efforts provided at their workplace. The OSHA Technical Manual recommends that workers handling hazardous drugs receive monitoring as part of the medical surveillance program that includes the taking of a medical and exposure history, physical examination, and some laboratory measures. Professional organizations also recommend medical surveillance as the recognized standard of occupational health practice for hazardous drug handlers.

AA.   Ventilated Cabinets

Conduct all tasks related to mixing, preparing, or manipulating hazardous drugs within a ventilated cabinet designed specifically to prevent hazardous drugs from being released into the surrounding environment. Follow aseptic requirements established by state boards of pharmacy. Recommend the use of ventilated cabinets when concerned about hazardous drug containment and aseptic processing. Use a Class I BSC or an isolator for containment when not required to process with asepsis. In some applications, a containment isolator may suffice. For mixing requiring an aseptic technique, use the Class II, Type B2 BSC. If possible, use Class III cabinets as isolators intended for asepsis and containment. Equip all ventilated cabinets with a continuous monitoring device to confirm adequate airflow prior to each use. Filter exhaust from these controls with a HEPA filter.

Exhaust 100 percent to the outside if feasible. Install an outside exhaust system to prevent entrainment by the building envelope or heating, ventilation, and air conditioning systems (HVAC). Place the fan downstream of the HEPA filter to ensure contaminated ducts remain under negative pressure. Never use a ventilated cabinet with air recirculation unless the hazardous drug in use will not volatilize during process manipulation or after capture by the HEPA filter. Use the information on volatilization provided by the drug manufacturer or determined from air sampling data. Refer to NSF/ANSI 49 for additional information regarding placement of the cabinet, exhaust system, and stack design. Never consider additional engineering or process controls, such as needleless systems, glove bags, and closed system drug transfer devices, as substitutions for ventilated cabinets. Clean the cabinet according to the manufacturer's instructions. Some manufacturers recommend weekly decontamination, as well as whenever spills occur, or when the cabinet requires moving, service, or certification. Decontamination should consist of surface cleaning with water and detergent followed by thorough rinsing. The use of detergent is recommended because there is no single accepted method of chemical deactivation for all agents involved. Avoid quaternary ammonium cleaners due to the possibility of vapor buildup in recirculated air. Use ethyl alcohol or 70 percent isopropyl alcohol if the contamination is soluble only in alcohol. Alcohol vapor buildup has also been a concern, so the use of alcohol should be avoided in BSCs where air is recirculated. Avoid spray cleaners due to the risk of spraying the HEPA filter. Never use ordinary decontamination procedures, which include fumigation with a germicidal agent, when handling dangerous drug waste.

AB. RECORDKEEPING

Maintain any workplace exposure records created in connection with hazard drug handling for at least 30 years. Maintain medical records for the duration of employment plus 30 years in accordance with the Access to Employee Exposure and Medical Records Standard (29 CFR 1910.1020). In addition, sound practice dictates that training records should include the following information:

- Dates of the training sessions
- Contents or a summary of the training sessions
- Names and qualifications of the persons conducting the training
- Names and job titles of all persons attending the training sessions

Maintain training records for three years from the date on which the training occurred.

XII. PROPOSED UNIVERSAL WASTE RULE FOR PHARMACEUTICALS

The proposed rule encourages generators to dispose of nonhazardous pharmaceutical waste as universal waste, thereby removing this unregulated waste from wastewater treatment plants and municipal solid waste landfills. The addition of hazardous pharmaceutical waste to the Universal Waste Rule will facilitate the collection of personal medications from the public at various facilities so that they can be more properly managed. This proposed rule applies to hazardous pharmaceutical wastes generated by the following types of facilities: pharmacies, hospitals, physicians' offices, dentists' offices, other healthcare practitioners, outpatient care centers, ambulatory healthcare services, residential care facilities, veterinary clinics, and reverse distributors. This rule does not apply to pharmaceutical manufacturing or production facilities. The EPA believes that hazardous pharmaceutical wastes meet the factors considered when determining if waste is appropriate for inclusion in the Universal Waste Rule. Specifically, most hazardous pharmaceutical waste presents a relatively low risk during accumulation and transport due to their form and packaging, which is typically in small, individually packaged dosages, such as pills or capsules.

A.   CURRENT RCRA REQUIREMENTS

Under current RCRA requirements, any facility that generates RCRA hazardous pharmaceutical waste falls under RCRA generator regulations. Under the universal waste program, generators of hazardous pharmaceutical waste may opt to manage this waste as *universal waste*. If a facility opts to manage its hazardous pharmaceutical waste under the universal waste option, then that facility will become a *handler* of pharmaceutical universal waste, rather than a *generator* of hazardous pharmaceutical waste. Compared to a generator of hazardous pharmaceutical waste, a handler of pharmaceutical universal waste can note the following benefits: (1) an increased accumulation threshold; (2) an increased onsite accumulation limit; (3) an increased storage time limit; (4) no manifest requirement; and (5) basic training requirements.

Approximately 31 commercial chemical products listed on RCRA's P and U lists have pharmaceutical uses. The EPA bases their P and U lists on chemical designations; this number does not completely represent the total number of brand-name pharmaceuticals that may actually be listed as hazardous waste. In addition, waste pharmaceuticals may also pose hazards because they exhibit one or more of the four characteristics of hazardous waste: ignitability, corrosivity, reactivity, and toxicity. Characteristic pharmaceutical wastes include those that exhibit the ignitability characteristic, such as solutions containing more than 24 percent alcohol. An example of a pharmaceutical that may exhibit the reactivity characteristic is nitroglycerine. Pharmaceuticals exhibiting a corrosive characteristic generally apply to compounding chemicals, including strong acids, such as glacial acetic acid, and strong bases, such as sodium hydroxide. Depending on the concentration in different preparations, pharmaceuticals may also exhibit the toxicity characteristic because of some contaminants such as arsenic, barium, cadmium, chromium, selenium, and silver.

B.   USP 797: PHARMACEUTICAL COMPOUNDING—STERILE PREPARATIONS

The Food and Drug Administration Modernization Act of 1997 (FDAMA) included a section on pharmacy compounding. The law introduced limits on pharmacy compounding and attempted to protect patients from unnecessary use of compounded drugs. The power of regulation granted to the FDA by the FDAMA was ruled unconstitutional by the Supreme Court in 2001. The new USP Chapter 797, Pharmaceutical Compounding—Sterile Preparations became enforceable by regulatory agencies on January 1, 2004. The provisions and requirements of USP 797 are designed to achieve compounding accuracy and sterility to ensure the safety of patients. USP 797 combines process and preparation quality controls with formal staff training and competency assessment guidelines. USP 797 addresses the responsibilities of compounding personnel, how to determine risk levels, and maintaining quality after a medication leaves the pharmacy. The development of USP 797 came after decades of increasing safety and quality consciousness.

Compounding medications continues as a high-risk and labor-intensive process. The implementation of USP 797 ensured that pharmacies would not overlook problems or evade compliance with necessary provisions. Pharmacies must comply with the USP 797 sterile compounding requirements during inspections, enforcement actions, and accreditation surveys. The ASHP Accreditation Services Division requires that a pharmacy comply with all federal, state, and local regulations concerning pharmacy practice.

C.   USP 797 2008 REVISION

The 2008 revision shifted emphasis to human factors and diminished mandates for primary engineering controls. New technological advances can supersede the written requirements of Chapter 797, *if* such advances can be shown to produce equivalent, or better, results than the ideas in Chapter 797. The revision added a new risk level category of "immediate use" to the previous categories of low, medium and high risk. An immediate-use compounded sterile product (CSP) is defined as a compound prepared with no more than three sterile, commercially supplied nonhazardous drugs; using commercial, sterile devices; for an infusion that will start within one hour of preparation and be completed within 12 hours. It should be emphasized here that this does not include any chemotherapeutic or other hazardous drug preparations. The revision also specified that this immediate-use classification is NOT intended to circumvent USP 797 intent. Never store immediate-use compounds for later use. Eliminating the requirement for an ISO 7 buffer basically eliminated the engineering control mandates that caused so much concern. The revision also increased beyond-use dating for medium risk compounding to a maximum of nine days refrigerated instead of seven. This allows alternate-site pharmacies to have a week's worth of in-date IVs, such as TPN and antibiotics. Multi-dose vials (MDVs) now contain specific maximum beyond-use dates of 28 days. Single-dose vials (SDVs) contain new guidelines that restrict beyond-use dating. Never store ampoules for any period of time. CSPs must develop adequate controls from preparation until the time of the administration to meet USP 797 requirements. The revision addressed radiopharmaceuticals, and also addressed clean rooms and environmental sampling. Air sampling is now required only monthly for low- and medium-risk certified compounding areas and weekly for high-risk areas. The revision addressed protective garb requirements and isopropyl alcohol glove washing. Under the revision, garbing order is more clearly defined to focus on moving from dirtiest to cleanest in the process. When dressing to work in clean areas begin with a hair net, beard mask, gown and finish with a gown, gloves, shoe covers and finish with a 70 percent isopropyl alcohol (IA) glove washing.

REVIEW EXERCISES

1. Describe the origin and characteristics of the following forms of ionizing radiation:
   a. Alpha particles
   b. Beta particles
   c. Gamma rays
   d. X-rays
2. Define the following radiation-related terms:
   a. Absorbed dose
   b. Ci
   c. RAD
   d. Radioactive half-life
   e. Rem
   f. ALARP
3. Describe the concept of radiation shielding.
4. What role does the NRC play in healthcare radiation safety?
5. Describe the key duties of a radiation safety committee.
6. What is the mission of the NCRP?
7. What is the key goal of the FDA CDRH in relation to radiation safety?
8. Define "nonionizing radiation" and provide several examples of generating devices.
9. List at least three safety concerns involving the use of MRI devices.
10. List and define at least five types of healthcare laboratories.
11. What was the purpose of the CLIA?
12. List at least five mandates of the OSHA Laboratory Standard.
13. What is formaldehyde and how do OSHA safety standards relate to worker exposures?
14. List and describe the four biosafety levels.
15. Describe investigational medications.
16. What nongovernmental organization sets guidelines on drug quality and specifications?
17. List the occupational exposure routes for hazardous medications.
18. List the five characteristics as defined by NIOSH that would classify a drug as hazardous.
19. What was the purpose for the development of the USP 797?

9 Agencies, Associations, and Organizations

I. ADMINISTRATIVE LAW

The U.S. Code (U.S.C.) consists of a compilation of laws in force from 1789 to the present. As prima facie or presumed to be law, it does not include repealed or expired acts. The Federal Register (FR) publication system established on July 26, 1935, provides the official legal information service of the United States government. It functions under the authority of the Administrative Committee of the Federal Register (ACFR). The FR operates through a statutory partnership with National Archives' Office of the Federal Register (OFR) and the U.S. Government Printing Office (GPO). The ACFR delegates its day-to-day authority to the Director of the OFR. The OFR administers programs under the Federal Register Act (44 U.S.C. Ch.15), the GPO Electronic Information Access Enhancement Act (44 U.S.C. 4101), the Freedom of Information Act (5 U.S.C. 551 et seq.), and other public information laws of the United States. The daily *Federal Register* posted on the GPO's Federal Digital System serves as the official daily publication for rules, proposed rules, and notices of federal agencies and organizations, including executive orders and other Presidential documents.

The Code of Federal Regulations (CFR) serves as the codification source for the general and permanent rules published previously in the *Federal Register*. Users may access the CFR free of charge on the Federal Digital System maintained by the GPO. The CFR contains 50 titles with each title assigned to specific agencies that issue regulations pertaining to broad subject areas. The CFR permits the arrangement of official text of agency regulations or standards into a single publication. This provides a comprehensive and convenient reference for anyone needing to access federal general and permanent regulations. The CFR works with the FR, which provides updates on a daily basis to keep the CFR current. A full set of the CFR consists of approximately 200 volumes, with revision occurring annually on a quarterly basis as follows:

- Titles 1–16 as of January 1
- Titles 17–27 as of April 1
- Titles 28–41 as of July 1
- Titles 42–50 as of October 1

The Administrative Procedure Act was passed in 1945 (5 U.S.C. 551) and added several provisions to the FR system. The Act gave the public, in most instances, the right to participate in the rule-making process by commenting on proposed standards. Informing the public about proposed rule-making actions remains the primary purpose of the FR. Governmental agencies can promulgate standards using the administrative law process. Most federal agencies publish a regulatory agenda in the FR to inform the public of proposed or expected actions.

BOX 9.1   SELECTED CFR TITLES

- Title 10, Energy & Radiation (NRC)
- Title 21, Food and Drugs (FDA)

(Continued)

BOX 9.1   SELECTED CFR TITLES (*Continued*)

- Title 29, Labor (OSHA)
- Title 40, Environmental Protection (EPA)
- Title 42, Health and Human Services (NIOSH, CMS)
- Title 44, Emergency Management (FEMA)
- Title 49, Transportation (DOT)

II. OCCUPATIONAL SAFETY AND HEALTH ADMINISTRATION

The OSH Act and standards issued by the Occupational Safety and Health Administration (OSHA) apply to every private employer with one or more employees except those covered by other federal legislation. Under the Act, employers have the general duty of providing their workers a place of employment free from recognized hazards to safety and health, and must comply with OSHA standards. When OSHA compliance officers discover hazards, employers can receive citations listing alleged violations including proposed penalties and abatement periods. Employers may contest these before the independent Occupational Safety and Health Review Commission (OSHRC). Employees must comply with standards and with job safety and health rules and regulations applying to their own conduct. Employees or their representatives have the right to file a complaint with OSHA requesting a workplace inspection. OSHA can withhold complainants' names from the employer. Employees have the right, on request, to be advised of OSHA actions regarding their complaint. They also possess the right to an informal review to be made of any OSHA decision not to inspect a workplace. Employees also may attend the employer's informal conference with OSHA to discuss any issues raised by inspection, citation, and notice of a proposed penalty or abatement period. In developing new or amended standards, OSHA invites full participation by employers and employees.

A. SUMMARY OF GENERAL DUTIES

- 5(a)(1): Each employer must provide each employee a place of employment free from recognized hazards that could cause or likely cause death or serious physical harm to employees.
- 5(a)(2): Each employer must comply with occupational safety and health standards promulgated under the OSH Act.
- 5(b): Each employee must comply with occupational safety and health standards and rules, regulations, and orders issued pursuant to this Act as applicable to his or her actions and conduct.

BOX 9.2   TYPES OF OSHA STANDARDS

- General Industry Standards, 29 CFR 1910
- Standards for Shipyard Employment, 29 CFR 1915
- Marine Terminals Standards, 29 CFR 1917
- Long Shoring Standards, 29 CFR 1918
- Construction Standards, 29 CFR 1926

Every workplace must display the OSHA Poster (Pub 3165), or the state plan equivalent. The poster explains worker rights to a safe workplace and how to file a complaint. Place the poster where employees will see it. You can order one free copy from OSHA or download a copy from the agency website at www.osha.gov. OSHA uses special criteria to conduct a programmed inspection. Criteria may include injury rates, death rates, exposure to toxic substances, or a high amount of lost workdays for the industry. Another nonprogrammed inspection can occur when an employee makes a formal complaint to OSHA regarding a possible unsafe working condition or imminent danger at the workplace. If the inspector does not have a warrant, employers do not have to provide access to the facility under the Fourth Amendment. OSHA regulations require employers to report deaths on the job within eight hours. The agency then investigates the circumstances of the death, usually onsite, to determine the cause of death and if violations of the OSH Act occurred. If the agency determines that the employer failed to follow safety and health requirements, it issues citations and proposed civil penalties. OSHA bases proposed penalties on the statutory factors of employer size, gravity of violation, good faith of the employer, and the history of previous violations. OSHA penalties do not correspond to or reflect the value of a worker's life or the cost of an injury or illness. OSHA instructs compliance officers to check the OSHA 300 Log and other documents to determine occupational health hazard trends. Inspectors can check healthcare facility safety records and Nuclear Regulatory Commission (NRC) radioisotope or radiation source licenses.

OSHA can refer a willful citation that resulted in a fatality to the Department of Justice for consideration for criminal prosecution. Any criminal prosecution by the Department of Justice does not impact OSHA authority to issue civil citations and penalties. OSHA can use criminal referral as an enforcement tool. However, most cases involving willful citations do not merit criminal prosecution.

B. PRIORITIES

Imminent danger situations receive top priority. An imminent danger refers to any condition with reasonable certainty that a danger exists that could cause immediate death or serious physical harm. If an imminent danger exists, the compliance officer will ask the employer to voluntarily abate the hazard and to remove endangered employees from exposure. Should the employer refuse, OSHA will apply to the nearest federal District Court for legal action to correct the situation. OSHA's second priority involves investigation of fatalities and catastrophes resulting in hospitalization of three or more employees. Employees retain a right to request an OSHA inspection when placed in imminent danger from a hazard or whenever there is a violation of an OSHA standard that threatens physical harm. If the employee so requests, OSHA will withhold the employee's name from the employer. OSHA establishes inspection priorities aimed at specific high hazard industries, occupations, or health hazards. Establishments cited for alleged serious violations may undergo a reinspection to determine whether the hazards are corrected. OSHA regulations require employers to report deaths on the job within eight hours. Employers may call their local office or the agency's toll-free number of 800-321- 6742. The agency investigates the circumstances of the death, usually onsite, to determine the cause of death and if violations of the OSH Act occurred. OSHA makes exceptions when the situation falls outside of the agency's jurisdiction.

C. CITATIONS

After the compliance officer reports the findings, the area director determines which citations warrant formal issuance and which penalties requirement assessment. An *other than serious* violation addresses issues that would not normally cause death or serious physical harm. OSHA can issue a *serious* violation if substantial probability exists that death or serious physical harm could result and the employer knew or should have known of the situation or hazard. OSHA cites imminent dangerous citations as serious violations. A *willful* violation refers to a situation that the employer intentionally and knowingly committed. The employer either knew that the operation constituted a violation, or was aware that a hazardous condition existed but made no reasonable effort to eliminate it. A *repeat* violation can address any standard, regulation, rule, or order where, upon reinspection, another violation of the previously cited section is found. Failure to correct any violations may bring civil penalties for every day the violation continues beyond the prescribed abatement date. Falsifying records, reports, or applications can bring a fine and/or six months in jail upon conviction. Assaulting a compliance officer, or otherwise resisting, opposing, intimidating, or interfering with a compliance officer in the performance of his or her duties is a criminal offense.

D. STATE-APPROVED PLANS

States and jurisdictions can operate their own occupational safety and health plans with OSHA approval. State plans must establish standards that meet federal requirements. Approved state plans must extend their coverage to state and local government workers. Alliances enable organizations committed to workplace safety and health to collaborate with OSHA to prevent injuries and illnesses in the workplace. OSHA and its allies work together to reach out to, educate, and lead the nation's employers and their employees in improving and advancing workplace safety and health.

E. RECORDKEEPING (29 CFR 1904)

Congress directed the Secretary of Labor through section 8(c)(2) of the OSH Act to prescribe regulations requiring employers to maintain accurate records and periodic reports on work-related deaths, injuries, and illnesses that do not involve medical treatment, loss of consciousness, restriction of work or motion, or transfer to another job. The Act requires OSHA to develop and maintain the effective collection, compilation, and analysis of occupational safety statistics. 29 CFR 1904, Recording and Reporting Occupational Injuries and Illnesses, requires employers to record information on the occurrence of injuries and illnesses in their workplaces. The employer must record work-related injuries and illnesses that meet one or more of published recording criteria. OSHA rules found in 29 CFR 1904 require all employers under OSHA jurisdiction with 11 or more employees to keep OSHA injury and illness records, unless the establishment is classified in a specific low-hazard retail, service, finance, insurance, or real estate industry. Employers with 10 or fewer employees must keep OSHA injury and illness records if OSHA or the Bureau of Labor Statistics informs them in writing that they must keep records under 29 CFR 1904.41.

F. RECORDING WORK-RELATED INJURIES AND ILLNESSES

The OSHA Log of Work-Related Injuries and Illnesses (Form 300) is used to document and classify work-related injuries and illnesses. The Log also documents the extent and severity of each case. Employers use the Log to record specific details about what happened and how it happened. The Summary (Form 300A) shows totals for the year in each category. At the end of the year, post the Summary in a visible location to make employees aware of the injuries and illnesses occurring in the workplace. Employers must keep a Log for each establishment or site. If an employer operates more than one establishment, each location must keep a separate Log and Summary. Keep all logs for five years.

1. Medical Treatment

Medical treatment includes managing and caring for a patient for the purpose of combating disease or disorder. Do not consider the following as medical treatments and do not record them: (1) visits to a doctor or healthcare professional solely for observation or counseling, (2) diagnostic procedures, including administering prescription medications used solely for diagnostic purposes, and (3) any procedure labeled as first aid.

2. Restricted Work

Restricted work activity occurs when an employer or healthcare professional recommends that the worker can't do the routine functions of his or her job for a full workday. Count the number of calendar days the employee was on restricted work activity or was away from work as a result of the recordable injury or illness. Do not count the day on which the injury or illness occurred in this number. Begin counting days from the day after the incident occurs. If a single injury or illness involved both days away from work and days of restricted work activity, enter the total number of days for each. You may stop counting days of restricted work activity or days away from work once the total of either or the combination of both reaches 180 days.

3. Classifying Injuries

An injury is a wound or damage to the body resulting from an event in the work environment. Examples include cuts, punctures, lacerations, abrasions, fractures, bruises, contusions, chipped teeth, amputations, insect bites, electrocution, or thermal, chemical, electrical, or radiation burns. Classify sprains and strain injuries to muscles, joints, and connective tissues as injuries when they result from a slip, trip, fall, or other similar accidents.

4. Classifying Illnesses

OSHA considers skin diseases as illnesses caused by exposure to chemicals, plants, or other hazardous substances. OSHA defines respiratory conditions or illnesses as breathing-related problems associated with pneumonitis, pharyngitis, rhinitis, farmer's lung, beryllium disease, tuberculosis, occupational asthma, reactive airways dysfunction syndrome, chronic obstructive pulmonary disease, and hypersensitivity. Examples can include heatstroke, hypothermia, decompression sickness, effects of ionizing radiation, exposure to ultraviolet rays, anthrax, and bloodborne pathogen diseases.

5. Posting the Summary

Post the Summary, OSHA Form 300A, only (not the Log) by February 1 of the year following the year covered by the form. Keep it posted until April 30 of that year. Retain the Log and Summary for five years following the year to which they pertain. Do not send the completed forms to OSHA unless specifically asked to do so.

6. Form 301 Injury and Illness Incident Report

Employers may use the OSHA 301 form or an equivalent form that documents the same information. Some state workers' compensation, insurance, or other reports may be acceptable substitutes, as long as they provide the same information as the OSHA 301 form.

G. ACCESS TO EMPLOYEE EXPOSURE AND MEDICAL RECORDS (29 CFR 1910.1020)

Employees potentially exposed to toxic substances or harmful physical agents in the workplace possess the right to access relevant exposure and medical records. Access includes current workers, former employees, and employees assigned or transferred to work involving toxic substances or harmful physical agents. Designated employee representatives may access employee medical or exposure records and analyses created from those records only in very specific circumstances. Designated employee representatives include any individual or organization employee with written authorization to exercise a right of access. Access gives the right to examine or copy medical and exposure records. Employees have the right to access records and analyses based on work concerns. Employers must permit access free of charge and within a reasonable period of time. Employees can access the material in one of three ways: (1) the employer may provide the employee a copy of the document, (2) the employer may provide facilities for employees to copy the document, or (3) the employer may loan a copy to the employee to copy offsite.

III. ENVIRONMENTAL PROTECTION AGENCY

The EPA was created in 1970 to protect the environment and exercise control over release of harmful substances that could threaten public health. Many EPA rules define which substances can be hazardous to human health and/or pose a threat to the environment. The EPA or state-approved agencies provide guidance for handling hazardous materials, regulate the operation of waste disposal sites, and establish procedures for dealing with environmental incidents such as leaks or spills. The EPA also has an interest in hazardous and infectious wastes produced by healthcare facilities. It publishes informative guides to assist risk managers in understanding and complying with a number of environmental laws and regulations. Environmental laws are published in 40 CFR.

A. RESOURCE CONSERVATION AND RECOVERY ACT

The Resource Conservation and Recovery Act (RCRA) is one of the EPA's main statutory weapons. The Act created a *cradle-to-grave* management system for current and future waste, while the EPA authorizes cleanup of released hazardous substances. Several statutes are media-specific and limit the amount of waste introduced into the air, waterways, oceans, and drinking water. Other statutes directly limit the production, rather than the release, of chemical substances and products that may contribute to the nation's waste. The RCRA is unique in that its primary purpose is to protect human health and the environment from the dangers of hazardous waste. The RCRA has a regulatory focus and authorizes control over the management of wastes from the moment of generation until final disposal. The RCRA was passed in 1976 as an amendment to the Solid Waste Disposal Act. The general RCRA objectives are to (1) protect human health and the environment, (2) reduce waste and conserve natural resources and energy, and (3) reduce/eliminate generation of hazardous waste as expeditiously as possible.

BOX 9.3   RCRA WASTE GENERATOR CLASSIFICATIONS

- Large quantity generators (over 1000 kg/month)
- Small quantity generators (100 to 1000 kg/month)
- Conditionally exempt generators (less than 100 kg/month) with no more than 1 kg acutely hazardous waste

BOX 9.4   KEY RCRA REGULATORY GENERATOR REQUIREMENTS

- Identify and label all waste and notify the EPA of hazardous waste operations
- Maintain secure storage areas, keep records, and train waste handlers
- Use permitted treatment, storage, and disposal facilities

The RCRA was amended in 1984 through passage of the Hazardous and Solid Waste Amendment. This action enabled the EPA to regulate underground storage tanks (USTs) to better control and prevent leaks. Title I of the RCRA regulated substances including petroleum products and CERCLA-regulated substances. Tanks with hazardous wastes are not regulated under Subtitle I but are regulated under Subtitle C of the RCRA. A UST has been defined as a container that has at least 10 percent of its contents underground. Facilities with underground tanks should take action to ensure piping does not fail, control corrosion of tanks and piping, and prevent spills and overflows. Refer to 40 CFR Part 280 for a listing of tanks excluded from the regulation.

B. COMPREHENSIVE ENVIRONMENTAL RESPONSE, COMPENSATION AND LIABILITY ACT

This law intends to remedy the mistakes in past hazardous waste management, whereas the RCRA is concerned with avoiding such mistakes through proper management in the present and future. The RCRA mainly regulates how waste should be managed. The Comprehensive Environmental Response, Compensation and Liability Act (CERCLA) authorized a number of government actions to remedy the conditions or the effects of a release. CERCLA, as originally enacted in 1980, authorized a five-year program by the federal government to perform cleanup activities. CERCLA authorized the EPA to identify those sites where release of hazardous substances had occurred or might occur and posed a serious threat to human health, welfare, or the environment. The parties responsible for the releases were required to fund the cleanup actions.

C. Superfund Amendments and Reauthorization Act of 1986

During a five-year period, it became clear that the problem of abandoned hazardous waste sites was more extensive than originally determined. The Superfund Amendments and Reauthorization Act of 1986 (SARA) established new standards and schedules for site cleanup and also created new programs for informing the public of risks from hazardous substances in the community and for preparing communities for hazardous substance emergencies. Under Public Law 99-499, SARA specified new requirements for state and local governments. It also has provisions for the private sector related to hazardous chemicals. The Emergency Planning and Community Right-To-Know Act specifically requires states to establish a State Emergency Response Commission (SERC). The SERC must designate emergency planning districts within the state. The SERC appoints a Local Emergency Planning Committee (LEPC) for each district. SERC and LEPC responsibilities include implementing various planning provisions of Title III and serve as points of contact for the community right-to-know reporting requirements. SARA Title III requires that the local committees must include, at a minimum, representatives from the following groups: state and local officials, law enforcement, civil defense, firefighting, environmental, hospital, media, first aid, health, transportation, and facility owners or operators subject to the emergency planning requirements.

D. Clean Air Act

The Clean Air Act (CAA) was passed to limit the emission of pollutants into the atmosphere; it protects human health and the environment from the effects of airborne pollution. The EPA established National Ambient Air Quality Standards (NAAQS, also known as KNACKS) for several substances. The KNACKS provide the public some protection from toxic air pollutants. Primary responsibility for meeting the requirements of the CAA rests with each state. States must submit plans for achieving KNACKS. Under Section 112 of the CAA, the EPA has the authority to designate hazardous air pollutants and set National Emission Standards for Hazardous Air Pollutants. Common air pollutants include the following: (1) ozone, (2) nitrogen dioxide, (3) carbon monoxide, (4) particulate matter, (5) sulfur dioxide, and (6) other sources including metal refineries, solvent usage, and manufacture of lead batteries. Air emissions from incinerators regulated by the RCRA must also comply with ambient air standards and emission limitations published under the provisions of the CAA. Extraction of pollutants from air emissions under CAA control using equipment such as scrubbers can create hazardous waste or sludge containing such waste. Disposal of incinerator materials must also comply with the RCRA.

E. Clean Water Act

The Clean Water Act (CWA) of 1977 strengthened and renamed the Federal Water Pollution Control Act of 1972. The Act has several major provisions, including the establishment of the National Pollutant Discharge Elimination Systems (NPDES) Permit Program to permit discharges into the nation's waterways. Any direct discharges into surface water requires a NPDES permit. An indirect discharge means that the waste is first sent to a publicly owned treatment works and then discharged pursuant to a permit. Sludge resulting from wastewater treatment should be handled as RCRA waste and disposed of at an RCRA facility if deemed hazardous. The act requires that certain industrial and municipal storm water discharges be regulated under the NPDES permit system.

F. FEDERAL INSECTICIDE, FUNGICIDE, AND RODENTICIDE ACT

The Federal Insecticide, Fungicide, and Rodenticide Act (FIFRA) was passed in 1947 and was initially administered by the U.S. Department of Agriculture (USDA). The EPA became responsible for the Act in 1970. A 1972 amendment included provisions to protect public health and the environment. FIFRA controls risks of pesticides through a registration system. No new pesticide can be marketed until it is registered with the EPA. The EPA can refuse to register a pesticide or to limit use if evidence indicates a threat to humans and the environment. All pesticides and general disinfectants used in healthcare facilities should be approved and registered by the EPA.

G. TOXIC SUBSTANCES CONTROL ACT

The Toxic Substances Control Act (TSCA) was enacted in 1976 to help control the risk of substances not regulated as drugs, food additives, cosmetics, or pesticides. Under this law, the EPA can regulate the manufacture, use, and distribution of chemical substances. The TSCA mandates that the EPA be notified prior to the manufacture of any new chemical substance. The EPA ensures that all chemicals are tested to determine risks to humans. The TSCA also allows the EPA to regulate polychlorinated biphenyls (PCBs) under 40 CFR 761.

IV. OTHER FEDERAL AGENCIES

A. NUCLEAR REGULATORY COMMISSION

This independent agency, established by the Energy Reorganization Act of 1974, regulates civilian use of nuclear materials. NRC employs a staff of 3000 with two-thirds of the employees working at the Rockville, Maryland headquarters location. NRC staffs four regional offices and assigns resident inspector offices at each commercial nuclear power plant and some fuel cycle locations. Headed by a five-member commission, the NRC regulates byproduct, source, and special nuclear materials to ensure adequate public health and safety, common defense and security, and protection of the environment. The NRC adopts and enforces standards for the departments of nuclear medicine in healthcare facilities. Some states have agreements with the government to assume these regulatory responsibilities. NRC issues five-year licenses to qualified healthcare organizations that follow prescribed safety precautions and standards. Types of regulated facilities include (1) nuclear power plants, (2) departments of nuclear medicine at hospitals, (3) academic activities at educational institutions, (4) research work in scientific organizations, and (5) industrial applications such as gauges and testing equipment.

B. NATIONAL INSTITUTE FOR OCCUPATIONAL SAFETY AND HEALTH

The National Institute for Occupational Safety and Health (NIOSH) conducts research and makes recommendations that help prevent work-related injury and illness. NIOSH, established by the OSH Act of 1970, operates under the administrative control of the Centers for Disease Control and Prevention (CDC). Although NIOSH and OSHA were created by the same Act of Congress, they operate as distinct agencies with separate responsibilities. However, NIOSH and OSHA often work together toward the common goal of protecting worker safety and health. NIOSH publishes educational resources and guidelines on a number of healthcare- and hospital-related topics.

C. CENTERS FOR DISEASE CONTROL AND PREVENTION

The CDC, located in Atlanta, Georgia, is an agency of the Department of Health and Human Services (DHHS). It works to protect the health of the American people by tracking, monitoring, preventing, and researching disease. It is also responsible for surveillance and investigation of infectious disease in healthcare facilities. The CDC conducts research and publishes results in its Morbidity and Mortality Weekly Report. This weekly publication provides healthcare facilities with timely information on topics such as infection control, isolation procedures, bloodborne pathogens, tuberculosis management, infectious waste disposal recommendations, and how to protect workers. CDC performs many of the administrative functions for the Agency for Toxic Substances and Disease Registry (ATSDR), a sister agency of the CDC, and is one of eight federal public health agencies within the DHHS. CDC seeks to accomplish its mission by working with partners throughout the nation and world to monitor health, detect and investigate health problems, conduct research to enhance prevention, develop and advocate sound public health policies, implement prevention strategies, promote healthy behaviors, foster safe and healthful environments, and provide leadership and training. CDC develops and sustains many vital partnerships with public and private entities that improve service to the American people.

D.   FOOD AND DRUG ADMINISTRATION

The FDA, created by the Appropriation Act of 1931, operates under the auspices of the DHHS. The FDA is responsible for protecting the public health by assuring the safety, efficacy, and security of human and veterinary drugs, biological products, medical devices, our nation's food supply, cosmetics, and products that emit radiation. FDA advances public health by helping to speed innovations that make medicines more effective, safer, and more affordable and by helping the public get the accurate, science-based information they need to use medicines and foods to maintain and improve their health. The FDA possesses the responsibility for regulating the manufacturing, marketing, and distribution of tobacco products to protect the public health and to reduce tobacco use by minors. It plays a significant role in the nation's counterterrorism capability. The FDA fulfills this responsibility by ensuring the security of the food supply and by fostering development of medical products to respond to deliberate and naturally emerging public health threats. It ensures the safety of food, human and veterinary drugs, biological products, medical devices, cosmetics, and any electronic products that emit radiation, and oversees biologic product manufacturing and the safety of the nation's blood supply. The FDA also conducts research to establish product standards and develop improved testing methods. It sets standards for drug approvals and over-the-counter and prescription drug labeling/manufacturing standards. The FDA also oversees the labeling and safety of all food products (except meat and poultry) and bottled water. It grants premarket approval of new devices, establishes manufacturing and performance standards, and tracks device malfunctioning and serious adverse reaction events. The FDA develops radiation safety performance standards for microwave ovens, television receivers, diagnostic x-ray equipment, cabinet x-ray systems, laser products, ultrasonic therapy equipment, mercury vapor lamps, and sunlamps. The FDA also accredits and inspects mammography facilities. In 2007, the President signed into law the Food and Drug Administration Amendments Act. The revised law represented a very significant addition to FDA authority. Among the many components of the law are the Prescription Drug User Fee Act (PDUFA) and the Medical Device User Fee and Modernization Act (MDUFMA).

E.   AGENCY FOR HEALTHCARE RESEARCH AND QUALITY

The Agency for Healthcare Research and Quality (AHRQ) serves as the health services research arm of the Department of Health and Human Services. It provides a major source of funding and technical assistance for health services research and research training at leading universities and other institutions. The Agency is a science partner, working with the public and private sectors to build the knowledge base for what works and does not work in health and healthcare and to translate this knowledge into everyday practice and policymaking. Health services research examines how people get access to healthcare, how much care costs, and what happens to patients as a result of this care. The main goals of health services research include identifying the most effective ways to organize, manage, finance, and deliver high quality care; reduce medical errors; and improve patient safety. The Agency's research findings help practitioners diagnose and treat patients more effectively. A computerized clinical information system developed with the Agency's support now helps healthcare professionals determine the most appropriate timing for giving antibiotics to surgical patients. A national clearinghouse gives clinicians, health plans, and healthcare delivery systems a web-based mechanism for obtaining detailed objectives on clinical practice. The Agency complements the biomedical research mission of its sister agency, the National Institutes of Health (NIH).

F. NATIONAL INSTITUTES OF HEALTH

NIH operates 27 centers and serves as one of the world's foremost medical research organizations. It pursues fundamental knowledge about the nature and behavior of living systems and the application of that knowledge to extend healthy life and reduce the burdens of illness and disability. NIH supports the health of the nation by conducting and supporting health related research. It places an emphasis on developing, maintaining, and renewing scientific evidence that will improve the nation's capability to prevent disease.

G. CENTERS FOR MEDICARE & MEDICAID SERVICES

The Centers for Medicare & Medicaid Services (CMS) operates 10 field offices that were reorganized in February 2007. The agency moved from a geography-based structure to consortia with key responsibilities focusing on survey and quality improvement. The agency also publishes guidelines governing long-term nursing facilities. The guidelines emphasize resident's rights and quality of care. The Omnibus Budget Reconciliation Act (OBRA) of 1987 gave the CMS the power to regulate facilities receiving federal funds. The Medicare Modernization Act or MMA became law in December 2003 and added outpatient prescription drug benefits to Medicare.

H. INSTITUTE OF MEDICINE

The Institute of Medicine (IOM) serves as an adviser to the nation for health improvement. As an independent, scientific adviser, the IOM strives to provide advice that is unbiased, based on evidence, and grounded in science. The mission of the IOM embraces the health of people everywhere. The Institute is a part of the National Academies and provides science-based advice on matters of biomedical science, medicine, and health. It is a nonprofit organization that provides a vital service by working outside the framework of government to ensure scientifically informed analysis and independent guidance. The Institute provides evidence-based and authoritative information/advice concerning health and science to policymakers, professionals, and leaders in every sector of society. Committees of volunteer scientists serve without compensation. Each report produced by committees goes through a review and evaluation process. The review is conducted by a panel of experts that remain unknown to the committee. The Institute's work centers principally on committee reports or studies on subjects ranging from quality of medical care to medical errors. The majority of the studies and other activities receive funds from the federal government. Other studies can be initiated by private industry, foundations, state or local governments, and the Institute.

V. ACCREDITATION ORGANIZATIONS

A. JOINT COMMISSION

The most well known of the accrediting bodies, the Joint Commission, impacts the operation of most hospitals and a good number of other healthcare organizations, including nursing homes and surgery centers. The Joint Commission operates as an independent, not-for-profit organization, governed by a board that includes physicians, nurses, and consumers. The mission to continuously improve safety and quality care through accreditation supports performance improvement in healthcare organizations. The Joint Commission sets the standards to measure many aspects of patient care and quality. To maintain and earn accreditation, a healthcare organization must undergo an extensive onsite review by a team of surveyors, at least once every three years. The review helps evaluate the organization's performance in areas that affect patient care. Organizations should provide a safe, functional, supportive, and effective environment for patients, staff, visitors, and contractors. Effective safety management provides guidance to achieve quality patient care, good outcomes, and continuous improvement. Some key environment and care requirements include long range and continuous planning by organizational leaders to ensure space requirements, proper equipment, and necessary resources remain available to support services offered. The planning and design of the environment should be consistent with the organizational mission to support proper care considering the patient's physical condition/health, cultural background, age, and cognitive abilities. Organizations should educate staff on their roles in the environment of care in safely, sensitively, and effectively supporting patient care. Organizations should also educate staff on physical requirements processes for monitoring, maintaining, and reporting on the organization's environment of care. Other requirements include developing standards to measure staff and organizational performance in managing and improving the environment of care and establishing an effective Information Collection and Evaluation System (ICES). Environment and care standards do not prescribe any particular safety-related structure. The standards don't address the specific type of safety committee or a specific individual to serve as safety officer. An organization with multiple sites may develop separate management plans for each location or choose to use a single comprehensive set of plans. The organization should address specific risks and unique conditions at each site. The six functional areas addressed in EOC management plans include: (1) Safety Management, (2) Security Management, (3) Fire Safety Management, (4) Hazardous Materials and Waste Management, (5) Medical Equipment Management, and (6) Utilities Management. Emergency Management and Life Safety are now standalone standards.

EOC management plans provide guidance for taking action to minimize risk and hazards. The organization can decide the format of the plan. The plan could exist as a set of plans, one for each functional area. Organizations could also develop a single document that covers all functional areas. The use of annexes, attachments, and action plans can provide details for the six functional areas. Maintaining a systematic approach when developing plans permits organizational members to better use the document or documents. Management plans should contain content that addresses performance objectives, compliance requirements, and performance evaluation specifics. Create a cross-checking process to document compliance with stricter regulatory requirements, other accreditation standards, and rules mandated by any authority with jurisdiction. The organization should identify risks and implement processes to minimize adverse impacts on buildings, grounds, equipment, occupants, and internal physical safety systems. Accomplish the following actions:

- Develop a written plan to address the management of the care environment
- Designate persons to coordinate safety functions
- Designate persons to intervene in events threatening life, health, or property
- Review general safety policies as often as necessary but at least every three years
- Respond to product safety recalls by taking appropriate actions
- Ensure proper maintenance of all facility grounds and equipment
- Conduct periodic evaluations to assess safety effectiveness
- Assess staff knowledge behaviors during period environmental tours
- Identify new or altered tasks that could pose risks in construction areas
- Evaluate areas with changes in services to identify improvement opportunities

- Conduct environmental tours at least every six months in all patient areas
- Conduct environmental tours at least annually in nonpatient areas

B. American Osteopathic Association

The American Osteopathic Association (AOA) represents more than 47,000 osteopathic physicians, promotes public health, encourages scientific research, and serves as the primary certifying body for osteopath doctors. The association serves as the accrediting agency for all osteopathic medical schools and healthcare facilities, including acute care hospitals. The osteopathic accreditation was developed in 1945. This enabled the association to assure that osteopathic students received their training through rotating internships and residencies in facilities that provided a high quality of patient care. The association also has deeming authority to accredit laboratories within accredited hospitals under the Clinical Laboratory Improvement Amendments of 1988. The association developed accreditation requirements for ambulatory care, surgery, mental health, substance abuse, and physical rehabilitation medicine facilities. The Healthcare Facilities Accreditation Program (HFAP) meets or exceeds the standards required by CMS to provide accreditation to all hospitals, ambulatory care/surgical facilities, mental health facilities, physical rehabilitation facilities, clinical laboratories, critical access hospitals, and stroke centers. HFAP's surveying process and standards benefit from oversight by a wide range of medical professionals, including both allopathic and osteopathic disciplines. HFAP accreditation requirements are clearly tied to the corresponding Medicare Conditions of Participation. Base successful accreditation on the facility's ability to correct deficiencies. Surveyors should possess experience and understand the many aspects of a healthcare facility and help make the survey process more realistic and educational. If a deficiency is identified, surveyors are able to draw from their experience and offer feasible solutions, usually on the spot. HFAP accreditation also is recognized by the federal government, state departments of public health, insurance carriers, and managed care organizations. Healthcare facilities seeking accreditation must comply with all the requirements listed in the latest edition of *Accreditation Requirements for Healthcare Facilities.*

C. Det Norske Veritas National Integrated Accreditation for Healthcare Organizations

The Det Norsek Veritas (DNV) National Integrated Accreditation for Healthcare Organizations (NIAHO) is designed from the ground up to drive quality transformation into the core processes of running a hospital. NIAHO helps healthcare organizations meet their national accreditation obligations and achieve ISO 9001 compliance. It compresses the survey cycle from every three years to annually, thereby ensuring continual quality improvement. DNV's goal is to offer healthcare organizations and companies a new alternative to hospital accreditation. DNV was established in 1864 as an independent foundation with a purpose to safeguard life, property, and the environment. Increasing patient safety and reducing errors in healthcare is an important part of that purpose. DNV issues ISO certificates to healthcare facilities worldwide, including hospitals, outpatient clinics, diagnostic centers, laboratories, nursing homes, and homecare centers.

The facility should take actions to ensure patient safety, provide areas for diagnosis and treatment, and provide services to meet the needs of the community. Organizations should maintain the condition of the physical plant and overall hospital environment to ensure the safety and well-being of patients, visitors, and staff. The hospital should maintain adequate facilities for its services. Locate diagnostic and therapeutic facilities to ensure the safety of patients. Maintain facilities, supplies, and equipment at an acceptable level to ensure safety and quality. Determine the extent and complexity of facilities by evaluating the services offered. The organization should implement processes to maintain a safe environment for the organization's patients, staff, and visitors. It should use documented process, policies, and procedures to define how unfavorable occurrences, incidents, or impairments impact the facility's infrastructure. Key accreditation functional areas include life safety, safety, security, hazardous material/waste, emergency management, medical equipment, and utilities management. Organizations should evaluate their physical environment management systems at least annually. Measure occurrences, incidents, or impairments and analyze them to identify any patterns or trends.

D. ACCREDITATION CANADA

Accreditation Canada is a not-for-profit, independent organization that provides health organizations with an external peer review to assess the quality of their services based on standards of excellence. Accreditation Canada, accredited by the International Society for Quality in Healthcare (ISQua), fosters quality in health services across Canada and internationally. Accreditation Canada's first survey with ISQua occurred in 1998. Three surveyors spent a week at Accreditation Canada interviewing teams and reviewing self-assessments and evidence. Accreditation Canada achieved three separate accreditation awards from ISQua—for the organization, the standards, and the surveyor. Accreditation Canada's clients include Regional Health Authorities, hospitals, and community-based services. The organization uses more than 600 surveyors or peer reviewers. These experienced professionals come from accredited health facilities. They are physicians, nurses, health executives, administrators, occupational therapists, laboratory scientists, respiratory therapists, psychologists, social workers, and addiction counselors. Accreditation standards are developed in close consultation with healthcare experts. The survey features customized processes geared to organizational priorities, comprehensive performance measures, and automated tools for efficient data exchange. Patient safety is an integral component of the accreditation process. Complying with Accreditation Canada standards and Required Organizational Practices reduces the potential for adverse events occurring within healthcare and service organizations. Accreditation standards assess governance, risk management, leadership, infection prevention and control, and medication management, as well as services in over 30 sectors, including acute care, home care, rehabilitation, community and public health, labs and blood banks, and diagnostic imaging.

E. COMMISSION ON ACCREDITATION OF REHABILITATION FACILITIES

The Commission on Accreditation of Rehabilitation Facilities (CARF) serves as an independent not-for-profit organization that provides accreditation in the human services field with focus on the areas of rehabilitation, employment, child and family, and aging services. The survey is a consultative process rather than an inspection. The survey team works with the provider to improve service resources and outcomes. CARF develops standards through a series of leadership panels, national advisory committees, focus groups, and field reviews. The standards development process provides opportunities for persons receiving services and other stakeholders to participate in developing CARF standards. A three-year accreditation, the highest level of accreditation, demonstrates substantial fulfillment of the CARF standards. A one-year accreditation indicates the existence of deficiencies in relation to the provider's conformance to the CARF standards, yet there is evidence of the provider's capability and commitment to correct the deficiencies or make progress toward their correction. Provisional accreditation indicates a provider still functions at the level of a one-year accreditation.

F. COLLEGE OF AMERICAN PATHOLOGISTS LABORATORY ACCREDITATION

The college serves as the principal organization of board-certified pathologists, and serves and represents the interest of patients, pathologists, and the public by fostering excellence in the practice of pathology and laboratory medicine. The college is the world's largest association composed exclusively of pathologists and is widely considered the leader in providing quality improvement to laboratories around the world. The College of American Pathologists (CAP) products include resources designed specifically for pathologists and laboratory professionals. CAP accreditation improves the quality of clinical laboratory services through voluntary participation, professional peer review, education, and compliance. Upon successful completion of the inspection process, the laboratory is awarded CAP accreditation and becomes part of an exclusive group of more than 6000 laboratories worldwide that meet the highest standards of excellence. CAP utilizes working and experienced laboratory professionals in their peer review process. This approach provides a laboratory with inspectors who bring first-hand knowledge of the most current laboratory techniques and processes. The college serves the broadest patient population by accommodating the full spectrum of laboratory disciplines under one accreditation process. No other accreditation process provides such a comprehensive offering. An accredited laboratory helps assure the facility meets federal requirements. CAP accreditation provides a laboratory with the assurance of meeting the highest standards of practice. The accreditation cycle includes the following:

- The lab submits an Application Request Form with a deposit.
- CAP forwards the application and checklists to the lab.
- The lab completes the application and reviews the checklists.
- CAP receives and reviews the application.
- An inspection team is assigned.
- A mutually acceptable date is set for the inspection.
- A team of inspectors arrives on the designated date.
- The team conducts a thorough inspection using checklists as guides.
- The team meets with the lab for a summation conference to review findings.
- The inspectors leave a copy of the final Summation Report.
- The lab corrects any deficiencies and provides documentation for CAP.
- The lab is accredited for a two-year cycle, but conducts a self-inspection at the one-year mark.

VI. STANDARDS ORGANIZATIONS

A. AMERICAN CONFERENCE OF GOVERNMENTAL INDUSTRIAL HYGIENISTS

The independent National Conference of Governmental Industrial Hygienists (NCGIH) convened on June 27, 1938, in Washington, D.C. In 1946, the organization changed its name to the American Conference of Governmental Industrial Hygienists (ACGIH). In September 2000, conference members approved an amendment of the Bylaws to permit members, not government or academic employees, greater voting rights and the opportunity to serve on the ACGIH Board. The amendment set new member categories, including the organizational member category. Today, nine committees focus their energies on a range of topics: agricultural safety and health, air sampling instruments, bio-aerosols, biological exposure indices (BEIs), industrial ventilation, international, and small business. ACGIH publishes Threshold Limit Values (TLVs) for chemical substances (TLVs-CS), and Threshold Limit Values for physical agents (TLVs-PA). The list of TLVs includes 642 chemical substances and physical agents, as well as 47 BEIs. ACGIH offers approximately 400 publication titles addressing industrial hygiene, environment, safety/health, toxicology, medical, hazardous materials/waste, indoor air quality, physical agents, ergonomics, distance learning, and computer resources.

B. AMERICAN NATIONAL STANDARDS INSTITUTE

This organization was founded in 1918 to consolidate voluntary standards. The American National Standards Institute (ANSI) is a federation of more than 1500 professional, trade, governmental, industrial, labor, and consumer organizations. It publishes national consensus standards developed by various technical, professional, trade, and consumer organizations. ANSI also serves as the coordinating agency for safety standards adopted for international implementation. It represents the United States as a member of the International Organization for Standardization and the International Electro-technical Commission. OSHA adopts many ANSI standards. ANSI provides members access to more than 9000 standards from around the world and publishes specifications for protective eyewear including safety glasses and goggles, hard hats, safety shoes, fall-protection equipment, eyewash stations, and emergency shower equipment.

C. NATIONAL COUNCIL ON RADIATION PROTECTION AND MEASUREMENTS

The National Council on Radiation Protection and Measurements (NCRP) works to promote the importance of radiation protection and measurements. The council, established to represent all of the national radiological organizations in the United States on a collective basis, focuses on the science of radiation protection. The International X-Ray and Radium Protection Committee, created in July 1928, evolved into the International Commission on Radiological Protection. The NCRP originally operated as an informal association of scientists seeking to make available information and recommendations on radiation protection and measurements. The NCRP was reorganized and chartered by the U.S. Congress in 1964 as the National Council on Radiation Protection and Measurements.

D. NATIONAL FIRE PROTECTION ASSOCIATION

The National Fire Protection Association (NFPA) serves as the world's leading advocate of fire prevention and publishes about 300 safety codes and standards. NFPA influence is present in every building, process, service, design, and installation. It encourages the broadest possible participation in its consensus code development process. NFPA relies on more than 6000 volunteers from diverse professional backgrounds to serve on over 200 technical code and standard development committees. NFPA code development processes are accredited by ANSI. Some examples of NFPA codes relevant to healthcare include NFPA 70, National Electrical Code, NFPA 99, Healthcare Facilities, and NFPA 101, Life Safety Code. NFPA offers excellent education and training that addresses the latest fire and life safety requirements, technologies, and practices. NFPA also administers several professional certifications including Certified Fire Protection Specialist, Certified Fire Inspector, and Certified Fire Plans Examiner. NFPA also develops dozens of texts, guides, and other materials to assist firefighters and first responders.

VII. VOLUNTARY ASSOCIATIONS

A. AMERICAN HEALTHCARE ASSOCIATION

The American Healthcare Association (AHCA) operates as a nonprofit federation of affiliated state health organizations, together representing nearly 12,000 nonprofit and for-profit assisted living, nursing facility, developmentally disabled, and subacute care providers that care for more than 1.5 million elderly and disabled individuals nationally. The Association represents the long-term care community to the nation at large and to government, business leaders, and the general public. It also serves as a force for change within the long-term care field, providing information, education, and administrative tools that enhance quality at every level. At its Washington, D.C., headquarters, the association maintains legislative, regulatory, and public affairs, as well as member services staffs that work both internally and externally to assist the interests of government and the general public, as well as member providers.

B. AMERICAN HOSPITAL ASSOCIATION

The American Hospital Association (AHA) is a national organization that represents and serves all types of hospitals, healthcare networks, and their patients and communities. Close to 5000 hospitals, healthcare systems, networks, other providers of care, and 37,000 individual members come together to form the AHA. Advocacy efforts include the legislative and executive branches and include the legislative and regulatory arenas. Founded in 1898, the AHA provides education for healthcare leaders and is a source of information on healthcare issues and trends.

C.   AMERICAN SOCIETY OF HEALTHCARE RISK MANAGEMENT

Established in 1980, the American Society of Healthcare Risk Management (ASHRM) is a personal membership group of the AHA with more than 4400 members representing healthcare, insurance, law, and other related professions. The Society promotes effective and innovative risk management strategies and professional leadership through education, recognition, advocacy, publications, networking, and interactions with leading healthcare organizations and government agencies. The Society initiatives focus on developing and implementing safe and effective patient care practices, the preservation of financial resources, and the maintenance of safe working environments.

D.   AMERICAN ASSOCIATION OF OCCUPATIONAL HEALTH NURSES

The American Association of Occupational Health Nurses (AAOHN) serves the largest group of healthcare professionals in the workplace. The vision is to create a positive economic impact through worker health and well-being leading to optimal performance. AAOHN is dedicated to advancing and maximizing the health, safety, and productivity of domestic and global workforces by providing education, research, public policy, and practice resources for occupational and environmental health nurses. The mission of AAOHN is to advance the profession of occupational and environmental health nursing through five pillars: (1) education and research, (2) professional practice/ethics, (3) communications, (4) governmental issues, and (5) establishment of alliances. Values include reflecting strategic and forward thinking while promoting excellence for the association and the profession and conducting business and interpersonal action ethically, honestly, and with respect. AAOHN also seeks to provide stewardship of fiscal responsibility to reflect credibility, accountability, and respect.

E.   ASSOCIATION FOR THE HEALTHCARE ENVIRONMENT

The Association for the Healthcare Environment (AHE) is the premier professional membership society for healthcare environmental services, housekeeping, waste management, textile care professionals, and related support services disciplines. AHE provides education, recognition for personal and professional achievements, and national networking as well as affiliation and collaboration with the AHA on public policy and advocacy issues related to healthcare environmental services. AHE serves as the association of choice for healthcare environmental services and textile care professionals. It is a recognized resource and catalyst in the general and regulatory communities. AHE strives to provide strong leadership and progressive thinking in the face of a changing healthcare field. AHE provides the following member benefits:

- Educational materials that can increase an individual's knowledge and skills
- Leadership that is accessible and responsible to the needs of the members
- Opportunities to network with peers on a national level
- Recognition for personal and professional achievements
- Collaboration with the AHA and other organizations on public policy and advocacy issues relating to environmental services

F.   ASSOCIATION OF OCCUPATIONAL HEALTH PROFESSIONALS

The Association of Occupational Health Professionals (AOHP) works to define employee health issues and serve as a leading advocate for occupational health professionals serving in healthcare organizations. The board uses monthly conference calls to coordinate positions on hot topics and strategic initiatives. The association participates in governmental affairs and meets with OSHA, NIOSH, and congressional representatives to address Association positions. The Association sponsors an annual national conference where members meet to share, network, and attend professional education sessions.

G. ECRI

This nonprofit health services research agency works to improve the safety, quality, and cost effectiveness of healthcare. It is widely recognized as one of the world's leading independent organizations committed to advancing the quality of healthcare. The agency focuses on healthcare technology, risk, quality, and environmental management. It provides information services and technical assistance to more than 5000 hospitals, healthcare organizations, ministries of health, government and planning agencies, voluntary sector organizations, associations, and accrediting agencies around the world. ECRI maintains over 30 databases, publications, information systems, and technical assistance services that set the standard for the healthcare community. The agency provides alerts related to technology hazards and the results of medical product or technology assessments.

H. AMERICAN ASSOCIATION OF COLLEGES OF NURSING

The American Association of Colleges of Nursing (AACN) is the national voice for baccalaureate and graduate nursing education. AACN's educational, research, federal advocacy, data collection, publications, and special programs work to establish quality standards for nursing education; assist deans and directors to implement those standards; influence the nursing profession to improve healthcare; and promote public support for professional nursing education, research, and practice. From an original 121-member institution in 1969, AACN today represents more than 725 member schools of nursing at public and private universities and senior colleges nationwide. These schools offer a mix of baccalaureate, graduate, and post-graduate programs. The dean or chief nurse administrator serves as the representative to AACN, though the association serves all members of the academic unit.

I. ACADEMY OF MEDICAL-SURGICAL NURSES

The Academy of Medical-Surgical Nurses (AMSN) is the only specialty nursing organization dedicated to the practice of medical-surgical nursing. AMSN is a vibrant community of more than 10,000 medical-surgical nurses who care about improving patient care, developing personally and professionally, advocating for the specialty of medical-surgical nursing, and connecting with others who share their compassion and commitment. The strategic goals focus on workplace advocacy; evidence-based practice, research and knowledge; professional development; national leadership and influence; and organizational health. Medical-surgical nurses use their powerful voice and focused action to continuously improve patient care. Medical-surgical nursing is a distinct specialty with its own body of knowledge, and is the foundation of all nursing practice. It has evolved from an entry-level position to a distinct specialty. Medical-surgical nurses are the largest group of practicing professionals. It is one of the most demanding nursing specialties.

J. AMERICAN ASSOCIATION OF CRITICAL-CARE NURSES

The American Association of Critical-Care Nurses (ACCN), the world's largest specialty nursing organization, has been serving the needs of nurses caring for acutely and critically ill patients since 1969. Representing the interests of more than 500,000 nurses who care for acutely and critically ill patients, AACN is dedicated to creating a healthcare system driven by the needs of patients and their families, where acute and critical care nurses make their optimal contribution. AACN defines acute and critical care nursing as the specialty within nursing that deals with human responses to potential and actual life-threatening health problems. The Association helps acute and critical care nurses stay up to date in technology and treatment techniques. AACN offers a wide array of opportunities through educational products, national conferences, and local and regional seminars, and strives to be a voice for acute and critical care nurses on congressional and regulatory issues, thereby ensuring that nurses play an active role in shaping the future of acute and critical care. It seeks to influence and shape health policy in a number of ways. In addition to developing its own position statements on a variety of issues, AACN works within healthcare coalitions to support legislative efforts and to bring attention at the national level to issues important to acute and critical care nurses, nursing, and the healthcare system. In addition, AACN participates in national forums that shape health policy, care delivery, and environments where nurses work and patients are cared for.

K.   American Nurses Association

The American Nurses Association (ANA) is the only full-service professional organization representing the nation's entire registered nurse population. From the halls of Congress and federal agencies to the boardrooms, hospitals and other healthcare facilities, ANA is the strongest voice for the nursing profession. It is headquartered in Silver Spring, Maryland. ANA represents the interests of the nation's 3.1 million registered nurses through its constituent and state nurses associations and its organizational affiliates. Dedicated to ensuring that an adequate supply of highly skilled and well-educated nurses is available, ANA is committed to meeting the needs of nurses as well as healthcare consumers. ANA advances the nursing profession by fostering high standards of nursing practice, promoting the economic and general welfare of nurses in the workplace, projecting a positive and realistic view of nursing, and lobbying the Congress and regulatory agencies on healthcare issues affecting nurses and the general public. ANA is at the forefront of policy initiatives pertaining to healthcare reform. Among the priority issues are a restructured healthcare system that delivers primary healthcare in community-based settings; an expanded role for registered nurses and advanced practice nurses in the delivery of basic and primary healthcare; obtaining federal funding for nurse education and training; and helping to change and improve the healthcare environment.

L.   American Academy of Ambulatory Care Nursing

AAACN (formerly the American Academy of Ambulatory Nursing Administration) was founded in 1978 as a not-for-profit, educational forum. The Academy adopted the current name of the organization in 1993. Membership was broadened to include nurses in direct practice, education, and research roles as well as those in management and administration. Today, membership is open to nurses and other professionals interested in ambulatory care and telehealth nursing. Corporations and individual corporate representatives are also welcomed as members. AAACN is the only specialty nursing association that focuses on excellence in ambulatory care. The Academy serves as a voice for ambulatory care nurses across the continuum of healthcare delivery and has membership in the Nursing Organizations Alliance (NOA). The Alliance provides a forum for nursing organizations to dialogue, collaborate, and facilitate policy formulation on professional practice and national health. AAACN provides education to ambulatory care professionals through annual conferences, webinars, and CNE articles in their *ViewPoint* newsletter and online library.

M.   American Association of Managed Care Nurses

The American Association of Managed Care Nurses (AAMCN) was founded in 1994 by a group of nurses who were dedicated to providing educational and networking opportunities for nurses working throughout the managed care industry. AAMCN is a nonprofit active membership association of registered nurses, nurse practitioners, licensed practical nurses, and other healthcare professionals committed to helping nurses become successful in their local marketplace through resourceful, responsive leading-edge member services, education, and communication. The vision of the AAMCN is to be an interrelated member of the managed care delivery team and systems for positive healthcare outcomes. The mission of the Academy is to be recognized as the expert and resource in managed care nursing, to establish standards for managed care nursing practice, to positively impact public policy regarding managed healthcare delivery, and to assist in educating the public on managed care.

N.   AMERICAN NEPHROLOGY NURSES' ASSOCIATION

The American Nephrology Nurses' Association (ANNA) is the professional association that represents nurses who work in all areas of nephrology. It promotes excellence in and appreciation of nephrology nursing in order to make a positive difference for people with kidney disease. ANNA's Strategic Plan articulates what the Association wants to achieve for its members and other stakeholders. It describes the Association's mission, core beliefs, goals, and strategic initiatives. Established as a nonprofit organization in 1969, ANNA has a membership of approximately 10,000 registered nurses in more than 100 local chapters across the United States. Members practice in all areas of nephrology, including hemodialysis, chronic kidney disease, peritoneal dialysis, acute care, and transplantation. Most members work in freestanding dialysis units, hospital outpatient units, and hospital inpatient units. ANNA develops and updates standards of clinical practice, educates practitioners, stimulates and supports research, disseminates knowledge and new ideas, promotes interdisciplinary communication and cooperation, and monitors and addresses issues encompassing the breadth of practice of nephrology nursing.

O.   AMERICAN ORGANIZATION OF NURSE EXECUTIVES

Since 1967, the American Organization of Nurse Executives (AONE) has provided leadership, professional development, advocacy, and research to advance nursing practice and patient care, promote nursing leadership excellence, and shape public policy for healthcare nationwide. AONE is a subsidiary of the AHA. The mission is to shape healthcare through innovative and expert nursing leadership. AONE serves its members by

- Providing vision and actions for nursing leadership to meet the healthcare needs of society
- Influencing legislation and public policy related to nursing and patient care issues
- Offering member services that support and enhance the management, leadership, educational, and professional development of nursing leaders
- Facilitating and supporting research and development efforts that advance nursing administration practice and quality patient care

P.   AMERICAN PSYCHIATRIC NURSES ASSOCIATION

The American Psychiatric Nurses Association (APNA) was founded in 1986. In the ensuing 25 years, APNA has grown to be the largest professional membership organization committed to the specialty practice of psychiatric-mental health (PMH) nursing and wellness promotion, prevention of mental health problems, and the care and treatment of persons with psychiatric disorders. APNA is the only PMH nursing organization whose membership is inclusive of all PMH-registered nurses (RN) including associate degree (ADN), baccalaureate (BSN), and advanced practice (APN) comprised of clinical nurse specialists (CNS), psychiatric nurse practitioners (NP), and nurse scientists and academicians (PhD). APNA membership totals more than 8000 psychiatric mental health nurses from all over the world. The Journal of the American Psychiatric Nurses Association provides quality, up-to-date information.

Q.   ASSOCIATION OF periOPERATIVE REGISTERED NURSES

The Association of periOperative Registered Nurses (AORN) represents the interests of more than 160,000 perioperative nurses and provides nursing education, standards, and services that enable optimal outcomes for patients undergoing operative and other invasive procedures. AORN's 40,000 registered nurse members facilitate the management, teaching, and practice of perioperative nursing education, or are engaged in perioperative research. Members also include perioperative nurses who work in related business and industry sectors. AORN collaborates with professional and regulatory organizations, industry leaders, and other healthcare partners who support their mission. Clinical and administrative staff serve on the committees and boards of approximately 30 professional associations to protect and improve safe perioperative practices. AORN monitors nursing and healthcare laws and regulations and, through their National Legislative Committee and AORN state legislative coordinators, coordinate and engage AORN members who are active grassroots advocates to participate in the legislative and regulatory processes. AORN's mission is to promote safety and optimal outcomes for patients undergoing operative and other invasive procedures by providing practice support and professional development opportunities to perioperative nurses.

R. NATIONAL STUDENT NURSING ASSOCIATION

The National Student Nursing Association's (NSNA) mission is to mentor students preparing for initial licensure as registered nurses, and to convey the standards, ethics, and skills that students will need as responsible and accountable leaders and members of the profession. Founded in 1952, NSNA is a nonprofit organization for students enrolled in associate, baccalaureate, diploma, and generic graduate nursing programs. It is dedicated to fostering the professional development of nursing students. The organization has over 60,000 members in 50 states, the District of Columbia, Guam, Puerto Rico, and the U.S. Virgin Islands. NSNA's Board of Directors is made up of 10 nursing students who are elected at the organization's Annual Convention. Two nonvoting consultants are appointed by the ANA and the National League for Nurses to provide guidance. NSNA also employs a full-time staff headquartered in Brooklyn, New York. Over 3000 nursing students participate in NSNA's Annual Convention, which features leadership and career development activities, opportunities to listen to renowned nursing leaders and hear about job opportunities, and the chance to network with hundreds of other students. The program includes a state board exam mini-review. NSNA's official magazine, *Imprint,* publishes five times a year and is mailed to the entire membership.

S. AMERICAN ASSOCIATION OF NURSE ANESTHETISTS

The American Association of Nurses Anesthetists (AANA) is the professional organization for more than 45,000 nurse anesthetists. As advanced practice nurses, certified registered nurse anesthetists (CRNAs) administer more than 34 million anesthetics in the United States each year. CRNAs practice in every setting where anesthesia is available and are the primary providers of anesthesia care in rural America. They administer every type of anesthetic, and provide care for every type of surgery or procedure, from open heart to cataract to pain management. Just as CRNAs are vigorous advocates for patient safety in the clinical setting, the AANA is an equally determined advocate for CRNAs concerning issues such as patient safety, access to quality healthcare services, scope of practice, educational funding, reimbursement, and many other legislative and regulatory matters in Washington, D.C., and across the country. As such, the AANA tracks state and federal legislation and regulation affecting nurse anesthesia practice, develops and carries out federal grassroots lobbying efforts, coordinates meetings with federal legislators and agency officials, testifies at federal and state legislative and regulatory hearings, and educates CRNAs and state associations regarding effective advocacy strategy and practice.

T. AMERICAN HOLISTIC NURSES ASSOCIATION

The American Holistic Nurses Association (AHNA) promotes the education of nurses, other healthcare professionals, and the public in all aspects of holistic caring and healing. AHNA is a nonprofit membership association serving more than 5400 members and 125 local chapters across the United States and abroad. In December 2006, due to the efforts of AHNA, holistic nursing was recognized as an "official nursing specialty" by the ANA. AHNA serves as a bridge between conventional healthcare and complementary/alternative healing practices. As healthcare professionals, holistic nurses may integrate complementary and alternative modalities (CAM) into clinical practice to treat the whole person and view healing as a partnership between a person seeking treatment and their practitioner. Holistic nursing is a specialty practice that draws on nursing knowledge, theories, expertise, and intuition to guide nurses in becoming therapeutic partners with people in their care. This practice recognizes the totality of the human being—the interconnectedness of body, mind, emotion, spirit, social/cultural, relationship, context, and environment.

U. EMERGENCY NURSES ASSOCIATION

The Association, founded in 1970, was initially known as the Emergency Department Nurses Association (EDNA). In 1985, the Association name was changed to Emergency Nurses Association (ENA), recognizing the practice of emergency nursing as role-specific rather than site-specific. Originally aimed at teaching and networking, the organization has evolved into an authority, advocate, lobbyist, and voice for emergency nursing. ENA has more than 40,000 members and continues to grow, with members representing over 35 countries around the world. The mission of ENA is to advocate for patient safety and excellence in emergency nursing practice. In order for ENA to fulfill its mission and to pursue its vision for the future, it should, like the members it represents, be flexible, dynamic, and adaptable to its complementary environments of professional practice and association business. ENA is committed to thoughtful environmental scanning and forecasting in order to best take advantage of opportunities and to respond to critical challenges that might impact the delivery of emergency healthcare.

V. AMERICAN NURSING INFORMATICS ASSOCIATION

The American Nursing Informatics Association (ANIA) is the association of professional nurses and associates who are committed to their specialty that integrates nursing science, computer science, and information science to manage and communicate data, information, knowledge, and wisdom in nursing and informatics practice. ANIA is an organization originally founded in 1982 as the Capital Area Roundtable on Informatics in Nursing (CARING). CARING was developed and organized by nurses in 1982 as a nonprofit undertaking to provide a forum for the advancement of automated healthcare information systems. In 1992, ANIA was formed in Southern California to provide networking, education, and information resources that enrich and strengthen the roles of nurses in the field of informatics. ANIA seeks to identify informatics practice as a specialty that is essential to the delivery of high quality and cost-effective healthcare. The mission is to advance the field of nursing informatics through communication, education, research, and professional activities.

W. ASSOCIATION OF REHABILITATION NURSES

The Association of Rehabilitation Nurses (ARN) recognizes that rehabilitation nurses go above and beyond in caring for their patients every day. Their membership includes more than 5600 professional nurses who together form a community singularly devoted to advancing and promoting rehabilitation nursing in a variety of practice settings, including nursing schools, acute care hospitals, inpatient rehabilitation units/hospitals, long-term care hospitals, skilled nursing facilities, and home health agencies. The wealth of knowledge and information their community of nurses share is extensive. ARN works closely with members to develop their rehabilitation nursing practice by providing supportive resources, in-depth educational programs, and opportunities to network with peers and to connect with leaders in the field, as well as providing tools to become active health policy advocates. ARN's mission is to promote and advance professional rehabilitation nursing practice through education, advocacy, collaboration, and research to enhance the quality of life for those affected by disability and chronic illness.

REVIEW EXERCISES

1. What does the federal government use to announce proposed administrative law regulations and standards?
2. Explain the purpose of the CFR.
3. List the type of standards published in the following CFR references:
   a. 29 CFR 1910
   b. 29 CFR 1915
   c. 29 CFR 1917
   d. 29 CFR 1918
   e. 29 CFR 1926
   f. 29 CFR 1904
4. How long must employers covered by the OSH Act maintain their OSHA 300 Logs?
5. Explain in your own words the key difference between an OSHA-defined injury and illness.
6. List the four general elements of the OSHA Management Guidelines.
7. Describe the purpose of the following environmental laws:
   a. RCRA
   b. CERCLA
   c. SARA
   d. CWA
   e. TSCA
8. What is the mission of NIOSH?
9. What three organizations possess legal "deemed status" to conduct hospital accreditation surveys?
10. What type of facilities does CARF accredit?

Bibliography

10 reasons to use microfiber mopping. 2003. Factsheet on SHP website. Available at http://www.sustainablehospitals.org/PDF/tenreasonsmop.pdf.

A Practical Guide to the Determination of Human Exposure to Radiofrequency Fields, NCRP Report No. 119, National Council on Radiation Protection and Measurements, 1993.

Abrahamson, E. Change without Pain. Boston: Harvard Business School Press, 2004.

Accident Prevention Manual for Business and Industry – Administration and Programs (10th ed.), National Safety Council, Itasca, Illinois, 1992.

Accountability, Culture & Behavior, Dan Petersen, ASSE Professional Safety, October 1997.

Adams, S.J. Benchmarks of Safety Quality, ASSE Professional Safety, November 1997.

The Agency for Toxic Substances and Disease Registry, Glossary, Accessed Online, 2010.

Allen, C.H. Maritime Counterproliferation Operations and the Rule of Law, Westport, CT: Praeger Security International, 2007.

Amar, A.D. Principled Versus Analytical Decision-Making: Definitive Optimization. Mid-Atlantic Journal of Business, June 1995, 119.

American Biological Safety Association, ABSA Biosecurity Task Force White Paper: Understanding Biosecurity, Illinois: The Association, 2003.

American Institute of Architects, AIA Guidelines for Design and Construction of Health Care Facilities, 2006.

American National Standards Institute, Safe Use of Lasers in Health Care Facilities, ANSI Z136.3.

American National Standards Institute, Standard Safety Levels with Respect to Human Exposure to Radiofrequency Fields, ANSI C95.1–1991.

American National Standards Institute, Safety Levels with Response to Human Exposure to Radiofrequency Electromagnetic Fields, 300 KHz to 100 KHz, ANSI C95.1–1982: New York.

American Society of Health System Pharmacists, ASHP guidelines on quality assurance for pharmacy-prepared sterile products. Am J Health-Syst Pharm. 2000;57:1150–69.

American Society of Health System Pharmacists Compounding Resource Center website, "Revisions Continue on USP Sterile-Compounding Chapter," Cheryl Thompson, AJHP News, Sept. 15, 2005.

American Society of Health System Pharmacists Compounding Resource Center website, "USP Chapter 797 Standards: An Expert Panel Offers Practical Strategies for Implementation," E. Clyde Buchanan, M.S., FASHP, Eric S. Kastango, M.B.A., R.Ph., FASHP, Darryl S. Rich, Pharm.D., M.B.A., FASHP.

American Society of Hospital Pharmacists, ASHP technical assistance bulletin on quality assurance for pharmacy prepared sterile products. Am J Hosp Pharm. 1993;50:2386–90.

Anderson, R. Security Engineering, Hoboken, NJ: Wiley, 2001.

ANSI, Safety Requirements for Workplace Walking Working Surfaces and Their Access: Floor, Wall and Roof Openings, Stairs and Guardrail Systems, ANSI/ASSE A1264.1–2007.

APIC Bioterrorism Working Group: April 2002 Interim Bioterrorism Readiness Planning Suggestions, Association for Professionals in Infection Control and Epidemiology.

Apple, J. Materials Handling Systems Design, New York: The Ronald Press, 1972.

Argyris, C, and D. Schon, Organizational Learning II, London: Addison-Wesley, 1996.

Arnold, H.J. 1989. Sanctions and rewards: Organizational perspectives. In Sanctions and Rewards in the Legal System: A Multidisciplinary Approach. Toronto: University of Toronto Press.

ASHE Electrical Standard Compendium, 1999 Edition, American Society of Healthcare Engineering, Chicago, IL.

ASHRAE Handbook 2003, HVAC Applications.

Aspden P, Corrigan J, Wolcott J, Eds. Patient Safety: Achieving a New Standard for Care. Washington, DC: National Academies Press, 2004.

ASSE, Return on Safety Investment White Paper, American Society of Safety Engineers, June 2002.

Athos, A.G. and R.C. Coffey, Time, Space and Things in Behavior in Organizations: A Multidimensional View, Englewood Cliffs, NJ: Prentice Hall, 1975.

Back Injury Prevention Guide for Health Care Providers, Cal/OSHA Consultation Programs, accessed online January 2006, http://www.dir.ca.gov/dosh/dosh_publications/backinj.pdf.

Bainbridge, L. 1983. Ironies of Automation, Automatica 19:775–779.

Baker, S.P., B. O'Neil, M.J. Ginsburg, and G. Li. Injury Fact Book, New York: Oxford University Press.

Bamber, L. 1979. Accident costing in industry, Health and Safety at Work (Croyden) 2/4:32–34.

Barbera J.A., A.G. Macintyre, and D. Sunday. Medical Surge Capacity and Capability: A Management System for Integrating Medical and Health Resources during Large Scale Emergencies. Alexandria, VA.: CNA Corporation, 2004, http://www.cna.org/documents/mscc_aug2004.

Bare, A.R. Pressurization Strategies for Biocontainment, R&D Laboratory Design Handbook (11):59B63, 2006.

Barnard, C.I. The Functions of the Executive, Cambridge: Harvard University Press, 1968.

Barnett, R. and D. Brickman. 1986. Safety Hierarchy, J Saf Res 17:49–55.

Beauchamp, T. and N. Bowie, Ethical Theory and Business, Englewood Cliffs, NJ: Prentice Hall, 1993.

Beckmann, U., D.M. Gillies, and S.M. Berenholtz et al. Incidents Relating To Intra-Hospital Transfer of Critically Ill Patients. An Analysis of The Reports Submitted to the Australian Incident Monitoring Study in Intensive Care. Intensive Care Med 2004 Feb; 30(8):1579–85.

Behm, M., A. Veltri, I. Kleinsorge. Cost Of Safety: Cost Analysis Model Helps Build Business Case for Safety, ASSE Professional Safety, April 2004.

Bell, C., S. Brener, and G. Gunraj. 2011. Association of ICU or Hospital Admission with Unintentional Discontinuation of Medications for Chronic Diseases, JAMA, 306:840–847.

Berry, M. Protecting the Built Environment: Cleaning for Health, TRICOM 21st Press, Chapel Hill, NC, 1994.

Biosafety in Microbiological and Biomedical Laboratories (BMBL), 5th Edition, Centers for Disease Control and Prevention (CDC) and National Institutes of Health (NIH), 2007.

Bioterrorism, Annals of Emergency Medicine, 34:2, August 1999.

Bird, F., and G. Germain. Practical Loss Control Leadership, International Loss Control Institute, Loganville, GA, 1985.

Birky, B., L. Slaback, and B. Schleien, Handbook of Health Physics and Radiological Health, 3rd ed, Lippincott Williams & Wilkins, Philadelphia, PA, 1998.

Blackwell, Daid S., Rajhans, Practical Guide to Respiratory Usage in Industry, Boston, 1985.

Block, S.S. (ed.). Disinfection, Sterilization, and Preservation, 4th Edition, Philadelphia: Lea & Febiger, 1991.

Borisoff, D. and D.A. Victor, Conflict Management: A Communication Skills Approach. Englewood Cliffs, NJ: Prentice Hall, 1989.

Bothe, K.R. World Class Quality. NY: AMACOM, 1991.

Brauer, R. Safety and Health for Engineers, New York: Van Nostrand Reinhold, 1990.

Brogmus, G., W. Leone, L. Butler, and E. Hernandez. Best Practices In OR Suite Layout And Equipment Choices to Reduce Slip, Trips, And Falls, AORN J 86:384–398, 2007.

Bronstein, D.A., Demystifying the Law, Chelsea, MI. Lewis Publishers, Inc., 1990.

Brookhuis, K., A. Hedge, H. Hendrick, E. Salas, and N. Stanton, Handbook of Human Factors and Ergonomics Models, London: CRC Press, 2005.

Building Air Quality: A Guide for Building Owners and Facility Managers, NIOSH/EPA, December 1991.

Bull, M., D. Lee, J. Stucky, Y.L. Chiu, A. Rubin, H. Horton, M.J. McElrath, 2007. Defining blood processing parameters for optimal detection of cryopreserved antigen-specific responses for HIV vaccine trials. J. Immunol. Methods 322: 57–69, http://www.ncbi.nlm.nih.gov/pmc/articles/PMC1904432/.

Bureau of Labor Statistics, U.S. Department Of Labor, Washington, DC: Occupational Injuries and Illnesses in the United States by Industry.

California Patient Safety Program Manual (PDF Version), Published By California Department of Health Services, accessed online January 2008. http://www.cdph.ca.gov/programs/Documents /PatientSafetyProgramManual12-12-2005.pdf.

Campanella, P. and M. Robinson. Patient Safety and Peak Performance, Pharmaceutical Formulation & Quality, 2006.

Carley, S. and K.M. Jones. Major Incident Medical Management and Support: The Practical Approach in the Hospital by the Advanced Life Support Group, Malden, MA: BMJ Books: Blackwell Pub., 2005.

Carroll, R., (ed.) Healthcare Risk Management,. 2nd ed., Chicago, IL: AHA Publishing, 1997.

Carroll, S. and D. Gillen, Are the Classical Management Functions Useful in Describing Managerial Work? Academy of Management Review 12, No. 1 (1980): 38–51.

Carson, H.T. and D.B. Cox. Handbook on Hazardous Materials Management, 4th edition. Institute of Hazardous Materials Management, Rockville, MD, 1992.

Casadevall, A., L. Pirofski. The weapon potential of a microbe. Bethesda, MD: The National Institutes of Health, 2005.

Case Study: Are Micro Fiber Mops Beneficial For Hospitals? Sustainable Hospitals Project, 2003.

Castro, K.G. S.W. Dooley, M.D. Hutton, R.J. Mullan, J.A. Polder, and D.E. Snider, Guidelines for Preventing the Transmission of Tuberculosis in Health-Care Settings, with Special Focus on HIV-Related Issues, http://www.cdc.gov/niosh.

CDC, Department Of Health And Human Services, Biosafety in Microbiological and Biomedical Laboratories, 3rd Edition, CDC/NIOSH, May 1993.

CDC Guidelines for Environmental Infection Control In Health-Care Facilities: Recommendations of CDC and the Healthcare Infection Control Advisory Committee (HICPAC), MMWR 2003; 52(NO. RR-10).

CDC website, NIOSH Alert, Preventing Occupational Exposures to Antineoplastic and Other Hazardous Drugs in Health Care Settings, http://www.cdc.gov/niosh/docs/2004-165/.

Center for Maximum Potential Building Systems, Practice Greenhealth. Green Guide for Health Care (Green Guide): Version 2.2, Operations Section. Green Guide for Health Care, 2008.

Chandler, A. D., Jr, Strategy and Structure, Cambridge, MA: M.I.T. Press, 1962.

Chandler, P, and J. Sweller. Cognitive Load Theory and the Format of Instruction, Cognition and Instruction 8(4): 293–332, 1991.

Charney, W. 1997. Lift Team Method for Reducing Back Injury: A Ten Hospital Study, Journal of the American Association of Occupational Health Nursing, 45(6).

Chemical Terrorism Agents and Syndromes, University Of North Carolina, Chapel Hill, NC, Has Developed a Wall Chart on Chemical Terrorism Agents, 2002.

Chen, A.Y.S. and J.L. Rodgers, Teaching the Teachers TQM, Management Accounting 76 (1995): 42–46.

Chesbrough, H.W. Open Innovation: The New Imperative for Creating and Profiting from Technology. Boston: Harvard Business School Press, 2003.

Chiou S., B. Evanoff, H. Wellman, M. Matz, and A. Nelson. 2008. Multi-Disciplinary Research to Prevent Slip, Trip, and fall (STF) Incidents among Hospital Workers, Contemporary Ergonomics, 2008.

Ciriello, V.M., P.G. Dempsey, R.V. Maikala, and N.V. O'Brien. 2007. "Revisited: Comparison of Two Techniques to Establish Maximum Acceptable Forces of Dynamic Pushing for Male Industrial Workers," International Journal of Industrial Ergonomics, Vol. 37, No. 11–12, pp. 877–892.

Clancy, C.M., M.B. Farquhar, and B.A. Sharp, Patient Safety in Nursing Practice, J Nurs Care Qual Jul–Sep 2005, 20(3):193–7.

Code of Federal Regulations, Titles 10, 21, 29, 40, 44, & 49, accessed online 2010–2013, http://cfr.regstoday.com/CFR.aspx.

Coe, C.P. The Elements of Quality in Pharmaceutical Care, Bethesda, MD: American Society of Hospital Pharmacists, 1992.

Cohen, S. 2006. Boom Boxes: Containers and Terrorism in Protecting the Nation's Seaports: Balancing Security and Cost. Accessed March 19, 2008 from http://www.ppic.org/content/pubs/report/R_606JHR.pdf#page = 179.

College of American Pathologists. Laboratory General Checklist College of American Pathologists. October 2006.

Committee on the Quality of Health Care in America, Crossing the Quality Chasm: A New Health System for the 21st Century. Washington, DC: National Academy Press, 2001.

Cooper, D. Safety Culture, ASSE Professional Safety, June 2002.

Corbett, J.M. 1988. Ergonomics in the Development of Human Centered AMT, Applied Ergonomics 19:35–39.

County Health Care Agency, Emergency Medical Services. Santa Ana, Calif.: Orange County Health Care Agency, 1993.

Courtney, T.K., D.A. Lombardi, G.S. Sorock, H.M. Wellman, S. Verma, M.J. Brennan, J. Collins, J. Bell, W.R. Chang, R. Grnqvist, L. Wolf, E. DeMaster, and M. Matz, Circumstances of Slips Trips and Falls among Hospital Workers, accessed online, http://injuryprevention.bmj.com/content/16/Suppl_1/A173.3.abstract, 2010.

Coverage of various types of laboratories by the Laboratory Standard. OSHA Standard Interpretation, February 8, 1991.

Crainer, S. 75 Greatest Management Decisions Ever Made, Management Review, November 1998.

Crawford S.Y., W.A. Narducci, and S.C. Augustine. National survey of quality assurance activities for pharmacy-prepared sterile products in hospitals. Am J Hosp Pharm. 1991;48:2398–413.

Creech, B. The Five Pillars of TQM. New York: Truman Talley Books/Dutton, 1995.

Crossing the Quality Chasm: A New Health System For The 21st Century, Washington, DC: National Academy Press, 2001.

Cutler, T.W. Medication Reconciliation Victory after an Avoidable Error, AHRQ Web M&M, February/March 2009.

Dancer, J. Mopping Up Hospital Infection, J Hosp Infect. 1999;43(2):85–100.

Daschner, F.D., MD. A. Schuster, Dipl. Biol., M. Dettenkofer, MD. K. Kümmerer, PhD Institute of Environmental Medicine and Hospital Epidemiology, Freiburg University Hospital, Freiburg, Germany the link is: http://www.ajicjournal.org/article/S0196-6553(04)00564-4/abstract.

David, M. and J. Wilbur. Fault Tree Analysis Application Guide, Report No. FTA, Reliability Analysis Center, Rome, NY, 1990.

Davids, M. W. Edwards Deming (1900–1993) Quality Controller. Journal of Business Strategy 20, no. 5 (1999): 31–32.

Dellinger, S. and B. Deane, Communicating Effectively – A Complete Guide for Better Managing, Chilton Book Company, Radnor, Pennsylvania, 1982.

DeLorenzo, R.A., Cyanide, Journal of Emergency Medical Services, October 1999.

Deming, W.E. Out of the Crisis. Cambridge, MA: MIT Press, 1986.

Department of Health and Human Services, Managing Hazardous Material Incidents: A Planning Guide for the Management of Contaminated Patients, 1998.

Department of Transportation, Research and Special Programs Administration. Emergency Response Guidebook. Washington, DC, 1993.

Desa, J., A. Bello, C. Galligan, T. Fuller,, and M. Quinn. Case Study: Are Microfiber Mops Beneficial for Hospitals? Sustainable Hospitals Project. Lowell, MA: University of Massachusetts, 2004.

Diamond, J. and G.M. Ciampanelli. Integrated Emergency Management: Hospital Disaster Planning, Topics in Emergency Medicine, 20:2, June 1998.

Dipilla, S., Slip and Fall Prevention: A Practical Handbook. 2nd Edition. Boca Raton, FL: CRC Press, 2010.

Disaster Mitigation Act of 2000. Federal Emergency Management Agency, http://www.fema.gov/library.

Donaldson, M., ed. To Err Is Human: Building a Safer Health System, National Academy Press, 2000.

Dougherty, T.M. Reinforcing Safety Values in People, ASSE Professional Safety, November 1997.

Douglas, M. and D. Wildavsky. Risk and Culture, Berkeley, CA, University of California, 1983.

Drucker, P.F. Management: Tasks, Responsibilities, Practices. New York: Harper & Row, 1974.

Drucker, P.F. What Makes an Effective Executive, Harvard Business Review, 1 June 2004.

Duflo, E., M. Greenstone, and R. Hanna. 2008. Indoor air pollution, health and economic well-being. S.A.P.I.E.N.S 1(1).

Dumas, J. and M. Salzman. Reviews of Human Factors and Ergonomics, Human Factors and Ergonomics Society, 2006.

Earley, P.C., S. Ang, and J-S Tan, CQ: Developing Cultural Intelligence in the Workplace. Stanford, CA: Stanford University Press, 2005.

Earnest, R.E. Characteristics of Proactive & Reactive Safety Systems, ASSE Professional Safety, November 1997.

ECRI, A Clinician's Guide to Surgical Fires: How They Occur, How to Prevent Them, How to Put Them Out. Health Devices 2003; 32(1):5–24.

ECRI Institute. 2008. Top ten technology hazards. Health Devices. November, 343–350. Plymouth Meeting, PA: ECRI Institute.

Emergency Management in Health Care: An All Hazards Approach. Oakbrook Terrace, IL: Joint Commission Resources, 2008.

Emergency Management Principles and Practices for Healthcare Systems, The Institute for Crisis, Disaster, and Risk Management (ICDRM) at the George Washington University (GWU); for the Veterans Health Administration (VHA)/US Department of Veterans Affairs (VA), June 2006.

Emergency Management Program Guidebook, Emergency Management Strategic Healthcare Group (EMSHG), Veterans Health Administration (VHA), March 2008.

Emergency Readiness, American Hospital Association (AHA), October 2001.

Enforcement Policy and Procedures for Occupational Exposure to Tuberculosis, Occupational Safety and Health Administration, October 8, 1993.

Environment of Care: Essentials for Health Care, 8th Edition, Joint Commission Resources, Oakbrook, IL, 2008.

Environmental Best Practices for Hospitals: Using Micro Fiber Mops in Hospitals, EPA Region 9, November 2002.

Environmental Protection Agency, Washington, DC: Asbestos Waste Management Guidance Publication, 1985.

Environmentally Preferable Purchasing Program, Office of Pollution Prevention and Toxics, Cleaning Products Pilot Project, Washington DC, 1997.

Erickson, J. The Relationship between Corporate Culture & Safety Performance, ASSE Professional Safety, May 1997.

Error Reduction in Healthcare: A Systems Approach to Improving Patient Safety. Chicago: Health Forum Inc. 2000:199–234.

Esmail, R., D. Banack, C. Cummings, et al. Is Your Patient Ready For Transport? Developing an ICU Patient Transport Decision Scorecard, Health Q 2006 Oct, 9 Spec. No. 80–6.

Essential Components of a Tuberculosis Prevention and Control Program. Morbidity and Mortality Weekly Report, Vol. 44, No. RR-11. September 8, 1995.

Ezzati, M. and D.M. Kammen, 2002. The Health Impacts of Exposure to Indoor Air Pollution from Solid Fuels in Developing Countries: Knowledge, Gaps, and Data Needs. Environmental Health Perspectives 110:1057–1068.

Farmer, J. and M.D. Christopher. Are You Prepared? Hospital Emergency Management Guidebook: Joint Commission Resources, Oakbrook, IL, 2006.

Favero, M. and W. Bond. Chemical disinfection of medical surgical material. In: Block, S., editor. Disinfection, Sterilization, and Preservation. 5th edition. Philadelphia: Lippencott, Williams and Wilkens. 2001, pp. 881–917.

Fayol, H. General and Industrial Administration, London: Sir Issac Pitman & Sons, 1949.

FDA Safety Alert: Entrapment Hazards Hospital Bed Side Rails. August 23, 1995.

Federal Emergency Management Agency, An Orientation to Hazardous Materials for Medical Personnel: A Self-Study Guide, 1998.

Federal Emergency Management Agency, Disaster Planning Guide for Business and Industry, 2004.

Federal Emergency Management Agency, Guide for All-Hazard Emergency Operations Planning: State and Local Guide 101, 1996.

Federal Emergency Management Agency: NRF Resource Center, National Response Framework, http://www.fema.gov

Feiner F., D. Miller, and F. Walker. Chart of the Nuclides, 13th ed., California, General Electric Corporation, 1983.

Felton, J. Violence Prevention at the Health Care Site, Occ Med, State Of The Art Reviews 12(4):701–715, 1997.

Ferrell, O.C., J. Fraedrich, and L. Ferrell. Business Ethics. Boston, MA: Houghton Mifflin Company, 2002.

Ferry, T. Modern Accident Investigation and Analysis, 2nd ed, New York: John Wiley & Sons, 1988.

Feyer, A.M. and A.M. Williamson. 1991. An Accident Classification System for Use in Preventive Strategies, Scand J Work Environ Health 17:302–311.

Fiedler, F.E. A Theory of Leadership Effectiveness, New York: McGraw-Hill, 1967.

Fink, R., D.F. Liberma, K. Murphy, et al. 1988. Biological safety cabinets, decontamination or sterilization with paraformaldehyde. Am Ind Hyg Assoc. 49:277–279.

Flood, R.L. and E.R. Carson, Dealing with Complexity: An Introduction to the Theory and Application of Systems Science. 2nd ed., New York: Plenum, 1993.

Food and Drug Administration. Public Health Service, U.S. Department of Health and Human Services, Washington, DC: FDA/CPSC Public Health Advisory: Hazards Associated with the Use of Electric Heating Pads. December 12, 1995.

Food and Drug Administration. Guidance for Industry and FDA Reviewers: Content and format of premarket notification [510(k)] submission chemical sterilants/high level disinfectants. Rockville, MD: U.S. Department of Health and Human Services, Food and Drug Administration, 2000.

Fragala, G., ed. The Epidemic of Health Care Worker Injury. Boca Raton, FL: CRC Press, 1999.

Framework for a Comprehensive Health and Safety Program in the Hospital Environment, OSHA, 1993.

Franz, D. 1997. Clinical Recognition and Management of Patients Exposed to Biological Warfare Agents, Journal of the American Medical Association, 278:5.

Frederiksen, L.W. Handbook of Organizational Behavior Management, Hoboken, NJ: Wiley, 1982.

Friend, M. Examine Your Safety Philosophy, ASSE Professional Safety, February 1997.

Friend, M.A. and Pagliari, L.R. Establishing a Safety Culture: Getting Started, ASSE Professional Safety, May 2000.

Friis, R. and T. Sellers. Epidemiology for Public Health Practice, 2nd ed. Gaithersburg, Maryland: Aspen Publishers, Inc., 1999.

Gaba, D.M. 1989. Human Error in Anesthetic Mishaps, Int Anesthesiology Clin. 27(3):137–47.

Gaba, D.M. and A. Deanda. 1988. Comprehensive Anesthesia Simulation Environment: Re-Creating the Operating Room for Research and Training. Anesthesiology 69(3):387–94.

Galloway, S. Critical Questions to Improve Behavior Based Safety, Safety Culture Excellence, 2008.

Garg, A. 1999. Long-Term Effectiveness of "Zero-Lift Program" in Seven Nursing Homes and One Hospital. U.S. Department of Health and Human Services, Centers for Disease Control and Prevention, National Institution for Occupational Safety and Health (NIOSH), Cincinnati, OH. August. Contract No. U60 /CCU512089-02.

Garg, A. and B. Owen. 1992. Reducing Back Stress to Nursing Personnel: An Ergonomic Intervention in a Nursing Home, Ergonomics, 35: 1353–1375.

Garner, J. and M. Favero. 1985. Handwashing and Hospital Environmental Control. HHS Publication #99-1117.

Garner, J.S. 1996. Guideline for isolation precautions in hospitals. Infect Control Hosp Epidemiol. 17:53–80.

Garvin, D.A. Building a Learning Organization, Harvard Business Review 71, 1993.

Gebrewold, F. and F. Sigwart. Performance Objectives: Key to Better Safety Instruction, ASSE Professional Safety, August 1997.

Geller, E.S. Working Safe: How to Help People Actively Care For Health and Safety, CRC Press, 1996.

Geller, S. Ten Leadership Qualities for a Total Safety Culture: Management Is Not Enough, ASSE Professional Safety, May 2000.

General Industry Digest, OSHA Pub 2201, OSHA 2201-05R, 2013.

Gido, J. and J. Clements. Successful Project Management, Cincinnati: South-Western College Publishing, 1999.

Gielen, A.C. 1992. Health education and injury control: Integrating approaches. Health Educ Q 19(2):203–218.

Gielo-Perczak, K., W.S. Maynard, and A. Didomenico. Multidimensional Aspects of Slips, Trips, and Falls. In R. Williges, Ed. Reviews of Human Factors and Ergonomics, HFES, January, 1, 2006, Vol. 2, No. 1, pp. 165–194(30), Santa Monica, CA.

Gillespie, U., A. Alassaad ,D. Henrohn. 2009. A Comprehensive Pharmacist Intervention to Reduce Morbidity in Patients 80 Years or Older: A Randomized Controlled Trial, Arch Intern Med. 169:894–900.

Gillman, L., G. Leslie, T. Williams, et al. 2006. Adverse Events Experienced While Transferring the Critically Ill Patient from the Emergency Department to the Intensive Care Unit. Emerg Med J 23(11):858–61.

Gleason K., H. Brake, V. Agramonte, and C. Perfetti. Medications at Transitions and Clinical Handoffs (MATCH) Toolkitf or Medication Reconciliation, Rockville, MD: Agency for Healthcare Research and Quality; Revised August 2012, AHRQ Publication No. 11(12)-0059.

Goals for Working Safely with Mycobacterium Tuberculosis in Clinical, Public Health, and Research Facilities, Centers for Disease Control and Prevention (CDC), Office of Health and Safety Information System (OHASIS).

Goodman, R.A., E.L. Jenkins, and J.A. Mercy. 1994. Workplace-Related Homicide among Health Care Workers in the United States, 1980 through 1990, JAMA 272(21):1686–1688.

Goodstein, L.P., H.B. Anderson, and S.E. Olsen. 1988. Tasks, Errors and Mental Models. London: Taylor & Francis.

Grimaldi, J. and R. Simonds. Safety Management, 5th ed. Homewood,IL: IRWIN, 1989.

Guastello, S.J. 1991. The Comparative Effectiveness of Occupational Accident Reduction Programs. Paper presented at the International Symposium Alcohol Related Accidents and Injuries. Yverdon-les-Bains, Switzerland, Dec. 2–5.

Guide for All Hazards Emergency Operations Planning, State and Local Guidance 101, Federal Emergency Management Agency, September 1996.

Guidelines for Preventing Workplace Violence for Health Care and Social Service Workers, OSHA 3148. 1996, 2003.

Haddon, W.J. 1972. A Logical Framework for Categorizing Highway Safety Phenomena and Activity, J Trauma 12:193–207.

Hale, A.R. and M. Hale, 1970. Accidents in perspective. Occup Psychol 44:115–122.

Hansen, D.J. The Work Environment. 1991. Chelsea: Occupational Health Fundamentals, pp. 304–308.

Hare, V.C. System Analysis: A Diagnostic Approach, New York: Harcourt Brace World, 1967.

Harrison, R. Diagnosing Organizational Culture, Trainer's Manual, San Francisco, CA: Jossey-Bass, 1993.

Harteloh, P.P.M., 2003. The Meaning of Quality in Health Care: A Conceptual Analysis. Health Care Analysis 11(3):259–67.

Harvard Business Review on Knowledge Management. Boston: Harvard Business School Press, 1998.

Hayes, A.W., ed. Principles and Methods of Toxicology. New York: Raven, 1989.

Hazards and Regulation of Cleaning Chemicals, Health Care without Harm, Health Facilities Management Magazine, Oct 2006.

Health and Safety Executive, Road Transport in Factories and Similar Workplaces, Guidance Note GS9(R) London: HMSO, 1992.

Healthcare Professionals: Guidelines on Prevention of and Response to Infant Abductions, 6th ed., March 2000, National Center for Missing & Exploited Children.

Heinrich, H. Industrial Accident Prevention: A Scientific Approach, 4th ed. New York: McGraw-Hill, 1959.

Helmreich, R.L. 1997. Managing Human Error in Aviation, Sci Am. 276(5):62–7.

Henriksen, K., J.B. Battles, M.A. Keyes, eds. Advances in Patient Safety: New Directions and Alternative Approaches (Vol. 1: Assessment). Rockville, MD: Agency for Healthcare Research and Quality, 2008.

Hinkel, J.M. 2008. Report on the NCCN Third Annual Patient Safety Summit. J Natl Compr Canc Netw 6(6):528–35.

Hoppin, P. and S. Donahue. Improving Asthma Management by Addressing Environmental Triggers: Challenges and Opportunities for Delivery and Financing. Symposium Edition. Asthma Regional Council of New England. 2004.

Hosmer, L. Ethics of Management, Homewood, IL: Irwin, 1991.

Hugentobler, M.K., B.A. Israel, and S.J. Schurman. 1992. An action research approach to workplace health: Intergrating methods. Health Educ Q 19(1):55–76.

Humbert, P Top Ten Tools For Effective Listening, accessed online, http://www.philiphumbert.com /Articles/10EffectiveListening.html, January 24, 2011.

Huotari, M.L. Trust in Knowledge Management and Systems in Organizations, Hershey, PA: Idea Group Publishing, 2003.

Incidence Rates for Nonfatal Occupational Injuries and Illnesses Involving Days Away from Work per 10,000 Full-Time Workers by Industry and Selected Events or Exposures Leading to Injury or Illness, Bureau of Labor Statistics (BLS), accessed online, http://www.bls.gov/news.release/pdf/osh2.pdf, January 2011.

Industry Standards for the Prevention of Work-Related Musculoskeletal Disorders in Sonography, Society of Diagnostic Medical Sonography, 2003.

Institute Of Medicine, Appendix K Glossary and Acronyms, Biological Threats and Terrorism: Assessing the Science and Response Capabilities: Workshop Summary, National Academy Press: 2002, pp. 275–286.

Institute Of Medicine, Committee on Quality of Health Care in America, Crossing the Quality Chasm: A New Health System for the 21st Century. Washington, DC: National Academy Press, 2001.

International Ergonomics Association, What Is Ergonomics? Accessed online, http://www.iea.cc/, December 2010.

International HACCP Alliance, International Organization for Standardization ISO 15189: Medical Laboratories, Particular Requirements for Quality and Competence, 2003.

International Organization for Standardization (ISO), 196, Symbols, Dimensions, and Layout for Safety Signs, ISO R557, Geneva.

Jervis, S. and T. Collins. Measuring Safety's Return on Investment, ASSE Professional Safety, September 2001.

Joint Commission Resources & Institute for Healthcare Improvement. The Essential Guide for Patient Safety Officers, JCR Publishing, 2009.

Joint Commission, Standing Together: An Emergency Planning Guide for America's Communities, 2005.

Kast, F.E. and J.E. Rosenzweig. 1972. General Systems Theory: Applications for Organizations and Management, Academy of Management Journal. 15(4):451.

Kastango, E. "Brutal Facts" from Overview USP 797 and Control of Contamination Presentation, Compliance Tools and Aseptic Certification for USP 797, STAR Center Training, 2007.

Katz, D. and R.L. Kahn. The Social Psychology of Organizations, New York: John Wiley & Sons, 1978.

Keim, M. and A. Kaufman. 1999. Principles for Emergency Response to Bioterrorism, www.sciencedirect .com/science/article/pii/.../pdf?...ScienceDirect Aug 2, 1999–August 1999. 34:2. ANNALS OF EMERGENCY MEDICINE. 1 7 7.

Kitzes, W.F. Safety Management and the Consumer Product Safety Commission, ASSE Professional Safety, April 1991.

Kjellén, U. 1984. The Deviation Concept in Occupational Accident Control, Part I: Definition and Classification and Part II: Data Collection and Assessment of Significance, Accident Anal Prev 16:289–323.

Kjellén, U. and T.J. Larsson, 1981. Investigating accidents and reducing risks – a dynamic approach, J Occupational Acc. 3:129–140.

Kleiman, L.S. Human Resource Management: A Managerial Tool for Competitive Advantage. Cincinnati: South-Western College Publishing, 2000.

Klein, E, D.L. Smith, and R. Laxminarayan. 2007. Hospitalizations and Deaths Caused By Methicillin-Resistant Staphylococcus Aureus, 1999–2005, Emerg Infect Dis. 13(12):1840–6.

Klevens, R.M. J.R. Edwards, C.L. Richards, Jr., T.C. Horan, R.P. Gaynes, D.A. Pollock, and D.M. Cardo. 2007. Estimating health care-associated infections and deaths in U.S. hospitals, 2002. Public Health Rep. 122(2):160–6.

Kline, P. and B. Saunders. Ten Steps to a Learning Organization, Arlington, VA: Green River Books, 1997.

Knapp, J.E. and D.L. Battisti. Chlorine Dioxide. In S. Block, ed. Disinfection, Sterilization, and Preservation. 5th ed. Philadelphia: Lippencott, Williams.

Koenig, K.L. New Standards in Emergency Management: Major Changes in JCAHO Requirements for Disasters. American College Emergency Physicians, Disaster Section Newsletter, May 2001.

Kohn, L.T., J.M. Corrigan, and M.S. Donaldso, eds. To Err Is Human: Building a Safer Health System. Washington, DC: National Academic Press, 2000.

Koontz, H. and C. O'Donnell, Principles of Management: An Analysis of Managerial Functions. New York: McGraw-Hill Book Co., 1955.

Kraus, J.F., K.A. Brown, D.L. McArther, C. Peek-Asa, L. Samaniego, C. Kraus, and L. Zhou. 1996. Reduction of Acute Low Back Injuries by Use of Back Supports, International Journal of Occupational and Environmental Health 2:263–273.

Krause, T. Behavior-Based Safety Process: Managing Involvement for an Injury-Free Culture. New York: Van Nostrand Reinhold, 1999.

Krause, T. Motivating Employees for Safety Success, ASSE Professional Safety, March 2000.

Krause, T. Leading with Safety, Hoboken, NJ: Wiley, 2005.

Kwan, J., L. Lo, M. Sampson, and K. Shojania. 2013. Medication Reconciliation during Transitions of Care as a Patient Safety Strategy: A Systematic Review. Ann Intern Med. 158(5 Pt. 2):397–403.

Labelle, J.E. What Do Accidents Truly Cost, ASSE Professional Safety, April 2000.

Lado, A.A. and M.C. Wilson. 1994. Human Resource Systems and Sustained Competitive Advantage: A Competency-Based Perspective, Academy of Management Review 19(4):699–727.

LaRue, B. and R.R. Ivany. Transform Your Culture, Executive Excellence, December 2004, 14–15.

Laser Safety Guide. Laser Institute of America. Fourth Edition. Cincinnati, OH, 1977.

Lawrence, P.R. and J.W. Lorsch,. Organization and Environment. Homewood, IL: Richard D. Irwin, Inc, 1969.

Lehto, M.R. and J.M. Miller. Warnings: Volume I: Fundamentals, Design, and Evaluation Methodologies. Ann Arbor, MI: Fuller Technical Publications, 1986.

Leonard, D., and W.C. Swap. Deep Smarts: How to Cultivate and Transfer Enduring Business Wisdom, Boston: Harvard Business School Press, 2005.

Leonard, M., S. Graham, and D. Bonacum. 2004. The Human Factor: The Critical Importance of Effective Teamwork and Communication in Providing Safe Care. Qual Safe Health Care 13(Suppl 1):I85–90.

Leplat, J. 1978. Accident Analyses and Work Analyses, Journal Occup Acc. 1:331–340.

Liberty Mutual Executive Survey of Workplace Safety, Liberty Mutual Insurance Company, Boston, 2001.

Lipscomb J. Violence in the Health Care Industry: Greater Recognition Prompting Occupational Health and Safety Interventions. In W. Charney, ed. Essentials of Modern Hospital Safety, Vol. 3. Boca Raton, FL: Lewis Publishers, 1994.

Lohr, K. Committee to Design a Strategy for Quality Review and Assurance in Medicare. Medicare: A Strategy for Quality Assurance, Vol. 1. Washington, DC: National Academy Press; 1990.

Lowrance, W. Of Acceptable Risk, Los Altos, CA: William Kaufmann, Inc, 1976.

Macintyre, A. 2000. Weapons of Mass Destruction Events with Contaminated Casualties: Effective Planning for Health Care Facilities, Journal of the American Medical Association, 283:2.

Malhotra, Y. 2005. Integrating Knowledge Management Technologies in Organizational Business Processes: Getting Real Time Enterprises to Deliver Real Business Performance. Journal of Knowledge Management 9(1):7–28.

Managing Hazardous Material Incidents: A Planning Guide for the Management of Contaminated Patients, DHHS, 1998.

Manuelle, F. Principles for the Practice of Safety, ASSE Professional Safety, July 1997.

Manzella, J.C. Measuring Safety Performance to Achieve Long-Term Improvement, ASSE Professional Safety, September 1999.

Martin, J.T. and M.A. Warner, eds. Anesthesiologic Considerations, Positioning in Anesthesia and Surgery. 3rd ed. Philadelphia, PA: WB Saunders; 1997:127.

Martins, S.B. and K.G. Shojania. Safety during Transport of Critically Ill Patients, Chapter 47, Agency for Healthcare Research and Quality. Making Health Care Safer: A Critical Analysis of Patient Safety Practices, available at http://www.ahrq.gov/.

Massawe, E, K. Geiser, M. Ellenbecker, and J. Marshall, 2007. Health, safety, and ecological implications of using biobased floor-stripping products. J Environ Health. 69(9):45–52, 76–7.

Mathis, T. May 2005. Lean Behavior-Based Safety – How the Process Is Evolving to Survive in Today's Economy, Occupational Hazards.

Matson, J.V. Effective Expert Witnessing, Chelsea, MI: Lewis Publishers, Inc., 1990.

Mayo, E. The Human Problems of Industrial Civilization. New York: Macmillan, 1933.

McAfee, R.B. and A.R. Winn. 1989. The Use of Incentives/Feedback to Enhance Work Place Safety: A Critique of the Literature. J Saf Res 20:7–19.

McAtamney, L. and E.N. Corlett. 1992. Ergonomic Workplace Assessment in a Health Care Context Ergonomics 35(9):965–978.

Mcintosh, I.B.D., C.B. Morgan, and C.E. Dorgan. ASHRAE Laboratory Design Guide. Atlanta: American Society of Heating, Refrigerating and Air-Conditioning Engineers, 2001.

Mclenon, M. 2004. Use of a Specialized Transport Team for Intrahospital Transport of Critically Ill Patients. Dimens Crit Care Nurs 23(5):225–9.

Medication Reconciliation Handbook, Oakbrook Terrace, IL, Joint Commission Resources and the American Society of Health-System Pharmacists, 2006.

Meredith, J.R. and S.J. Mantel, Jr. Project Management: A Managerial Approach, 5th ed, Hoboken, NJ: John Wiley & Sons, 2002.

Meyer, M.W. Theory of Organizational Structure, Indianapolis: Bobbs-Merrill, 1977.

Miller, J.M., M.R. Lehto, and J.P. Frantz. Warnings and Safety Instructions: Annotated and Indexed, Ann Arbor, MI: Fuller Technical Publications, 1994.

Mintzberg, H. The Nature of Managerial Work, New York: Harper & Row, 1973.

Mitchell, P.H. and N.M. Lang. 2004. Framing the Problem of Measuring and Improving Healthcare Quality: Has the Quality Health Outcomes Model Been Useful? Med Care 42:II4–11.

Model Plans for Bloodborne Pathogens and Hazard Communications, OSHA Pub 3186.

Mooney, J. and A. Reiley. Onward Industry, New York: Harper & Row, 1931.

Moran, G. Biological Terrorism Part 1: Are We Prepared? Emergency Medicine, February 2000.

Morath, J. and J. Turnbull. To Do No Harm. San Francisco: Jossey Bass, 2005.

Mueller, S., K. Sponsler, S. Kripalani, and J. Schnipper. 2012. Hospital-Based Medication Reconciliation Practices: A Systematic Review, Arch Intern Med. 172:1057–1069.

NAS, Reliability Centered Maintenance Guide for Facilities and Collateral Equipment, February 2000.

National Council on Radiation Protection. Report Number 48: Radiation Protection for Medical and Allied Health Personnel. Washington, DC, 1976.

National Institute for Occupational Safety and Health (NIOSH), 1997. Musculoskeletal Disorders and Workplace Factors–A Critical Review of Epidemiologic Evidence for Work-Related Musculoskeletal Disorders of the Neck, Upper Extremity, and Low Back.

National Institute of Occupational Safety and Health, U.S. Department of Health and Human Services, Washington, DC: A Guide to Safety in Confined Spaces, NIOSH Publication 87–113. 1987.

National Quality Forum, National Consensus Standards for Nursing-Sensitive Care: An Initial Performance Measure Set. Washington, DC: National Quality Forum, 2004.

National Research Council and Institute Of Medicine, Musculoskeletal Disorders and the Workplace – Low Back and Upper Extremities, National Academy Press, 2001.

National Safety Council, Accident Prevention Manual for Business and Industry, Administration and Programs, 10th Edition, 1992.

Nestor, D. Hospital Bed Design and Operation Effect on Incidence of Low Back Injury among Nursing Personnel. In Hazadeh, F., ed., Trends in Ergonomics/Human Factors V, North-Holland: Elsevier Science Publishers, B. V., 1988.

No Routine Surface Disinfection, American Journal of Infection Control 32(8): 513–515 (December 2004).

Noe, R.A., et al. Human Resource Management: Gaining a Competitive Advantage. 5th ed., Boston: McGraw-Hill, 2006.

Noise and Hearing Conservation Manual, American Industrial Hygiene Association, 1986.

Noji, E. and G. Kelen. Manual of Toxicologic Emergencies. Chicago: Year Book Medical Publishers, 1989.

Noll, G, M. Hildebrand, and J. Yvorra. Hazardous Materials: Managing the Incident, Fire Protection Publications, Oklahoma State University, 1988.

Nursing Homes Safety and Health Training Course, OSHA Office of Training and Education, Oakland, CA.

Occupational Health and Safety, 2nd edition, Chicago, IL: National Safety Council, 1994.

Occupational Safety and Health Administration, U.S. Department of Labor, Washington, DC: Hospital and Nursing Home e-Tools, Online, May 2011.

Oden, H.W. Managing Corporate Culture, Innovation, and Intrapreneurship. Westport, CT: Quorum Books, 1997.

Organization for Economic Cooperation and Development. Behavioral Adaptation to Changes in the Road Transport System, Paris: OECD, 1990.

Orr, G. Identifying Risk Factors for Work Related Musculoskeletal Disorders in Nursing Homes, (OSHA Directorate of Technical Support), Nursing Homes Safety and Health Training Course, OSHA Office of Training and Education, USDL, July 1996.

Ouchi, W.G. Theory Z: How American Business Can Meet the Japanese Challenge, Reading, MA: Addison-Wesley Publishing, 1982.

Overview of Recent Back Support Studies: Evidence Favorable to the Use of Elastic Lumbar Back Supports in Industry, Chase Ergonomics: 1998 Edition.

Owen, B., Garg, A., Back Stress Isn't Part of the Job, American Journal of Nursing, 48–51, February 1993.

Oxenrider, J. Creative Root Cause Analysis. Middlebury, IN: Center for Creative Teamwork, 1998.

Panico, C.R. 2004. Culture's Competitive Advantage, Global Cosmetic Industry 172(12) 58–60.

Pascale, R. Managing on the Edge, New York: Simon & Schuster, 1990.

Pesanka, D.A., P.K. Greenhouse , L.L. Rack, et al. 2008. Ticket to Ride: Reducing Handoff Risk during Hospital Patient Transport. J Nurs Care Qual 26:1–7.

Petersen, D. Safety Management: A Human Approach, 2nd edition. Goshen, NY: Aloray, 1988.

Pfaff, B.L. Emergen cy Department Management of Nerve Agent Exposure, International Journal of Trauma Nursing, July–September, 1998.

Poon, E. Medication Reconciliation: Whose Job Is It? AHRQ Web M&M, September 2007.

Pope, W. Managing for Performance Perfection. Weaverville, NC: Bonnie Brae Publications, 1990.

Porter, E. Manpower Development: The System Training Concept, New York: Harper and Row, 1964.

Practical Guide to the Determination of Human Exposure to Radiofrequency Fields, NCRP Report No. 119, National Council on Radiation Protection and Measurements, 1993.

Preventing Bed Falls. Nursing Update, 4(1), Spring 1993. Arcadia, CA: J.T. Posey Company.

Preventing Work-Related Musculoskeletal Disorders in Sonography, NIOSH Publication No. 2006–148.

Principal Emergency Response and Preparedness Requirements, OSHA Pub 3122, 2004.

Proceedings from the Economic Evaluation of Health and Safety Interventions at the Company Level, Journal of Safety Research, 36(3), 2005.

Project (SHP), Lowell Center for Sustainable Production,

Proper Patient Positioning in Wheelchairs, Nursing Update, 5(1), Winter 1994. Arcadia, CA: J.T. Posey Company.

Protect Yourself Against Tuberculosis – A Respiratory Protection Guide for Health Care Workers, DHHS (NIOSH) Publication No. 96–102. December 1995.

Pugliese, G. Green Link: News and Success Stories about Green Purchasing and Environmental Health. January 2006, available at https://www.premierinc.com/quality-safety/tools-services/safety/topics/epp/downloads/green-link-01-web.pdf

Purchasers' Group 'Leapfrogs' to Quality, Healthcare Benchmarks 8(4):44–5, 2001.

Rasmussen, J. 1983. Skills, Rules, and Knowledge: Signals, Signs, and Symbols, and Other Distinctions in Human Performance Models, IEEE T Syst Man Cyb 13(3):257–266.

Reason, J. Human Error, New York: Cambridge University Press, 1990.

Reason, J. Managing the Risks of Organizational Accidents, Hampshire, England: Ashgate, 1997.

Reason, J. Human Error: Models and Management. BMJ. 2000, 320:768–770, http://www.bmj.com/content/320/7237/768.

Redley B. and M. Botti. 2013. Reported Medication Errors after Introducing an Electronic Medication Management System, Clinical Nursing 22:579–589.

Reinhardt, P. and J. Gordon Infectious Waste Management. Chelsea, MI: Lewis Publishers, 1991.

Reopening Shuttered Hospitals to Expand Surge Capacity, Rockville, MD.: U.S. Dept. of Health and Human Services, http://archive.ahrq.gov/research/shuttered/.

Rhame, F.S. The inanimate environment. In J.V. Bennett and P.S. Brachmann, eds. Hospital infections. 4th ed. Philadelphia: Lippincott-Raven, 1998, pp. 299–324.

Richmond, J.Y. HIV biosafety: Guidelines and regulations. In G. Schochetman and J.R. George, eds. AIDS testing. 2nd ed. New York: Springer-Verlag, 1994, pp. 346–60.

Richmond, J.Y. and S.L. Nesby-O'Dell. 2002 Laboratory security and emergency response guidance for laboratories working with select agents. MMWR Recomm Rep. 51:(RR-19):1–6.

Riser, D.T., M.M. Rice, M.L. Salisbury, et al. 1999. Potential for Improved Teamwork to Reduce Medical Errors in the Emergency Department. Emergency Med. 34(3):373–383.

Robbins, S.P. and M. Coulter. Management. Upper Saddle River, NJ: Prentice Hall, 1999.

Robertson, L.S. Injury Epidemiology. New York: Oxford University Press, 1992.

Robustelli, P. and J. Kullmann, Implementing Six Sigma to Affect Lasting Change. Qualtec, 2006.

Romano, M. Back breaks, Contemporary Long Term Care, 45–51, February 1996.

Ross, J.E. Total Quality Management. Delray Beach, FL: St. Lucie Press, 1995.

Ross, T., Ed. Surgical Technology for the Surgical Technologist: A Positive Care Approach, Clifton Park, NY: Delmar, 2008.

Rubin, C., T. Renda, and W. Cumming, Disaster Timeline Series, http://www.disastertimeline.com.

Russell, N.D., M.G. Hudgens, Ha, R. Havenar-Daughton, C. McElrath, M.J. Moving to human immunodeficiency virus type 1 vaccine efficacy trials: defining T cell responses as potential correlates of immunity. J Infect Dis. 2003;187:226–242. [PubMed].

Rutala, W.A. 1990. APIC Guideline for Selection and Use of Disinfectants, American Journal for Infection Control 18:99–117.

Rutala, W.A. (ed.). Chemicals Germicides in Health Care, Polyscience Publications, Morin Heights, Quebec, Canada, 1994.

Rutala, W.A., M.F. Gergen, D.J. Weber. 2007 Microbiologic evaluation of microfiber mops for surface disinfection. Am J Infect Control. 35(9):569–73.

Salthammer, T., ed. Organic Indoor Air Pollutants: Occurrence, Measurement, Evaluation Germany: Wiley-VCH, 1999.

Santell, J.P. and R.F. Kamalich. 1996. National survey of quality assurance activities for pharmacy-prepared sterile products in hospitals and home infusion facilities. Am J Health-Syst Pharm. 53:2591–603.

Schneier, Bruce, Beyond Fear: Thinking Sensibly about Security in an Uncertain World, New York: Copernicus Books, pp. 26–27.

Schein, E.H. Organizational Culture and Leadership: A Dynamic View. San Francisco: Jossey-Bass, 1995.

Schelp, L. 1988. The Role of Organizations in Community Participation – Prevention of Accidental Injuries in a Rural Swedish Municipality, Soc Sci Med 26(11):1087–1093.

Schneider, B., ed., Organizational Climate and Culture, San Francisco: Jossey-Bass, 1990.

School of Law, Code of Medical Ethics, Annotated Current Opinions. Chicago, IL: AMA, 1994.

Scott, W.R. Organizations: Rational, Natural, and Open Systems. Englewood Cliffs, NJ: Prentice-Hall, 1981

Screening and Surveillance: A Guide to OSHA Standards, OSHA Pub 3162.

Selection of Chemical Protective Clothing, Third Edition, American Conference of Governmental Industrial Hygienists, Cincinnati, OH, 1988.

Senge, P. The Art & Practice of the Learning Organization (1990). In M. Ray and A. Rinzer, eds., The New Paradigm in Business: Emerging Strategies for Leadership and Organizational Change, World Business Academy, 1993.

Senge, P.M. The Fifth Discipline: The Art and Practice of Learning Organization. London: Century Business, 1993.

Shirley P.J. and J.F. Bion. 2004 Intra-Hospital Transport of Critically Ill Patients: Minimizing Risk. Intensive Care Med 30(8):1508–10.

Shojania, K.G., B.W. Duncan, K.M. McDonald et al., eds. Making Health Care Safer: A Critical Analysis of Patient Safety Practices. Evidence Report/Technology Assessment No. 43, AHRQ.

Siegel, L.C. and T.K. Gandhi. Hospital Admission due to High-Dose Methotrexate Drug Interaction, AHRQ Web M&M, January 2009.

Simonowitz, J.A. Health Care Workers and Workplace Violence, Occ Med: State Of The Art Reviews 11(2):277–291, 1996.

Small Business Handbook, OSHA Pub 2209, 2005.

Small Entity Compliance Guide for Respiratory Protection Standard, OSHA Pub 9071, 2011.

Smeltzer, LR. and J.L. Waltman, et al. Managerial Communication: A Strategic Approach, Needham, MA: Ginn Press, 1991.

Smith, G.S. and H. Falk. 1987. Unintentional Injuries, Am J. Prev Medicine 5, Sup:143–163.

Sotter, G. Stop Slip and Fall Accidents. Mission Viejo, CA: Sotter Engineering Company, 2000.

Spaulding, E.H. 1972. Chemical Disinfection and Antisepsis in the Hospital, J Hosp Res. 9:5–31.

Speir, R. Punishment in Accident Investigation, ASSE Professional Safety, August 1998.

Spengler, J.D. and J.M. Samet. Indoor Air Pollution: A Health Perspective. Baltimore, MD: Johns Hopkins University Press, 1991.

Standard for the Provision of Slip Resistance on Walking/Working Surfaces, ANSI/ASSE A1264.2–2006.

Standardizing a Patient Safety Taxonomy: A Consensus Report, Washington, DC: National Quality Forum, 2006.

Stanton, N., P. Salmon, G. Walker, C. Baber, and D. Jenkins, Human Factors Methods: A Practical Guide for Engineering and Design, Aldershot, England: Ashgate Publishing Limited, 2005.

Steers, R.M. and L.W. Porter. Motivation and Work Behavior, 5th Ed. New York: McGraw-Hill, 1991.

Supervisors Safety Manual, 8th edition, Chicago, IL: National Safety Council, 1993.

Survey Results: Community Liaison Programs to Decrease Hospital Readmissions, ISMP Medication Safety Alert! Acute Care Edition, March 7, 2013; 18:1–3.

Sweller, J. 1988. Cognitive Load during Problem Solving: Effects on Learning, Cognitive Science 12(1): 257–285.

Tam, V., S. Knowles, P. Cornish, N. Fine, R. Marchesano, and E. Etchells, 2005. Unintended Medication Discrepancies at the Time of Hospital Admission, CMAJ 173:510–515.

Taylor, F.W. The Principles of Scientific Management, New York: Harper, 1917.

The Compass: Management Practice Specialty News. Why You've Been Handed Responsibility for Safety, J.J. Keller and Associates, Winter 2000, pp. 1, 4.

The Public Health Security and Bioterrorism Preparedness and Response Act of 2002, http://www.fda.gov/oc /bioterrorism.

Thrall, J.H. 2006. Education and Cultural Development of the Health Care Workforce, Part II. Opportunities for Nonprofessional Workers, Radiology 240(2):311–4.

Threats, and Terrorism: Assessing the Science and Response Capabilities: Workshop Summary. Washington, DC: National Academy Press, 2002.

Training Requirements and Guidelines, OSHA Pub 2254, 1998.

Travers, K., and P.J. Kanki. In A. Aldovini and B.D. Walker, eds. Techniques in HIV Research. New York:. Stockton Press, 1990, pp. 6–10.

Turner, J. ed. Violence in the Medical Care Setting: A Survival Guide. Rockville, MD: Aspen Systems Corporation.

Tweedy, J. Leading Safety & Health: You Make It Happen, Helena, AL: TLC Services, 2000.

Tweedy, J. Healthcare Hazard Control and Safety Management, 3rd edition. Boca Raton, FL: CRC Press, 2014.

Vesley, D., J. Lauer, and R. Hawley. Decontamination, sterilization, disinfection, and antisepsis. In D.O. Fleming, and D.L. Hunt, eds. Laboratory Safety: Principles and Practices, 3rd ed. Washington, DC: ASM Press, 2001, pp. 383–402.

Vinas, T. Best Practices – Dupont: Safety Starts at the Top, Industryweek.Com, July 2002.

Vincoli, J. Basic Guide to System Safety, New York: Wiley & Sons, 2006.

Vojtecky, M.A., P. Harber, J.W. Sayre, E. Billet, and S. Shimozaki. 1987. The Use of Assistance While Lifting, Journal of Safety Research, 18: 49–56.

Von Bertalanffy, L., General System Theory: Foundations, Developments, Applications. New York: Braziller, 1968.

Vouros, G.A. 2003. Technological Issues towards Knowledge-Powered Organizations, Journal of Knowledge Management 7(2): 114–127.

Waganaar, W.A., P.T. Hudson, and J.T. Reason. 1990. Cognitive Failures and Accidents, Appl Cogn Psychol 4:273–294.

Wallerstein, N. and R. Baker. 1994. Labor Education Programs in Health And Safety, Occup Med State Art Rev 9(2):305–320.

Walters, H. Identifying & Removing Barriers to Safe Behaviors, ASSE Professional Safety, January 1998.

Wandelt, M.A. Definitions of Words Germane to Evaluation of Healthcare, NLN Pub 1976(15–1611):57–8.

Warren, J., R.E. Fromm Jr., R.A. Orr, et al. 2004. Guidelines for the Inter and Intrahospital Transport of Critically Ill Patients. Crit Care Med 32(1):256–62.

Weber, D.J. and W.A. Rutala. Occupational risks associated with the use of selected disinfectants and sterilants. In W.A. Rutala, ed. Disinfection, Sterilization and Antisepsis in Health Care. Champlain, NY: Polyscience Publications, 1998, pp. 211–26.

Weber, R. Medication Reconciliation Pitfalls, AHRQ Web M&M, February 2010.

Weeks, J.L. 1991. Occupational Health and Safety Regulation in The Coal Mining Industry: Public Health at The Workplace, Annu Rev Publ Health 12:195–207.

Weick, K.E. Sensemaking in Organizations, Thousand Oaks, CA: Sage, 1995.

Wickens, C. and J. Hollands. Engineering Psychology and Human Performance, New York: Prentice Hall, 1999.

Wickens, C., J. Lee, Y. Liu, and B. Gorden. An Introduction to Human Factors Engineering, 2nd ed., Prentice Hall, 1997.

Williams, J. Improving Safety Leadership, ASSE Professional Safety, April 2002.

Youngberg, B. and M. Halite, M., eds. The Patient Safety Handbook, Jones and Bartlett Publishers, Sudbury, MA, 2003.

Appendix 1
Nurses Safety Perception Survey

1. Did you receive adequate job-related training prior to assuming your current position?
2. Do supervisors discuss accidents, incidents, and injury events with involved workers?
3. Do supervisors enforce hazard control and safety rules fairly and correct unsafe behaviors?
4. Do supervisors take appropriate disciplinary action when work rules are not followed?
5. Do you perceive the major cause of accidents to be unsafe work conditions?
6. Does the organization actively promote and encourage employees to work safely?
7. Do you believe senior leaders view hazard control and safety as an important organizational priority?
8. Do supervisors seem more concerned with their personal records than accidents?
9. Would some type of "safety incentive" program motivate you to work more safely?
10. Does the organization conduct required hazard surveys thoroughly?
11. Do you feel that supervisors receive adequate safety and health training?
12. Do supervisors provide sufficient training on the proper selection and use of personal protective equipment?
13. Do you understand what a performance-based OSHA standard means?
14. Have you received any safety-related training since your orientation or follow-up training?
15. Does the organization keep records of safety inspections and identified hazards?
16. Do you feel that employees are influenced by the organizational hazard control efforts?
17. Does the organization provide you information about the costs, trends, types, and causes of accidents?
18. Do you feel the organization conducts accident investigations to assign blame?
19. Do you feel that the organization deals with problems caused by alcohol or substance abuse?
20. Does the organization conduct post-accident drug testing for all involved workers?
21. Do safety leadership personnel and supervisors conduct unscheduled hazard surveys and inspections?
22. Do you understand how the workers' compensation system works?
23. Does the organization provide special education and training for all shift workers?
24. Does the organization make safety a part of all job reviews or evaluations?
25. Do you think injured workers should participate in an "early return to work" initiative?
26. Is off-the-job safety an integral part of the overall hazard control and safety efforts?
27. Do supervisors report accidents promptly?
28. Do you wish to view the facility accident and injury record and compare it to similar facilities?
29. Do you feel your coworkers support the organization hazard control management efforts?
30. Do supervisors take hazard control and job safety seriously?
31. Does the organization take a proactive or reactive role when promoting hazard control?
32. Does the organization encourage supervisors to recognize those that work safely?
33. Do workers participate in the development of safe work practices and hazard control policies?
34. Do you feel workers play an important role in making hazard control and safety decisions?

35. Does senior management support supervisors when they make decisions affecting hazard control?
36. Do coworkers understand the relationship between their job tasks and safety?
37. Do you know where to access the written emergency action plan?
38. Do you know where to access the hazard communication program?
39. Do you feel that you received appropriate hazard control orientation and training before beginning this job?
40. Do you feel that the company has too many rules governing safety and health issues?
41. Do supervisors enforce hazard control and safety rules in the same manner as other job-related policies?
42. Does the organization set hazard control or safety-related goals and objectives?
43. Does senior management communicate goals to all workers in the organization?
44. What role do employees take in the goal-setting process?
45. Who serves as the key person in the hazard control management function?
46. Do you feel the organization quickly evaluates hazards and takes appropriate actions?
47. Can supervisors reward workers for good safety performance?
48. Do you think alcohol and drugs increase the risk of an accident?
49. Do workers caution others about unsafe conditions and behaviors?
50. Can you initiate actions to correct an unsafe situation?
51. Do you know how to report unsafe conditions, hazards, or behaviors?
52. Do workers fear the threat of reprisal when reporting safety deficiencies?
53. Do you feel hazard control and safety issues receive the same priority as other organizational issues?
54. Does the organization value its good compliance record over other hazard control objectives?
55. Do supervisors model safe behaviors to their workers?
56. Do supervisors promote safety with statements such as, "This is management's idea"?
57. Do all employees receive an adequate safety orientation?
58. Do you feel the safety orientation program adequately prepares you to work safely?
59. Does senior management recognize safety-related work behaviors?
60. Do safety meetings directly impact safety performance on the job?
61. Do workers have the opportunity to attend safety meetings and training classes?
62. Do supervisors handle workers with personal problems in an effective manner?
63. Do you view your job as stressful?
64. Do you know your organizational or departmental hazard control and safety goals?
65. Do supervisors consistently require the use of personal protective equipment?
66. Can workers use alcohol or drugs on the job without detection?
67. Do supervisors sometimes overlook risks and hazards to get the job done?
68. Do hourly workers serve on the hazard control or safety committee?
69. Do you know the name of the organizational hazard control manager?
70. Does adherence with some established safety rules hinder job accomplishment?
71. Do you feel overworked?
72. Do you feel pushed on the job?
73. Does the organization mandate overtime work?
74. Do you feel satisfied with your job?
75. Do you feel that you can achieve the goals of your job?
76. Does the organization recognize you for doing a good job?
77. Do superiors assign you responsibility without delegating authority?
78. Do you feel overwhelmed with too much responsibility?
79. Do you enjoy your job?
80. Does your immediate supervisor ask you for input?
81. Do you consider yourself loyal to the organization?

82. Do you feel any organizational loyalty directed at you?
83. Do you believe in the importance of teamwork in your work area?
84. Does the organization provide you with adequate job-related training?
85. Do superiors ever judge you by things beyond your control?
86. Do you consider job security an important issue?
87. Do organizational hazard control plans, policies, and procedures protect the employee?
88. Do you believe that accidents will just happen?
89. Can the organization truly prevent accidents?
90. Do workers feel free to discuss accident causal factors with investigators?
91. Do you feel that your job exposes you to more hazards than most other workers?
92. Do you understand the purpose of job safety analysis?
93. Do you understand your responsibilities during an emergency or disaster?
94. Do you have sufficient education and training to accomplish your job in a safe manner?
95. Do supervisors use safety practices and set an example for subordinates?
96. Do senior managers visit job areas and discuss the importance of working safely?
97. How would you classify hazard control efforts in your facility? Reactive _____
    Proactive _____
98. Do leaders and supervisors promote hazard control as the right thing to do or do they promote compliance?
99. Does the organization require supervisors to conduct job-related safety training?
100. Do you know the most hazardous substance used in the workplace?
101. Do you know the two safest egress routes from your work area to a safe place?
102. Does the organization conduct realistic emergency egress drills on a regular basis?
103. What hazard or safety issue causes you the most concern?

_____

_____

_____

104. How would you improve the effectiveness of hazard control management?

_____

_____

_____

105. What actions would you suggest for improving senior management involvement and employee participation in the organizational hazard control efforts?

_____

_____

Appendix 2
Safety Improvement Principles

Organizations can apply the principles of quality improvement to safety efforts in a number of ways.

Develop a Policy – The organization should publish a hazard control policy that outlines objectives and goals. Leaders must communicate the policy to all members of the organization. Require everyone to participate and support hazard control efforts.

Promote the Importance of Inspections – Members at every organizational level must understand the purpose of hazard control inspections, audits, and surveys.

Constantly Improve the Hazard Control Functions – Leaders, managers, and hazard control personnel must endorse the "philosophy" of continuous improvement.

Leadership and Management – Managers and supervisors must use effective management principles and leadership concepts to improve organizational efficiency. Top management must provide staff members with the tools and the time to pursue improvement ideas.

Promote Trust and Innovation – Provide a climate where organizational members at all levels can identify problems and make suggestions to improve operations. Trust creates an atmosphere that promotes innovation. Encourage coordination and open communication among all operational departments.

Eliminate Meaningless Slogans – Hazard control and safety messages must clearly define safe work practices, rules, or job procedures.

Promote Pride in Quality Work – Hazard control objectives must stress the importance of accomplishing jobs or tasks safely and correctly. Most individuals take pride in doing quality work. Provide organizational members with good equipment, a safe work area, and effective training.

Encourage Members to Work on Self-Improvement – Enhance organizational performance improvement efforts by affording everyone the opportunity to engage in educational and other personal development opportunities.

Facilitate Change – Managers and supervisors must change or modify their management style to benefit the improvement of the organization. Leaders must not only accept change, but promote the need for change when warranted.

Appendix 3
Accident Causal Factor Chart

Causal Factors	Identification of Factors	Possible Corrective Actions
Environmental: Unsafe procedure or process	Hazardous processes; management failed to adequately plan	Job hazard analysis or formulation of safe job practices
Defective, overused	Buildings, machines, or equipment worn, cracked, broken, or defective	Inspection; replacement; proper maintenance
Improperly guarded	Work areas, machines, or equipment that are unguarded or inadequately guarded	Inspection; check plans, blueprints, purchase orders, contracts, and materials; provide guards for existing hazards
Defective design	Failure to consider safety in design, construction, or installation of buildings, machinery, and equipment; too large, too small, or not strong enough	Unreliable source of supply; check plans, blueprints, purchase orders, contracts, and materials; provide guards for existing hazards
Unsafe dress or apparel	Management failed to provide/specify the use of goggles, respirators, safety shoes, hard hats, or safe apparel	Provide proper apparel or personal protective equipment; specify acceptable dress, apparel, or protective equipment
Unsafe housekeeping	Poor job area layout; lack of required equipment for good housekeeping—shelves, boxes, bins, aisle markers, etc.	Provide suitable layout and equipment necessary for good housekeeping
Improper ventilation	Poorly ventilated or unventilated work areas	Improve the ventilation
Improper illumination	Poorly illuminated or no illumination at all	Improve the illumination
Behavior related: Lack of knowledge or skill	Unaware of safe practice; unskilled; not properly instructed or trained	Job training
Improper attitude	Worker properly trained but failed to follow instructions due to one of the following: willful, reckless, absent-minded, emotional, or angry	Supervisor; discipline; behavior correction; human resources documentation
Physical deficiencies	Poor eyesight, defective hearing, heart trouble, hernia, etc.	Pre-placement physical examination; periodic physical examination; proper placement of employees; identification of workers with temporary bodily defects

Appendix 4
Ergonomic Symptoms Report

NAME _____ DEPARTMENT _____

JOB TITLE _____ SHIFT _____

Identify All Affected Areas

• **Forearm**	• **Wrist**	• **Knee**	• **Elbow**	• **Upper Back**
• **Lower Back**	• **Hand**	• **Fingers**	• **Ankle**	• **Foot**
• **Shoulder**	• **Thigh**	• **Lower Leg**	• **Neck**	

Check Terms That Best Describe Your Problem

• **Aching**	• **Numbness**	• **Tingling**	• **Loss of Color**
• **Burning**	• **Swelling**	• **Weakness**	• **Stiffness**
• **Cramping**	• **Other**		

1. List how and when the problem first occurred. _____
   _____

2. List the length and description of each episode. _____
   _____

3. How often did the problem occur during the past year? _____
   _____

4. What do you think caused the problem? _____
   _____

5. Describe the problem at its worst. _____
   _____

6. Describe any medical treatment for this problem._____
   _____

7. What medical treatment helps the problem? _____
   _____

8. Describe the frequency and type of treatment during the past year._____
   _____

9. How many workdays did you lose because of this problem during the past year? _____
   _____

10. Have you been placed in another job or a modified duty status because of the problem?

_____

11. What would help the problem?_____

_____

12. Do you work another job in addition to this one? Yes No

13. If you answered yes to Item 12 above, please describe the requirements of the job.

_____

14. List off-the-job activities, hobbies, or interests._____

_____

15. Describe any other jobs you accomplished for more than two weeks during the past year.

_____

16. Provide any other helpful information. _____

_____

_____   _____

Name/Signature   Date

Appendix 5
Sample Personal Protective Equipment Hazard Assessment Form

Department/Facility _____ **Date** _____

A. Eye and Face
  1) Airborne Particles_____
  2) Biohazards _____
  3) Hazardous Chemicals_____
  4) Caustic Substances _____
  5) Gas/Vapor Exposures _____
  6) Flammable Liquids _____
  7) Laser Hazards _____
  8) Other _____

B. Head
  1) Falling or Flying Objects _____
  2) Work Being Performed Overhead_____
  3) Electrical Hazards_____
  4) Construction Hazards _____
  5) Other Hazards_____

C. Other
  1) Lifting _____
  2) Repetitive Motion or Prolonged Standing _____
  3) Bloodborne Pathogens Exposures_____
  4) Infection Risks _____

D. Foot
  1) Falling and Rolling Objects _____
  2) Slip and Trip Hazards _____
  3) Electrical Hazards_____
  4) Wet or Slippery Surfaces _____
  5) Chemical Exposures _____
  6) Environmental_____
  7) Other _____

E. Hand
  1) Skin Absorption _____
  2) Cuts or Lacerations _____
  3) Abrasions _____
  4) Punctures _____

    5) Chemical Exposures _____

    6) Thermal Burns _____

    7) Harmful Temperature Extremes_____

    8) Machine Hazards _____

    9) Other _____

F. Respiratory Exposures

    1) Harmful Dusts _____

    2) Hazardous Drugs _____

    3) Fumes or Mists _____

    4) Smoke Hazards _____

    5) Vapors _____

    6) Biohazards _____

    7) Asbestos _____

    8) Other _____

G. Torso

    1) Hot Hazards _____

    2) Cuts _____

    3) Acid Exposures_____

    4) Ionizing Radiation _____

    5) Other _____

H. Comments

_____

_____

_____

_____

I. Certification

The assessment used the following methods to determine the existence of workplace hazards requiring personal protective equipment (PPE):

- Walk-through survey
- Specific job analysis
- Review of accident statistics
- Review of safety equipment selection guideline materials
- Selection of appropriate or required PPE

Assessment Conducted By _____ Date _____

Assessment Certified By _____ Date _____

Appendix 6
Workplace Violence Prevention Policy

A. POLICY STATEMENT

The safety and security of organization personnel, patients, and visitors is of vital importance. Threats, threatening behavior, or acts of violence against personnel, patients, visitors, or contractors will not be tolerated. It is the policy of the organization to provide a safe environment in order to conduct the mission of the organization in the most effective manner possible.

A.1. A safe environment will be attained by the following:

1. Appropriate employee screening
2. Employee education and training
3. Surveillance of the work area
4. Effective management of situations involving violence or threats of violence

A.2. This policy supports the written procedures set forth by the organization's safety and health workplace violence plan, which is incorporated into this policy by reference.

B. DEFINITIONS

B.1. Workplace: Any location, either permanent or temporary, wherein an employee performs any work-related duty. This includes, but is not limited to, the building or facility and the surrounding perimeters, including the parking lots, field locations, alternate work locations, and travel to and from work assignments.

B.2. Workplace violence: Any physical assault, threatening behavior, or verbal abuse by employees or third parties that occurs in the workplace. It includes, but is not limited to, beating; stabbing; suicide; attempted suicide; shooting; rape; psychological trauma such as threats, obscene phone calls, and intimidating presence; and harassment of any nature such as stalking, shouting, or swearing.

C. PROCEDURES

C.1. The organization will not tolerate the following conduct or behavior:

1. Threats, direct or implied
2. Physical conduct that results in harm to people or property
3. Possession of weapons on any property or workplace of the organization
4. Intimidating conduct or harassment that results in fear for personal safety

C.2. Inappropriate and threatening behaviors include, but are not limited to, the following:

1. Unwelcome name-calling, obscene language, and other verbally abusive behavior
2. Throwing objects, regardless of the type or whether a person is a target of a thrown object
3. Touching a person in an intimidating, malicious, or sexually harassing manner
4. Acts such as hitting, slapping, poking, kicking, pinching, grabbing, and pushing
5. Physical and intimidating acts such as obscene gestures, getting in your face, and fist shaking

D. REPORTING AND INVESTIGATING

D.1. Any employee who experiences, observes, or has knowledge of actual or threatened workplace intimidation or violence has a responsibility to report the situation as soon as possible.

1. In the case of an actual or imminent act or threat of violent behavior, immediately report the situation to senior leadership or organizational security
2. In all cases, the report should be immediately made to the employee's supervisor or department head, and to the compliance committee

D.2. All reports of workplace intimidation or violence will be investigated impartially, and as confidentially as possible.

D.3. Employees are required to cooperate in any investigations. A timely resolution of each report should be reached and communicated to all parties involved as soon as possible.

D.4. Any form of retaliation against employees for making a bona fide report concerning workplace intimidation or violence is prohibited, and must be immediately reported to the compliance committee.

E. REPORTING NON-WORK-RELATED VIOLENCE

E.1. Employees who are victims of domestic or non-work-related violence, or who believe they are potential victims of such violence, or who believe they are potential victims of such violence and fear it may enter the workplace, are encouraged to promptly notify their supervisor or department head. All such reports will be investigated as described above.

F. NONDISCIPLINARY AND DISCIPLINARY ACTION

F.1. Upon completion of an investigation, incidents will be reviewed before proceeding with nondisciplinary or disciplinary action, according to the provisions of the performance and corrective action policy.

F.2. Examples of actions that might be taken when an employee has been found to have violated the policy include, but are not limited to, the following:

1. Mandatory participation and counseling
2. Corrective action up to and including termination
3. Criminal arrest and prosecution
4. Initiation of a court order

F.2.a. Those who believe they are victims of intimidation or violence, whether workplace or non-work-related, should contact their supervisor or department head to inquire about available employee assistance and to obtain advice about dealing with the situation.

G. WEAPONS POLICY

G.1. The possession, carrying, or use of weapons on the organization's property is strictly prohibited. This includes firearms, edged weapons, illegal knives, martial arts weapons, clubs, and any device capable of projecting a ball, pellet, arrow, bullet, shell, or other similar devices. Violation of this policy is grounds for immediate termination. Refer to the firearms and other weapons policy, which is incorporated into this policy by reference.

Appendix 7
Bloodborne Training Requirements

Employers must be sure that all employees with occupational exposure participate in a training program that is provided at no cost to the employee and during working hours. Training shall be provided at the time of the initial assignment for tasks where occupational exposure may take place and at least annually thereafter. For employees who have received training on bloodborne pathogens in the year preceding the effective date of the standard, only training with respect to the provisions of the standard that were not included need be provided. Annual training for all employees must be provided within one year of their previous training. Employers will provide additional training when changes, such as modification of tasks or procedures or institution of new tasks or procedures, affect the employee's occupational exposure. The additional training may be limited to addressing the new exposures created. Material appropriate in content and vocabulary to educational level, literacy, and language of employees shall be used. The training program should contain at a minimum the following elements:

- An accessible copy of the regulatory text of this standard and an explanation of its contents
- A general explanation of the epidemiology and symptoms of bloodborne diseases
- An explanation of the modes of transmission of bloodborne pathogens
- An explanation of the employer's exposure control plan and the means by which the employee can obtain a copy of the written plan
- An explanation of the appropriate methods for recognizing tasks and other activities that may involve exposure to blood and other potentially infectious materials
- An explanation of the use and limitations of methods that will prevent or reduce exposure, including appropriate engineering controls, work practices, and personal protective equipment
- Information on the types, proper use, location, removal, handling, decontamination, and disposal of personal protective equipment
- An explanation of the basis for selection of personal protective equipment
- Information on the hepatitis B vaccine, including information on its efficacy, safety, method of administration, the benefits of being vaccinated, and that the vaccination will be offered free of charge
- Information on the appropriate actions to take and persons to contact in an emergency involving blood or other potentially infectious materials
- An explanation of the procedures to follow if an exposure incident occurs, including the method of reporting the incident and the medical follow-up that will be made available
- Information on the post-exposure evaluation and follow-up that the employer is required to provide for the employee following an exposure incident
- An explanation of the signs and labels or color coding required
- An opportunity for interactive questions and answers with the person conducting the training session

The person conducting the training should be knowledgeable in the subject matter covered by the training program as it relates to the workplace.

Sample Bloodborne Training Program
Learning Objective A
Participants will be able to

- Describe the Occupational Safety and Health Administration (OSHA) Bloodborne Pathogens standard
- State which employees are covered under the standard
- Describe what is required to meet the standard

Discuss the following:

- Provisions of the OSHA Bloodborne Pathogens standard
- Workers covered by the standard
- How to meet the OSHA standards
- Training program requirements

Additional training is necessary if new equipment is introduced to the work area. Standards include the *potential* for exposure, not just exposure. A written exposure control plan is necessary for the safety and health of workers. Participants should be able to

- Identify job classifications where there is exposure to blood or other potentially infectious materials.
- Explain the protective measures currently in effect in the acute-care facility and methods of compliance to be implemented, including hepatitis B vaccination and post-exposure follow-up procedures, how hazards are communicated to employees, personal protective equipment, housekeeping, and recordkeeping.
- Establish procedures for evaluating the circumstances of an exposure incident.

Learning Objective B
Participants will be able to explain the transmission, course, and effects of bloodborne pathogens by acquiring an understanding of the following:

- Definition of bloodborne pathogens
- Types of bloodborne pathogens
- Hepatitis B virus (HBV)—definition of, symptoms, course, and effects
- Human immunodeficiency virus (HIV)— definition of, symptoms, course, and effects
- Mode of transmission of bloodborne pathogens (including modes of transmission other than occupational transmission)

Employees should be reminded that

- HBV is more persistent than HIV.
- Nonintact skin makes employees vulnerable to infection.
- Clean work surfaces decrease the risk of infection.
- You cannot identify whether someone is HBV or HIV positive just by appearances.
- Due to the length of time before symptoms appear, employees may not be aware that they have contracted a disease.
- All exposures must be reported.
- Handwashing is important for preventing transmission of bloodborne pathogens.
- Work surfaces must be cleaned with an appropriate disinfectant.
- Personal protective equipment must be used as per the employer.

With regard to workplace transmission, employees should be made aware that HBV, HIV, and other pathogens may be present in

- Body fluids such as saliva, semen, vaginal secretions, cerebrospinal fluid, synovial fluid, pleural fluid, peritoneal fluid, pericardial fluid, amniotic fluid, and any other body fluids visibly contaminated with blood
- Saliva and blood contacted during dental procedures
- Unfixed tissue or organs other than intact skins from living or dead humans
- Organ cultures, culture media, or similar solutions
- Blood, organs, and tissues from experimental animals infected with HIV or HBV

Employees should learn that the means of transmission include

- An accidental injury by a sharp object contaminated with infectious material, such as
  - Needles
  - Scalpels
  - Broken glass
  - Exposed ends of dental wires
- Anything that can pierce, puncture, or cut your skin
- Open cuts, nicks, and skin abrasions, and even dermatitis and acne, as well as the mucous membranes of the mouth, eyes, or nose
- Indirect transmission, such as touching a contaminated object or surface and transferring the infectious material to the mouth, eyes, nose, or open skin

Learning Objective C
Participants will be able to describe the types of hepatitis B vaccines and their usage and contraindications and should be able to identify who should receive the vaccine. The following points should be addressed:

- Types of hepatitis B vaccines available.
- What workers should receive the Hepatitis B vaccine?
- When and how the vaccine is given:
  - Vaccine must be given to employee during work hours.
  - Vaccine is given in three doses according to a dose schedule.
  - Vaccine is given in the upper arm area (deltoid muscle).
- Vaccine is provided by the employer at no out-of-pocket expense to the employee.
- Employers cannot require employees to bill their insurance plans for the cost of the vaccine.
- Employers cannot require that employees stay at a specific job for a specific amount of time to receive the vaccine free of charge.
- No prescreening is required, and the employer cannot make this a requirement for receiving the vaccine.
- To whom the vaccine does not have to be offered:
  - Employees who have previously completed the hepatitis B vaccination series
  - When immunity is confirmed through antibody testing
  - When contraindications to receiving the vaccine exist
- Side effects of the vaccine.
- Dosage (10 g or 1.0 mL):
  - Dose 1—at target date
  - Dose 2—30 days later
  - Dose 3—6 months after first dose

Learning Objective D
The participants will be able to describe the controls that reduce exposure to bloodborne pathogens:

- Provide an overview of four prevention strategies: engineering controls, work practice controls, personal protective equipment (PPE), and universal precautions.
- Differentiate between engineering and work practice controls.
- Discuss handwashing and location of handwashing facilities.
- Go over work practice requirements.
- Discuss sharps, including how to handle contaminated sharps (disposable/reusable).
- Describe the PPE used in the facility and give examples.
- Emphasize that PPE must be readily available and sized appropriately.
- Make it clear that PPE must be provided and maintained at no cost to the employee.
- Advise employees what to do if
  - PPE becomes contaminated
  - Personal clothing becomes contaminated
- Discuss gloves and activities that may alter the integrity of gloves (e.g., cleaning with surfactants).
- Give examples of the limited exceptions to using PPE.
- Describe the concept of universal precautions.
- Describe acceptable containers.
- Describe biohazard containers.
- Describe surfaces in the facility that could be contaminated with blood or other potentially infectious materials.
- Explain how to clean work surfaces in the facility (e.g., what cleaners are used).
- Explain the cleaning schedule established in the facility.
- Explain how to clean up broken glass or a contaminated spill.
- Explain how to clean reusable sharps (if appropriate to the facility).
- Explain how laundry is handled in the facility.
- Caution employees to be careful when handling contaminated laundry.
- Advise employees to watch for hidden sharps.

Ask participants if they have any questions about this material.
Summary:

- Use puncture-resistant, leak-proof containers, color-coded red or labeled depending on the standard, to discard contaminated items such as needles, broken glass, scalpels, or other items that could cause a cut or puncture wound.
- Use puncture-resistant, leak-proof containers, color-coded red or labeled, to store contaminated reusable sharps until they are properly reprocessed.
- Store and process reusable contaminated sharps in a way that ensures safe handling; for example, use a mechanical device to retrieve used instruments from soaking pans in decontamination areas.
- Use puncture-resistant, leak-proof containers to collect, handle, process, store, transport, or ship blood specimens and potentially infectious materials.
- Label these specimens if they are shipped outside the facility.
- Labeling is not required when specimens are handled by employees trained to use universal precautions with all specimens and when these specimens are kept within the facility.
- Wash hands when gloves are removed and as soon as possible after contact with blood or other potentially infectious materials.
- Provide and make available a mechanism for immediate eye irrigation in the event of an exposure incident.

- Do not bend, recap, or remove contaminated needles unless required to do so by specific medical procedures or the employer can demonstrate that no alternative is feasible.
- In these instances, use mechanical means such as forceps or a one-handed technique to recap or remove contaminated needles.
- Do not shear or break contaminated needles.
- Discard contaminated needles and sharp instruments in puncture-resistant, leak-proof, red or biohazard-labeled containers that are readily accessible, maintained upright, and not allowed to be overfilled.
- Do not eat, drink, smoke, apply cosmetics, or handle contact lenses in areas of potential occupational exposure.
- Do not store food or drink in refrigerators or on shelves where blood or potentially infectious materials are present.
- Use red labels or affix biohazard labels to containers when storing, transporting, or shipping blood or other potentially infectious materials.

Universal precautions:

- Workers should protect their skin from contact with any body fluids.
- Workers should wear gloves when handling anything that contains these fluids.
- Workers should use such a barrier when handling blood, body fluids (e.g., urine, vomit), mucous membranes, or skin that is not attached.
- If the splatter of blood is anticipated, then workers should wear masks, protective eyewear and aprons.
- Workers should wash hands and other skin surfaces immediately and thoroughly if they become contaminated with blood or other body fluids.
- Workers should wash their hands thoroughly after removing protective gloves.
- All healthcare workers should be careful not to be injured by needles, scalpels, and other sharp instruments or devices during procedures.
- Workers should take special care when cleaning up after surgical procedures or disposing of any needles or instruments so other workers do not come into contact with these instruments.
- Workers should use puncture-resistant containers to collect and dispose of these objects.
- Workers should take care at all times when working near sharp instruments and the disposal containers.
- Workers can minimize the need for emergency mouth-to-mouth resuscitation by using other ventilation devices to perform these operations where necessary; this precaution is taken even though saliva has not been implicated in HIV transmission.
- Healthcare workers who have exudative lesions or weeping dermatitis should refrain from all direct patient care and handling patient equipment until the condition is resolved.
- Pregnant healthcare workers should be completely familiar with and particularly careful to observe the precautions to minimize the risk of HIV transmission to their infants.

Learning Objective E
Participants will be able to determine whether an exposure incident could occur within their worksite and describe steps to take after an exposure incident. Discussion should address the following topics:

- Local exposure control plan
- Defining an occupational exposure incident
- Information given to healthcare professionals
- Steps to minimize exposure
- Potential exposure incidents applicable to the work site

Sessions must be comprehensive in nature and include information on bloodborne pathogens as well as on OSHA regulations and the employer's exposure plan. The person conducting the training must be knowledgeable in the subject matter. The training program must accomplish the following:

- Explain the regulatory text and make a copy available.
- Explain the epidemiology and symptoms of bloodborne diseases.
- Explain the modes of transmission of bloodborne pathogens.
- Explain the employer's written exposure control plan.
- Describe the methods to control transmission of HBV and HIV.
- Explain how to recognize occupational exposure.
- Inform workers about the availability of free hepatitis B vaccinations, vaccine efficacy, safety, benefits, and administration.
- Explain the emergency procedures and reporting of exposure incidents.
- Inform workers of the post-exposure evaluation and follow-up available from healthcare professionals.
- Describe how to select, use, remove, handle, decontaminate, and dispose of personal protective clothing and equipment.
- Explain the use and limitations of safe work practices, engineering controls, and personal protective equipment.
- Explain the use of label, signs, and color coding required by the standard.
- Provide an interactive question and answer session on the training.

Appendix 8
Patient Handling Guidance

Guideline 1. Lifting and Lateral Transfers Lifting—Use upright, neutral working postures and proper body mechanics:

- Bend your legs, not your back. Use your legs to do the work.
- When lifting or moving people, always face them.
- Do not twist when turning. Pick up your feet and pivot your whole body in the direction of the move.
- Try to keep the person you are moving, equipment, and supplies close to the body. Keep handholds between your waist and shoulders.
- Move the person toward you, not away from you.
- Use slides and lateral transfers instead of manual lifting.
- Use a wide, balanced stance with one foot slightly ahead of the other.
- Lower the person slowly by bending your legs, not your back. Return to an erect position as soon as possible.
- Use smooth movements and do not jerk. When lifting with others, coordinate lifts by counting down and synchronizing the lift.

Lateral Transfers

- Position surfaces (e.g., bed and gurney, bed and cardiac chair) as close as possible to each other. Surfaces should be at approximately waist height, with the receiving surface slightly lower to take advantage of gravity.
- Lower the rails on both surfaces (e.g., beds and gurneys).
- Use draw sheets or incontinence pads in combination with friction-reducing devices (e.g., slide boards, slippery sheets, plastic bags, low-friction mattress covers, etc.).
- Get a good hand-hold by rolling up draw sheets and incontinence pads or use other assist equipment such as slippery sheets with handles.
- Kneel on the bed or gurney to avoid extended reaches and bending of the back.
- Have team members on both sides of the bed or other surfaces. Count down and synchronize the lift. Use a smooth, coordinated push-pull motion. Do not reach across the person you are moving.

Guideline 2. Ambulating, Repositioning, and Manipulating—When using gait or transfer belts with handles, observe the following:

- Keep the individual as close as possible.
- Avoid bending, reaching, or twisting your back when
  - Attaching or removing belts (e.g., raise or lower beds, bend at the knees)
  - Lowering the individual down
  - Assisting with ambulation
- Pivot with your feet to turn.

- Use a gentle rocking motion to take advantage of momentum.
- Stand/pivot type transfers are used for transferring an individual from bed to chair, for example, or to help an individual get up from a sitting position.
- Use transfer discs or other assists when available. If using a gait or transfer belt with handles, follow the above guidelines.
- Keep feet at least a shoulder width apart.
- If the patient or resident is on a bed, lower the bed so the individual can place his or her feet on the floor to stand.
- Place the receiving surface (e.g., wheelchairs) on the individual's strong side (e.g., for stroke or hemiparalysis conditions) so the individual can help in the transfer.
- Get the person closer to the edge of bed or chair and ask him or her to lean forward when trying to stand (if medically appropriate).
- Block the individual's weak leg with your legs or knees (this may place your leg in an awkward, unstable position; an alternative is to use a transfer belt with handles and straddle your legs around the weak leg of the patient or resident).
- Bend your legs, not your back.

Lifting or Moving Tasks with the Patient or Resident in Bed—Some common methods include scooting up or repositioning individuals using draw sheets and incontinence pads in combination with a log roll or other techniques:

- Adjust beds, gurneys, or other surfaces to waist height and as close to you as possible.
- Lower the rails on the bed, gurney, etc., and work on the side where the individual is closest.
- Place equipment or items close to you and at waist height.
- Get help and use teamwork.

Guideline 3. Transporting Patients, Residents, and Equipment—It is often necessary to transport people in gurneys, wheelchairs, or beds or handle various types of carts, monitors, instrument sets, and other medical equipment:

- Decrease the load or weight of carts, instrument trays, etc.
- Store items and equipment between waist and shoulder height.
- Use sliding motions or lateral transfers instead of lifting.
- Push, don't pull. Keep loads close to your body. Use an upright, neutral posture and push with your whole body, not just your arms.
- Move down the center of corridors to prevent collisions.
- Watch out for door handles and high thresholds that can cause abrupt stops.

Guideline 4. Performing Activities of Daily Living—Cramped showers, bathrooms, or other facilities in combination with poor work practices may cause providers to assume awkward positions or postures or use forceful exertions when performing activities of daily living (ADLs):

- Use upright, neutral working postures and proper body mechanics. Bend your legs, not your back.
- Eliminate bending, twisting and long reaches by
  - Using long-handled extension tools (e.g., handheld shower heads, wash and scrub brushes)
  - Wheeling people out of showers or bathrooms and turning them around to wash hard-to-reach places

- Use shower-toilet chairs that are high enough to fit over toilets. This eliminates additional transfers to and from wheelchairs, toilets, etc.
- Use shower carts or gurneys, bath boards, pelvic lift devices, bathtub and shower lifts, and other helpful equipment.
- When providing in-bed medical care or other services, follow the guidelines listed previously.

Guideline 5. Transferring from the Floor—When it is medically appropriate, use a mechanical assist device to lift people from the floor. If assist devices are not readily available or appropriate, you may have to perform a manual lift. When placing slings, blankets, draw sheets, or cots under the person, observe the following:

- Position at least two providers on each side of the person. Get additional help for large patients or residents.
- Bend at your knees, not your back. Do not twist.
- Roll the person onto his or her side without reaching across the person.
- If using hoists, lower the hoist enough to attach slings without strain.
- If manually lifting, kneel on one knee, and grasp the blanket, draw sheet, or cot.
- Count down and synchronize the lift. Perform a smooth lift with your legs as you stand up. Do not bend your back.

Guideline 6. Assisting in Surgery

- Use retractor rings instead of prolonged manual holding of retractors.
- Position operating tables or other surfaces at waist height.
- Stand on lifts or stools to reduce reaching.
- Frequently shift position or stretch during long operations.
- Avoid prolonged or repeated bending of the neck or the waist.
- Stand with one foot on a lift and frequently alternate feet to reduce pressure on the back.
- Reduce the number of instrument sets (trays) on a case cart.
- Store instrument sets (trays) in racks between the waist and shoulders.
- Use stands or fixtures to hold extremities.
- Get help from coworkers as needed to
  - Position legs or extremities in stirrups.
  - Move heavy carts, microscopes, monitors, alternate operating tables, equipment, or fixtures.

Appendix 9
Patient Safety Plan Development Considerations

Purpose—Patient safety plans must direct actions to improve healthcare safety and reduce risk to patients through a culture that encourages

- Recognizing and reporting risks to patient safety and medical errors
- Reviewing reported risks to identify causal factors and process changes necessary to improve systems
- Determining and initiating the best solutions for eliminating and reducing identified risks, hazards, or behaviors
- Internal reporting of identified risks and details of all corrective and improvement actions
- A continuing focus on all procedures, processes, and systems
- Minimizing individual blame or retribution for involvement in a medical or care error
- Analyzing selected healthcare services prior to an adverse event occurrence to identify system or processes that require redesign
- Ongoing organizational learning about medical and health care errors
- Sharing that knowledge and information to impact behavioral changes within all levels of the organizational structure

Approach—The patient safety plan must provide a systematic, coordinated, and continuous approach to all improvement efforts. The plan must encourage establishment of mechanisms to promote effective responses to actual events. The plan should also encourage an ongoing proactive reduction in medical, medication, and care errors by integrating patient safety objectives into all new and existing processes. Maintaining and improving patient safety must be coordinated and a collaborative effort. Achieving optimal results depends on the existence of a total safety culture. Patient safety must be an integral part of all departments and disciplines within the organization. It is important to develop plans, processes, and mechanisms for all activities identified by the interdisciplinary team.

Adverse Events—Adverse events are untoward incidents, therapeutic misadventures, iatrogenic injuries, or other adverse occurrences directly associated with care or services provided. Adverse events may result from acts of commission or omission. Adverse events can include patient falls, medication errors, procedural errors or complications, suicides, and missing patients. An adverse event can also be categorized as either a sentinel event or near miss.

Sentinel Event—A sentinel event is an unexpected occurrence involving death or serious physical or psychological injury or the risk thereof. Serious injury specifically includes loss of limb or function. The phrase "or the risk thereof" includes any process variation for which a recurrence would carry a significant chance of serious adverse outcome.

Near Misses—A near miss is an event or situation that could have resulted in an accident, injury or illness, but did not, either by chance or through timely intervention. Near misses are opportunities for learning and afford the chance to develop preventive strategies and actions. Near misses should receive the same level of scrutiny as adverse events that result in actual injury.

Root Cause Analysis—A root cause analysis (RCA) process can identify causal factors that feed adverse or near-miss events. Such an analysis focuses on improving systems and processes, not on individuals. A good analysis must go beneath the surface by asking probing questions regarding the how and why of the situation or incident. This approach permits the discovery of immediate and contributing causal factors. The analysis permits the team to identify problems and develop changes for systems and processes. Improvements through redesign or development of new processes will help reduce the risk of event recurrence. An RCA looks at human and other factors most directly associated with the event, as well as all causal factors related to the event. Analysis of the contributing causes helps identify risks and their potential contributions to the event. A credible RCA must include participation by the leadership of the organization, be internally consistent, and include consideration of relevant literature. The root cause team must have members that understand the event being analyzed.

Patient Safety Program Scope—The patient safety program must include the development of a continuous assessment process to prevent error occurrence. Event information from data and incident occurrence reports should be reviewed by the multidisciplinary team to prioritize organizational efforts. Sources of data could include incident reports, medication errors, adverse drug reactions, transfusion reactions, sentinel events, and other adverse events. The patient safety program encompasses the patient population, visitors, volunteers, physicians, and staff and addresses improvement issues in every department. Senior leadership must be responsible for ensuring full implementation of the program, with an emphasis on the following functions:

- Patient rights and assessment
- Patient and continuum of care
- Patient/family education
- Leadership and improvement of organization performance
- Information and human resource management
- Environment of care management and infection control

Methodology—The multidisciplinary team should represent leadership from throughout the organization and provide oversight for the patient safety program. All patient care and nonpatient care departments must report safety events and potential occurrences. The report will contain aggregated information related to the type of occurrence, severity of occurrence, number and type of occurrences per department, and impact on the patient. The team will

- Analyze this information to determine the need for further safety activities.
- Conduct a data review of all internal and external reports, including Joint Commission on Accreditation of Healthcare Organizations (JCAHO) sentinel event reports, ORYX and Core Measure performance data, and any reporting information from state and federal sources, including current literature.
- Select at least one high-risk safety process for proactive risk assessment annually.
- The proactive risk assessment requires the team to
  - Assess the intended and actual implementation of the process to identify the steps in the process where there is, or may be, undesirable variation.
  - Identify the possible effects of the undesirable variation on patients and how serious the possible effect on the patient could be.
  - For the most critical effects, conduct an RCA to determine why the undesirable variation leading to that effect may occur.

- Redesign the process or underlying systems to minimize the risk of that undesirable variation or to protect patients from the effects of that undesirable variation.
- Test and implement the redesigned process.
- Identify and implement measures of the effectiveness of the redesigned process.
- Implement a strategy for maintaining the effectiveness of the redesigned process over time.
- Describe the mechanisms necessary to ensure that all components of the healthcare organization are integrated into and participate in the organization-wide program.

Patient Care Provider Actions—In the event of a medical error, observe the following:

- Perform necessary healthcare interventions to protect and support the patient's clinical condition.
- Perform necessary healthcare interventions to contain the risk to others.
- Contact the patient's attending physician and other physicians, as appropriate, to report the error, carrying out any physician orders as necessary.
- Preserve any information related to the error, including physical evidence; preservation of information includes documenting the facts regarding the error on an adverse drug event or occurrence report.
- Report the medical error to the staff member's immediate supervisor.
- Submit an occurrence or adverse drug event report per organizational policy.

Reporting of Events—An effective patient safety program cannot exist without optimal reporting of medical or healthcare errors and occurrences. The organization should adopt a nonpunitive approach in its management of errors and occurrences. Personnel must be able to report suspected or identified medical or healthcare errors without the fear of reprisal in relationship to their employment. Organizations must support the concept that errors occur due to a breakdown in systems and processes. Improvement will be achieved by focusing on systems and processes rather than disciplining those involved in adverse events. Focusing on remedial actions and individual development assists rather than punishes organizational members.

Informing Patient and Family Members—The patient safety program should include an annual survey of patients, their families, and staff about their perceptions of risks to patients. The survey should solicit opinions and suggestions for improvement. It is important to make sure that patients, and when appropriate, their families, are informed about the outcomes of care; brief patients about unanticipated outcomes or when the outcomes differ significantly from the anticipated outcomes. When a healthcare error leads to injury, the patient and family should receive a truthful and compassionate explanation about the error, including remedies available to the patient. They should be informed that the factors involved in the injury are being investigated so steps can be taken to reduce the likelihood of similar injury to other patients. Staff should educate patients and their families about their role in helping to facilitate the safe delivery of care.

Appendix 10
Sample TB Exposure Control Plan

(Name) _____ maintains, reviews, and updates the Exposure Control Plan (ECP) at least annually, and whenever necessary to reflect new or modified tasks, procedures, and engineering controls that affect occupational exposure. The ECP is also updated to reflect new or revised employee positions with occupational exposure. This facility has had _____ cases of confirmed tuberculosis (TB) in the past 12 months. This facility is located in _____ County, which has reported _____ cases of TB in the last 12-month reporting period. The facility also considers the TB reported in the following surrounding counties that provide patients to this medical practice:
_____

These counties reported _____ cases of TB in the past 12 months.

TB OVERVIEW

Transmission of TB is a recognized risk in healthcare facilities and medical practices and clinics. Transmission is most likely to occur by contact with patients who have unrecognized pulmonary or laryngeal TB and are not on effective anti-tuberculosis therapy. Increases in TB in many areas are related to the high risk of TB among immune-compromised persons, particularly those infected with the human immunodeficiency virus (HIV). Healthcare and medical facilities should be alert to the need for preventing TB transmission in settings in which persons with HIV infection receive care or work. Tuberculosis is a bacterial disease caused by Mycobacterium tuberculosis. The bacteria are spread through the air by droplet nuclei when an infected person coughs, speaks, sneezes, or sings, or during procedures that cause droplets to be aerosolized. Droplet nuclei are one to five microns in size. Air currents can keep the nuclei airborne for long periods of time. Droplet nuclei must be inhaled and reach the alveoli to cause infection.

TB EXPOSURE CONTROL PLAN

This Tuberculosis Exposure Control Plan applies to all areas of this practice where exposure to pulmonary or laryngeal tuberculosis may occur. It is intended to prevent transmission of pulmonary Mycobacterium TB from infected individuals to susceptible hosts. All employees must comply with this plan. TB precautions are not necessary if the patient is on anti-tuberculosis medications (and compliant) and has no symptoms such as coughing, night sweats, weight loss, or fever. Person(s) responsible for this plan are listed below:

Person(s) responsible for implementation and evaluation of this plan and annual review of program with revisions as necessary: _____.

Person(s) responsible for employee TB skin testing, interpretation and follow-up procedures, and reporting skin test results: _____.

Person(s) responsible for limiting employee exposures, ensuring proper education and training of staff, and monitoring staff compliance with this plan: _____.

Person(s) responsible for properly evaluating suspected and confirmed tuberculosis in patients:
_____

EMPLOYER PROVISIONS

The employer will provide proper personal protective equipment (PPE), skin testing, education, and training at no cost to employees. Employees are responsible for the information contained in this plan and are expected to wear proper respiratory protection when directed and trained by the employer. Prevention of TB exposure will be based on adherence to the recommendations by the Centers for Disease Control and Prevention (CDC), and the Occupational Safety and Health Administration (OSHA). The employer will ensure employees use provided guidelines for patient assessment and evaluation of TB.

EMPLOYEE EXPOSURE DETERMINATION POTENTIAL

With respect to potential TB exposure, employees and students will be identified and placed into one of two categories:

Category 1: Job classifications in which employees have occupational exposure to tuberculosis, regardless of frequency.

Category 2: Job classifications in which the employees do not have occupational exposure to tuberculosis.

Note: The list of job classifications will be located in _____.

PATIENT ASSESSMENT FOR DETECTION OF TUBERCULOSIS

The first step in preventing exposure is to identify potentially infectious individuals. Triage of patients will include vigorous efforts to detect patients with active TB promptly and to minimize the time spent in contact with other patients, visitors, and employees and students.

Signs/Symptoms

- Persistent cough (greater than 3 weeks' duration)
- Coughing up blood
- Weight loss
- Loss of appetite
- Lethargy/weakness
- Night sweats
- Fever

Medical/Social History

- Recent immigrant (especially high risk, e.g., Asia, Africa, or Latin America)
- Known immune-suppressed
- Known previous positive purified protein derivative (PPD) and/or chest x-ray
- Resides in shelter, prison, or long-term care facility
- Known exposure to TB
- Known history of TB, did not complete therapy
- Alcohol or drug use

EXPOSURE INCIDENT REPORTING

All employees must report exposure incidents immediately to the responsible person(s). _____ is responsible for investigating, evaluating, and documenting the circumstances surrounding the exposure incident and for instituting changes to prevent similar occurrences.

PROCEDURES TO ISOLATE AND MANAGE CARE

Establish procedures to isolate individuals with suspected or confirmed infectious TB. All individuals with suspected or confirmed infectious TB are placed in a room away from other individuals. High-hazard procedures (where TB may be aerosolized) require precautions to prevent/minimize occupational exposure to infectious TB. The following high-hazard procedures are performed at this facility: _____. The policy of this organization is to transfer TB infected or suspected patients to a treatment or isolation facility. While awaiting transfer, the individual is masked or segregated to protect employees who are without respiratory protection. The organization uses the following procedures and equipment for masking an individual with suspected/confirmed infectious TB and isolating the patient from others until the patient can be transported to an acceptable care or treatment facility: _____.

TRANSPORTING OF PATIENTS WITH TUBERCULOSIS

The patient with suspected or confirmed infectious TB will wear a mask (surgical mask) if transported or sent to another department outside of the exam room (e.g., lab, hospital, rehab, prison).

EMPLOYEE NOTIFICATION OF TB RISKS

The organization uses the local risk assessment process to determine the potential for TB exposure. The organization assesses the medical procedures performed at the practice to ensure that all employees with job tasks that offer potential for occupational exposure are informed of the hazard and take proper precautions against exposure to TB.

WARNING SIGNS

Signs are posted at the entrance to rooms or areas used to isolate an individual with suspected or confirmed infectious TB. Signs also required in areas where procedures or services are being performed on an individual with suspected/confirmed infectious TB. The sign must include a signal word (e.g. "STOP," "HALT," or "NO ADMITTANCE") or biological hazard symbol and a descriptive message (e.g., "Respiratory Isolation, No Admittance Without Wearing a Type N95 or More Protective Respirator," or "See nurses' station before entering this room") [1910.145(f)(4)] - Specifications for Accident Prevention Signs and Tags.

PERSONS RESPONSIBLE FOR TB CONTROLS

Person(s) responsible for identifying and making available respiratory controls (masks for patients and respirators for exposed employees): _____.

Person(s) responsible for administrative work practice controls (identification, triage, and isolation) to reduce atmospheric contamination: _____.

ENGINEERING CONTROLS

Whenever possible, place a surgical or procedural mask on the patient. If the patient wears a mask, employees need not wear a mask. After placing the patient in an exam room that is away from other individuals, the door will remain closed at all times. Employees will wear respiratory protection (N-95 respirators) at all times while in the room if the patient cannot or will not wear a mask properly and at any time that there is potential for exposure.

WORK PRACTICES

Nonclinical support personnel will be excluded from any contact with known or suspected TB-infected patients. Prompt identification of patients with history or symptoms of TB will be done. Tissues and instructions for the patient to cough and sneeze into tissues and discard them in a waste container will be provided. The patient will be asked to wear a mask (regular surgical) until he/she is placed in a room with door closed. If the patient cannot wear a mask, employees/students who interview or examine a patient with confirmed or suspected TB will wear a mask (N-95) at all times while in contact with patient or in the exam room with patient. Patients with active TB who must come into the clinic will have appointments scheduled to avoid exposing others. Designated times of the day for TB appointments will be observed to avoid interaction with immune-compromised individuals. OSHA and the CDC have defined the following as "high-hazard procedures" when performed on patients with known TB or those who are at high risk of having infectious TB. These procedures are likely to produce bursts of aerosolized infectious particles or to result in copious coughing or sputum production. Other high-risk situations include (1) aerosolized medication treatment (including pentamidine) and (2) diagnostic sputum induction.

RESPIRATORY CONTROLS

Respirators will be provided to exposed employees at no cost. Respirators must be at a minimum a NIOSH-approved N-95. Employees will wear a respirator in the following circumstances: (1) when entering a room occupied by a known or suspected infectious tuberculosis patient or (2) while performing high-hazard procedures (if any). Employees and students will be instructed in the correct use and limitations of the N-95 mask in the prevention of exposure to TB. Employees and students are accountable to evaluate the mask for proper fit and condition at each usage. OSHA requirements state that a NIOSH-approved respirator (N-95 is NIOSH approved) is the minimum acceptable level of protection for staff exposed to TB. Respirators are issued and fit tested according to the provisions of OSHA Standard 29 CFR 1910.134. Responsible person(s) for respirator fit testing and program management: _____.

MEDICAL SCREENING OF EMPLOYEES

At the time of hire, employees, including those with a history of taking the Bacillus Calmettee–Guérin (BCG) vaccination, will receive a two-step PPD to reduce the likelihood of interpreting a boosted reaction as representing a recent infection. Individuals who have a history of a positive PPD reaction, documentation of completion of preventive therapy, or documentation of adequate therapy for active TB disease will not receive a skin test. Routine chest x-ray is NOT recommended for those who have tested positive in the past or have received treatment for active TB. Chest x-rays should not be used to screen for TB. Chest x-rays may be used detect active disease but may not detect latent infection that may be treated to prevent active disease. Employees will be tested annually thereafter. Employees/students with a positive history (+PPD or disease) will address the signs and symptoms portion of the PPD Skin Testing Form – Section III. Employees who have positive skin test results will be sent to the health department for further evaluation and treatment. There are no work restrictions for a positive skin test that is not accompanied by symptoms of the disease and/or a positive chest x-ray. Employees with active tuberculosis (chest x-ray and symptomatic) will be sent to _____ immediately for further evaluation and treatment. Work restriction is required. The healthcare provider must provide documentation that infection no longer exists before the employee can return to work. The cost of initial and annual testing and follow-up of positive PPDs is the responsibility of the employer.

EXPOSURE FOLLOW-UP PROCEDURES

Employees exposed to a confirmed or suspected source of TB will notify the employer and the following testing will be done:

- Baseline skin test
- Monitoring of the individual for development of symptoms of TB
- Follow-up skin test at 12 weeks

Note: If follow-up skin test converts to positive or if symptoms develop, the individual will be referred to _____ for follow-up and treatment. The PPD Skin Test Form will be utilized to record results.

EDUCATION AND TRAINING

Contents for initial training will include the basic concepts of TB transmission, pathogenesis, and diagnosis, including the difference between latent TB infection and active disease. Training will also address the signs and symptoms of TB and the possibility of reinfection and/or reactivation in persons with a positive TB skin test. Training must address potential for exposure to persons with infectious TB, including prevalence of TB, and the situations with increased risk of exposure to TB. Employees must be taught about the principles and practices of infection control that reduce the risk of transmission of TB, including the infection control measures and written policies/procedures. Site-specific education will be provided in the areas of controls, purpose of skin testing, the significance of positive skin testing results, and the importance of participating in the skin test program. Education includes basics of drug therapy for latent or active TB infection, indications, use, and effectiveness. Employees are educated on their responsibility to seek medical evaluation promptly if (1) symptoms develop that may be indicative of TB or (2) if skin test conversion occurs.

ANNUAL TB REQUIREMENTS

- Annual retraining including new knowledge, incidence statistics, and a forum for questions
- TB skin testing

SKIN TEST ADMINISTRATION

Employees must read and sign appropriate consent forms. Employer will obtain medical history PRIOR to testing. History will include signs and symptoms of TB and socioeconomic history. Consider the following:

- Hemoptysis, coughing >3 weeks, night sweats, unintentional weight loss, loss of appetite, malaise, weakness
- Close contact with anyone with active TB
- Nationality (foreign-born persons from areas where TB is endemic)
- Drug use
- Medical conditions that may cause immune-suppressed conditions (AIDS, cancer, prolonged steroid use, etc.)
- Medically underserved (migrant farm workers, homeless, etc.)
- Members of a high-risk racial group (e.g., Asians, Pacific Islanders, African Americans, Hispanics, and Native Americans)

Appendix 11
Model Respirator Plan for Small Organizations

GENERAL INFORMATION

The purpose of this Respirator Plan is to protect all employees from respiratory hazards through the effective use of respirators. The Respirator Plan Administrator (RPA) appointed at this location is _____. The employer has expressly authorized the RPA to audit and change respirator usage procedures whenever there is a chance of exposure to an air contaminant or airborne disease at the work site. This authority includes designating mandatory respirator usage areas and/or job-related tasks. The RPA is solely responsible for all aspects of this plan and has full authority to make decisions relevant to respirator usage. This authority includes training workers, purchasing the necessary equipment to implement the program, and developing local respiratory protection procedures.

LOCAL WRITTEN OPERATING PRACTICES

The RPA will develop written "operating practices" that detail specific instructions covering the basic elements in this plan. These local operating practices will be attached to the plan. These practices can only be amended by the RPA.

The RPA will develop detailed written standard operating procedures governing the selection and use of respirators, using the Occupational Safety and Health Administration (OSHA) standard and the National Institute of Occupational Safety and Health (NIOSH) Respirator Decision Logic as guidelines. Outside consultation, manufacturer assistance, and other recognized authorities will be consulted if there is any doubt regarding proper selection and use of respirators. **These detailed operating practices will be included as attachments to this respirator plan**.

RESPIRATOR SELECTION

Respirators will be selected on the basis of exposures. All selections will be made by the RPA and only NIOSH-certified respirators will be selected and used.

TRAINING AND USE

Respirator users will be instructed and trained in the proper use of respirators and their limitations. Both supervisors and workers will be trained by the RPA. The training should provide the employee an opportunity to handle the respirator, have it fitted properly, test its face piece-to-face seal, wear it in normal air during a familiarity period, and finally to wear it in a test atmosphere. Fit testing will be accomplished for all tight-fitting respirators used at this location.

FITTING INSTRUCTIONS

Every respirator wearer will receive fitting instructions, including demonstrations and practice in how the respirator should be worn, how to adjust it, and how to determine if it fits properly.

Respirators should not be worn when conditions prevent a good face seal. Such conditions may be a growth of beard, sideburns, a skull cap that projects under the face piece, or temple pieces on glasses. No employees of this facility who are required to wear tight fitting respirators may wear beards. Also, the absence of one or both dentures can seriously affect the fit of a face piece. The worker's diligence in observing these factors will be evaluated by periodic checks. To assure proper protection, the user seal check will be done by the wearer each time she/he puts on the respirator. The manufacturer's instructions will be followed.

ASSIGNMENT OF RESPIRATORS

When practicable, the respirators will be assigned to individual workers for their exclusive use. Nondisposable respirators will be regularly cleaned and disinfected. Those issued for the exclusive use of one worker will be cleaned after each day's use, or more often if necessary. Those used by more than one worker will be thoroughly cleaned and disinfected after each use. The RPA will establish a respirator cleaning and maintenance facility and develop detailed written cleaning instructions. Disposable respirators will be discarded if they are soiled or are no longer functional. For additional information, refer to the manufacturer's instructions. Respirators used routinely will be inspected during cleaning. Worn or deteriorated parts will be replaced.

EMPLOYEE SCREENING

Persons will not be assigned to tasks requiring use of respirators unless it has been determined that they are physically able to perform the work and use the respirator. The employer will designate a healthcare professional to determine and assess worker health and physical ability to wear a respirator. The respirator user's medical status will be reviewed annually.

EVALUATIONS

The employer will ensure that an annual inspection/evaluation of the program is conducted to determine the continued effectiveness of the program. The RPA will make frequent inspections of all areas where respirators are used to ensure compliance with the respiratory protection requirements.

EVALUATION CHECKLIST

In general, the respiratory protection plan should be evaluated for each job, or at least annually, with program adjustments, as appropriate, made to reflect the evaluation results. Functions can be separated into administration and operation.

ADMINISTRATION

1. Is there a written policy that acknowledges employer responsibility for providing a safe and healthful workplace, and assigns program responsibility, accountability, and authority?
2. Is program responsibility vested in one individual who is knowledgeable and who can coordinate all aspects of the program at the healthcare facility?
3. Can administrative and engineering controls eliminate the need for respirators?
4. Are there written procedures/statements covering the following topics?
   _____ a. Designation of an administrator
   _____ b. Respirator selection
   _____ c. Purchase of NIOSH-certified respirators
   _____ d. Medical aspects of respirator usage
   _____ e. Issuance of equipment

_____ f. Fitting
_____ g. Training
_____ h. Maintenance, storage, and repair
_____ i. Inspection
_____ j. Use under special conditions
_____ k. Work area surveillance

OPERATION

5. Equipment selection
   a. Are work area conditions and worker exposures properly surveyed?
   b. Are respirators selected on the basis of the hazard of exposure?
   c. Are selections made by individuals knowledgeable in selection procedures?
6. Are only NIOSH-certified respirators purchased and used? Do they provide adequate protection for the specific hazard?
7. Has a medical evaluation of the prospective user been made to determine physical and psychological ability to wear the selected respiratory protective equipment?
8. Where practical, have respirators been issued to the users for their exclusive use, and are there records covering issuance?
9. Respiratory protective equipment fitting
   a. Are the users given the opportunity to try on several respirators to determine whether the respirator they will be subsequently wearing is the best fitting one?
   b. Is the fit tested at appropriate intervals?
   c. Are those users who require corrective lenses properly fitted?
   d Is the face piece tested in a test atmosphere?
   e. Are workers prohibited from wearing respirators in contaminated work areas when they have facial hair or other characteristics that may cause face seal leakage?
10. Respirator use in the work area
    a. Are respirators being worn correctly (i.e., head covering over respirator straps)?
    b. Are workers keeping respirators on all the time while in the designated areas?
11. Maintenance of respiratory protective equipment
    a. Are nondisposable respirators cleaned and disinfected after each use when different people use the same device, or as frequently as necessary for devices issued to individual users?
    b. Are proper methods of cleaning and disinfecting utilized?
    c. Are respirators stored in a manner so as to protect them from dust, sunlight, heat, damaging chemicals, or excessive cold or moisture?
    d. Are respirators stored in a storage facility so as to prevent them from deforming?
    e. Is storage in lockers permitted only if the respirator is in a carrying case or carton?
12. Inspection
    a. Are respirators inspected before and after each use and during cleaning?
    b. Are qualified individuals/users instructed in inspection techniques?
    c. Are records kept of the inspection of respiratory protective equipment?
13. Repair
    a. Are replacement parts used in repair those of the manufacturer of the respirator?
    b. Are repairs made by trained individuals?
14. Training and feedback
    a. Are users trained in proper respirator use, cleaning, and inspection?
    b. Are users trained in the basis for selection of respirators?
    c. Are users evaluated, using competency-based evaluation, before and after training?
    d. Are users periodically consulted about program issues such as discomfort, fatigue, etc.?

SAMPLE RESPIRATOR INSPECTION RECORD

TYPE_____

NO._____ DATE: _____

    A. Face Piece
    B. Inhalation Valve
    C. Exhalation Valve Assembly
    D. Headbands/Straps
    E. Filter Cartridge
    F. Cartridge/Canister
    G. Harness Assembly
    H. Hose Assembly
    I. Speaking Diaphragm
    J. Gaskets
    K. Connections
    L. Other Defects

DEFECTS FOUND:
CORRECTIVE ACTION:

MEDICALLY SCREEN ALL USERS

Conduct a medical evaluation of workers to determine fitness to wear respirators. The use of respirators can place several physiological stresses on wearers—stresses that particularly involve the pulmonary and cardiac systems. However, respirators typically used by healthcare workers are generally lightweight, and the physiological stresses they create are usually small. Therefore, most workers can safely wear respirators. Current OSHA regulations (29 CFR 1910.139) state that workers should not be assigned tasks requiring respirators unless they have been determined to be physically able to perform the work while using the equipment. The regulations also note that a physician should determine the criteria on which to base this determination.

No general consensus exists about what elements to include in medical evaluations for respirator use in general industry. Some institutions use only a questionnaire as a screening tool; others routinely include a physical examination and spirometry; and some include a chest x-ray. No generally accepted criteria exist for excluding workers from wearing respirators. Specifically, no spirometric criteria exist for exclusion. However, several studies have shown that most workers with mild pulmonary function impairment can safely wear respirators. There are some restrictions, such as the type of respirator or workload, for those with moderate impairment. There should be no respirator wear for individuals with severe impairment. Some respirators have a latex component and should not be worn by those who are allergic to latex.

Because most healthcare workers wear the very light, disposable half-mask respirator, recommend that a health questionnaire be the initial step in the evaluation process. Refer to OSHA 29 CFR 1910.134 paragraph "e" for guidance on medical evaluation. Appendix B of the Standard contains a sample medical questionnaire. If results from this evaluation are essentially normal, the employee can be cleared for respirator wear. Further evaluation, possibly including a directed physical examination and/or spirometry, should be considered in cases in which potential problems are suggested on the basis of the questionnaire results.

Appendix 12
Glossary of Terms

A

Abatement: the process of minimizing public health dangers and nuisances, usually supported by regulation or legislation, i.e., noise abatement, smoke abatement

Absorbed dose: the amount of energy deposited by ionizing radiation in a unit mass of tissue

Absorption: transformation of radiant energy into a different form of energy by the interaction of matter, depending on temperature and wavelength or the process by which a liquid penetrates the solid structure of absorbents, fibers, or particles or the process of an agent being taken in by a surface much like a sponge and water

Access control: a method of restricting the movement of persons into or within a protected area by manual (guards), hardware (locks and keys), or software (electronic card or biometric readers)

Accessible: an ADA term about having the legally required features and/or qualities that ensure entrance, participation, and usability of places, programs, services, and activities by individuals with a wide variety of disabilities

Accessible emission level: the magnitude of accessible laser or collateral radiation of a specific wavelength or emission duration at a particular point as measured by appropriate methods and devices or the radiation to which human access is possible in accordance with the definitions of the laser's hazard classification

Accessible emission limit: the maximum accessible emission level permitted within a particular laser class

Accident type: the description and classification of a mishap

Acid: a compound either inorganic or organic that (1) reacts with a metal to evolve hydrogen, (2) reacts with a base to form a salt, (3) dissociates in water solution to yield hydrogen ions, (4) has a pH of less than seven, and (5) neutralizes bases or alkaline media by receiving a pair of electrons from the base so that a covalent bond is formed between the acid and the base

Action level: the amount of a material in air at which certain OSHA regulations to protect employees take effect; exposure at or above the action level is termed occupational exposure

Action plan: documented outline of specific projected activities to be accomplished within a specified period, to meet a defined need

Active electrode: electrosurgical accessory that directs current flow to the surgical site, also called a cautery tip

Activity (radioactivity): the rate of decay of radioactive material expressed as the number of atoms breaking down per second measured in units called Becquerels or Curies

Actual breakthrough time: the average time elapsed between initial contact of the chemical with the outside surface of the fabric and the detection time

Actuator: a power mechanism used to effect motion of a robot or a device that converts electrical, hydraulic, or pneumatic energy into robot motion

Acute effect: an adverse effect on humans or animals, with symptoms developing rapidly and quickly becoming a crisis resulting from a short-term exposure

Acute radiation exposure: an exposure to radiation that occurred in a matter of minutes rather than in longer or continuing exposure over a period of time

Acute radiation syndrome: a serious illness caused by receiving a dose of more than 50 rads of penetrating radiation to the body in a short time (usually minutes); the early symptoms include nausea, fatigue, vomiting, diarrhea with loss of hair, and swelling of the mouth and throat, and general loss of energy may follow

Adaptation: a change in the structure of an organism that results in its adjustment to its surroundings

Adequate: denotes the quality or quantity of a system, process, procedure, or resource that will achieve the relevant incident response objective

Adsorption: attachment of the molecules of a gas or liquid to the surface of another substance, normally a solid, or the process by which a liquid adheres to the surface of a material but does not penetrate the fibers of the material

Advanced life support: medical procedures performed by emergency medical technicians-paramedics that include the advanced diagnosis and protocol-driven treatment of a patient in the field

Aerobic: requiring the presence of air or oxygen to live, grow, and reproduce

Aerosol: liquid droplets or solid particles dispersed in air

Agency: a division of government with a specific function offering a particular kind of assistance or in the emergency incident command system

Aggressor: any person seeking to compromise a function or structure

Air blast: an airborne shock wave resulting from the detonation of explosives

Air exchange rate: speed at which outside air replaces air inside a building or the number of times the ventilation system replaces air within a room or built structure

Airborne radioactive material: any radioactive material dispersed in the air in the form of a dust, fumes, mists, aerosols, vapors, or gases

Alarm procedure: a means of alerting concerned parties to a disaster; various optical and audible means of alarm are available, including flags, lights, sirens, radio, and telephone

Alcohol-based hand sanitizers: a gel or rub that contains alcohol to reduce the number of viable microorganisms on the hands (60 to 95 percent ethanol or isopropanol)

Alkali: a term normally used to refer to hydroxides and carbonates of the metals of group 1a of the periodic table or ammonium hydroxide

All hazards: emergency management term referring to a natural or man-made event that would require actions to protect life, property, environment, public health, safety, and/or minimize disruptions to government, social, or economic activities

Allergen: a substance or particle that causes an allergic reaction

Alloy: a mixture or solution of metals, either solid or liquid, which may or may not include a nonmetal

Alpha particle: positively charged particle emitted by certain radioactive materials; identical to the nucleus of the helium atom, and the least penetrating form of radiation

Alternate site burn: a patient burn resulting from electricity exiting the body by unintended means

Ambient air: outside or surrounding air

Americium (am): a silvery metal that is a man-made element whose isotopes am-237 through am-246 are all radioactive; trace quantities of americium are widely used in smoke detectors, and as neutron sources in neutron moisture gauges

Analysis approach: selecting one of two primary approaches for FEMA; one is the hardware approach that lists individual hardware items and analyzes their possible failure modes and the second is the functional approach that recognizes that every item is designed to perform a number of outputs within a system

Andon cord policy: empowerment of healthcare personnel to act to ensure safety regardless of hierarchy and without risk or retaliation

Anemometer: a rotating vane, swinging vane, or hot-wire device used to measure air velocity

Anhydrous: a substance in which no water is present in the form of a hydrate or water of crystallization

Anion: a negatively charged ion

Annual summary: the occupational injury and illness totals for the year as reflected by OSHA Form 300 Log entries

Annual survey: survey conducted each year by the Bureau of Labor Statistics to produce national data on occupational injury and illness rates in various industries

Anode: the positive electrode in an electrolytic cell

Anosmia: reduced sensitivity to odor detection

Antidote: an agent that neutralizes or counteracts the effects of a poison

Antimicrobial: any agent that destroys microbial organisms

Antiseptics: substances applied to skin to reduce microbial flora such as alcohols, chlorine, iodine, etc.

Anti-terrorism: preventative in nature, it entails using passive and defensive measures such as education, foreign liaison training, surveillance, and counter surveillance which is designed to deter terrorist activities

Aperture: an opening through which radiation can pass

Application program: the set of instructions that defines the specific intended tasks of robots and robot systems

Approved: a method, procedure, equipment, or tool that has been determined to be satisfactory for a particular purpose

Aqueous: a solution or suspension in which the solvent is water

Area command: an organization established to oversee the management of multiple incidents that are each being handled by separate incident command systems or organizations; an area command is activated only if necessary, depending on the complexity of the incident and incident management span-of-control considerations

Argon: a gas used as a laser medium that emits blue-green light

Aromatic: term applied to a group of hydrocarbons characterized by the presence of the benzene nucleus; a major series of unsaturated cyclic hydrocarbons whose carbon atoms are arranged in closed rings

Asbestos: fibrous magnesium silicate

Asphyxiant: a chemical gas or vapor that can cause unconsciousness or death by suffocation

Assessment: the evaluation and interpretation of measurements and other information to provide a basis for decision-making

Assigned protection factor: a rating assigned to a respirator style by OSHA or NIOSH; this rating indicates the level of protection most workers can expect from the properly worn, maintained, and fitted respirator used under actual workplace conditions

Assignment: a task given to a resource to perform within a given operational period that is based on operational objectives defined in the incident action plan

Atmosphere supplying respirator: any respirator that provides clean air from an uncontaminated source to the face piece; examples include supplied-air (airline) respirators, self-contained breathing apparatus, and combination supplied air/SCBA devices

Atom: smallest particle of an element that can enter into a chemical reaction

Atomic mass number: the total number of protons and neutrons in the nucleus of an atom

Atomic mass unit: one unit is equal to one-twelfth of the mass of a carbon-12 atom

Atomic weight: the mass of an atom, expressed in atomic mass units

Atropine: an anti-cholinergic used as an antidote for nerve agents to counteract excessive amounts of acetylcholine

Attended continuous operation: the time when robots are performing (production) tasks at a speed no greater than slow speed through attended program execution

Attended program verification: the time when a person within the restricted envelope (space) verifies the robot's programmed tasks at programmed speed

Attenuated vaccine: a vaccine that has been weakened but is still required to be controlled as infectious by some regulatory programs

Attenuation: the decrease in energy (or power) as a beam passes through an absorbing or scattering medium

Authority gradient: an unwillingness to be truthful to those in power or authority

Autoclave: device used to sterilize medical instruments and equipment by using steam under pressure

Auto-ignition point: the lowest temperature at which a material will catch fire without the aid of a flame or spark

Auto-ignition temperatures: temperature at which a material will self-ignite and maintain combustion without a fire source

Automatic contour: a feature of a bed where the thigh section of the sleep surface articulates upward as the head section travels upward, thereby reducing the likelihood of patient/resident mattress from migrating toward the foot end of the bed

Automatic conveyor and shuttle systems: devices comprised of various types of conveying systems linked together with various shuttle mechanisms for the prime purpose of conveying materials or parts to prepositioned and predetermined locations automatically

Automatic guided vehicle system: advanced material-handling or conveying systems that involve a driverless vehicle which follows a guide path

Automatic mode: robot state in which automatic operation can be initiated

Automatic operation: time during which robots are performing programmed tasks through unattended program execution

Automatic sprinklers: system built in or added to a structure that automatically delivers water in case of fire

Automatic storage and retrieval systems: storage racks linked through automatically controlled conveyors and an automatic storage and retrieval machine or machines that ride on floor-mounted guide rails and power-driven wheels

Awareness barrier: physical and/or visual means that warns a person of an approaching or present hazard

Awareness signal: a device that warns a person of an approaching or present hazard by means of audible sound or visible light

Axis: the line about which a rotating body, such as a tool, turns

B

Background radiation: the radiation in man's natural environment, including cosmic rays and radiation from naturally occurring radioactive elements

Badging: process of providing outside personnel with identification that gives them access to the designated facilities of the organization requesting assistance

Ballistic protection: techniques for the protection of personnel (and material) against projectiles of all kinds, such as protective blankets for vehicles or protective gear (jackets, helmets, trousers, etc.)

Barricade: an intervening barrier (natural or artificial) of such type, size, and construction as to limit the effects of low angle high velocity fragments

Barrier: a physical means of separating persons from the restricted envelope or space

Base: substance that (1) liberates hydroxyl ions when dissolved in water; (2) liberates negative ions of various kinds in any solvent; (3) receives a hydrogen ion from a strong acid to form a weaker acid; (4) gives up two electrons an acid, forming a covalent bond with the acid

Basic human needs: physical, safety, social, self-esteem, and self-actualization

Basic life support: noninvasive first aid procedures and techniques utilized by most all trained medical personnel, including first responders, to stabilize critically sick and injured people

Beam: a collection of rays that may be parallel, convergent, or divergent

Becquerel: the amount of a radioactive material that will undergo one decay (disintegration) per second

Bed alarms: any alarm intended to notify caregivers of either an unwanted patient/resident egress or that the patient/resident is near the edge of the mattress

Bed rail extender: detachable device intended to bridge the space between the head and foot bed rail

Bed rails: adjustable metal or rigid plastic bars that attach to the bed; synonymous terms are side rails, bed side rails, and safety rails

Beta particle: a particle emitted from a nucleus during radioactive decay; can be stopped by a sheet of metal or acrylic plastic depending on the emitted energy level of a particular isotope

Bioassay: an assessment of radioactive materials that may be present inside a person's body through analysis of the person's blood, urine, feces, or sweat

Biocide: a substance that can kill living organisms

Biodegradable: a substance with the ability to decompose or break down into natural components

Biological half-life: the time required for one half of the amount of a substance, such as a radionuclide, to be expelled from the body by natural metabolic processes, not counting radioactive decay, once it has been taken in through inhalation, ingestion, or absorption

Biological threat: a threat that consists of biological material planned to be deployed to produce casualties in personnel or animals or damage plants

Bioremediation: the management of microorganisms

Bipolar: forceps-shaped active electrode; current flows through the tissue from one tip to the other

Blacklist: a counterintelligence agency listing of actual or potential hostile collaborators, sympathizers, or other persons viewed as threatening to friendly military forces

Blameless reporting: accountable, with emphasis on learning, not blame

Blast: the brief and rapid movement of air, vapor, or fluid away from a center of outward pressure, as in an explosion or in the combustion of rocket fuel; the pressure accompanying this movement

Blast containment: containing an explosive force so the blast wave and fragmentation materials are contained within a border made by barriers, walls, revetments, or other materials or objects

Blast effect: destruction of or damage to personnel, vehicles or structures from an explosive force by a weapon designed to explode on contact with or above the ground

Blast mitigation: various physical measures that can be used to lessen the damage of a blast wave on critical assets; these measures include, but are not limited to, blast walls, blast barriers, standoff distance, and structural hardening

Blast wave: a sharply defined wave of increased pressure rapidly propagated through a surrounding medium from a center of detonation or similar disturbance; a sharp jump in pressure is known as a shock wave and a slow rise is known as a compression wave

Blister agent: a chemical warfare agent that produces local irritation and damage to the skin and mucous membranes, pain and injury to the eyes, reddening and blistering of the skin, and when inhaled, damage to the respiratory tract

Block diagrams: diagrams that illustrate the operation, interrelationships, and interdependencies of the functions of a system are required to show the sequence and the series dependence or independence of functions and operations; block diagrams may be constructed in conjunction with or after defining the system and shall present the system breakdown of its major functions

Blood agent: a chemical warfare agent that is inhaled and absorbed into the blood; the blood (cyanogens) carries the agent to all body tissues where it interferes with the tissue oxygenation process

Bloodborne pathogens: pathogenic microorganisms that are present in human blood and can cause disease in humans; these pathogens include, but are not limited to, the hepatitis B virus and human immunodeficiency virus

Bloodborne pathogens engineering controls: sharps disposal containers, self-sheathing needles, and safer medical devices, such as sharps with engineered sharps injury protections and needleless systems that isolate or remove the bloodborne pathogens hazards from the workplace

Bloodborne pathogens exposure incident: means a specific eye, mouth, other mucous membrane, nonintact skin, or parenteral contact with blood or other potentially infectious materials that results from the performance of an employee's duties

Boiling point: the temperature at which a liquid changes to a vapor at sea level pressure

Bollard: any object used to confine traffic within or from a given area; they are vertical members made of wood, steel, or concrete that are permanently placed

Bolt ring: closing device used to secure a cover to the body of an open head drum; this ring requires a bolt and nut to secure the closure

Bonding: the interconnecting of two objects such as a tank or cylinder with clamps and wire as a safety practice to equalize the electrical potential between the objects and help prevent static sparks that could ignite flammable material; dispensing/receiving a flammable liquid requires dissipating the static charge by bonding between containers

Boundaries of excediency: these are margins of safety when errors become known that signal the need to reassess, slow down, or ask for help

Branch: organizational level having functional or geographical responsibility for major aspects of incident operations, organizationally situated between the section chief and the division or group in the operations section

Breakthrough time: the time from initial chemical contact to detection

Buffer: an acid-based balancing or control reaction where the pH of a solution is protected from major changes when acids or bases are added to it

Building-related illness: diagnosable illnesses with identifiable symptoms that can be attributed to building contaminants

Bung: a threaded closure located in the head or body of a drum

Bunker: a fortified structure, but primarily a buried or semi-buried structure, offering a high degree of protection to personnel, defended gun positions, or a defensive position, from enemy attack

Bureaucratic organization: line organization with an established hierarchy

Burn back: the distance a flame will travel from the ignition source back to the aerosol container

C

Cache: a predetermined complement of tools, equipment, and/or supplies stored in a designated location, available for incident use

Canister or cartridge: a container with a filter, sorbent, or catalyst, or a combination of these items, that removes specific contaminants from the air passed through the container

Carbon dioxide: a heavy, colorless, non-flammable, relatively nontoxic gas produced by the combustion and decomposition of organic substances and as a byproduct of many chemical processes

Carbon monoxide: a colorless, odorless, toxic gas generated by the combustion of common fuels in the presence of insufficient air or where combustion is incomplete

Carbonate: a compound formed by the reaction of carbonic acid with either a metal or an organic compound

Carcinogen: any substance that has been found to induce the formation of cancerous tissue in experimental animals

Carpal tunnel syndrome: a common affliction caused by compression of the median nerve in the carpal tunnel, often associated with tingling, pain, or numbness in the thumb and first three fingers

CAS: chemical abstracts service

CAS number: a number assigned to identify a chemical substance; the chemical abstracts service indexes information that appears in chemical abstracts, published by the American Chemical Society

Catalyst: an element or compound that accelerates the rate of a chemical reaction but is neither changed nor consumed by it

Catastrophic incident: any natural or man-made incident, including terrorism, that results in extraordinary levels of mass casualties, damage, or disruption severely affecting the population, infrastructure, environment, economy, national morale, and/or government functions

Catastrophic loss: a loss of huge and extraordinary proportion

Cathode: the negative electrode of an electrolytic cell

Causation: factors that come together and result in adverse events

Caustic material: that which is able to burn, corrode, dissolve, or eat away another substance

Caustic substances: strong alkalis; their solutions are corrosive to the skin and other tissues

CBRN: chemical, biological, radiological, or nuclear agent or substance

Ceiling concentration: maximum concentration of a toxic substance allowed at any time or during a specific sampling period

Ceiling limit: normally expressed as threshold limit value (TLV) and permissible exposure limit (PEL), ceiling limit is the maximum allowable concentration to which an employee may be exposed in a given time period

Ceiling maximum: allowable exposure limit not to be exceeded for an airborne substance

Ceiling value: concentration that should not be exceeded during the working exposure; exposure should at no time exceed the ceiling value

Cell: the smallest unit within a guerrilla or terrorist group, a cell generally consists of two to five people dedicated to a terrorist cause; the formation of cells is born of the concept that an apparent "leaderless resistance" makes it hard for counterterrorists to penetrate

CFC: chlorofluorocarbons, being phased out worldwide because of their detrimental effect on the ozone layer

CFM: cubic feet per minute, a unit measuring airflow when evacuating ventilation systems

Chain of command: a series of command, control, executive, or management positions in hierarchical order of authority

Chain reaction: a process that initiates its own repetition; in a fission chain reaction a fissile nucleus absorbs a neutron and fissions spontaneously releasing additional neutrons

Characteristic waste: hazardous waste that exhibits one of four characteristics: ignitability, reactivity, toxicity, or corrosivity

Chemical agent: any chemical substance that is intended for use in military operations to kill, seriously injure, or incapacitate humans because of its physiological effects

Chemical contamination: The presence of a chemical agent on a person, object, or area

Chemical disinfection: use of formulated chemical solutions to treat and decontaminate infectious waste

Chemical family: a group of compounds with related chemical and physical properties

Chemical hygiene plan: a written plan that addresses job procedures, work equipment, protective clothing, and training necessary to protect employees from chemical and toxic hazards, required by OSHA under its Laboratory Standard

Chemical name: scientific designation of a chemical substance

Chemical transportation emergency center: an organization that provides immediate information for members on what to do in case of spills, leaks, fires, or exposures

Chemical warfare agent: a chemical substance, which, because of its physiological, psychological, or pharmacological effects, is intended for use in military operations to kill, seriously injure, or incapacitate humans (or animals) through its toxicological effects

Chemotherapy: development and use of chemical compounds that are specific for the treatment of diseases

Chief: the incident command system title for individuals responsible for management of functional sections: operations, planning, logistics, finance/administration, and intelligence/investigations, if established as a separate section

Chlorinated solvent: organic solvent that contains chlorine atoms

Choking agents: agents that exert their effects solely on the lungs and result in the irritation of the alveoli of the lungs

Chronic effect: adverse effect on animals or humans in which symptoms develops slowly over a long period of time or recurs frequently

Chronic exposure: exposure to a substance over a long period of time, possibly resulting in adverse health effects

Citizen Corps: a community-level program, administered by the Department of Homeland Security, that brings government and private-sector groups together and coordinates the emergency preparedness and response activities of community members

Civil support: DOD support to U.S. civil authorities for domestic emergencies, and for designated law enforcement and other activities

Class A fire: that which involves wood, paper, cloth, trash, or other ordinary materials

Class B fire: that which involves grease, paint, or other flammable liquids

Class C fire: that which involes live electrical or energized equipment

Class D fire: that which involves flammable metals

Class K fire: that which involves kitchen oils used for frying

Clean Air Act: Public Law Pl 91-604, where the EPA set national ambient air quality standards and enforcement/discharge permits are carried out by the states under implementation plans

Clear zone: an area that is clear of visual obstructions and landscape materials that could conceal a threat or perpetrator

Clinical laboratory: workplace where diagnostic or other screening procedures are performed on blood or other potentially infectious materials

Clinicians: physicians, nurses, nurse practitioners, physicians' assistants, and others

Cobalt (co): gray, hard, magnetic, and somewhat malleable metal, cobalt is relatively rare and generally obtained as a byproduct of other metals, such as copper; most common radioisotope is cobalt-60, used in radiography and medical applications

Cold zone: the support zone is the area outside the warm zone where there is no contamination present

Collateral damage: unintended damages, beyond the destruction of the enemy forces or installations specifically targeted, to surrounding human and nonhuman resources, either military or nonmilitary, caused by the spillover of weapons effect, as opposed to the damage caused by aiming errors

Collective dose: the estimated dose for an area or region multiplied by the estimated population in that area or region

Colorimetry: an analytical method by which the amount of a compound in solution can be determined by measuring the strength of its color by either visual or photometric methods

Combustible: term used to classify certain liquids that will burn on the basis of flash point; NFPA and DOT classify combustible liquids as having a flash point of 100 degrees F (38 degrees C) or higher

Command: the act of directing, ordering, or controlling by virtue of explicit statutory, regulatory, or delegated authority

Command staff: an incident command element consisting of the following functions: public information officer, safety officer, liaison officer, and other positions as required (all report to the incident commander)

Committed dose: a dose that accounts for continuing exposures expected to be received over a long period of time from radioactive materials that were deposited inside the body

Common operating picture: an overview of an incident by all relevant parties that provides incident information enabling the incident commander/unified command and any supporting agencies and organizations to make effective, consistent, and timely decisions

Communications/dispatch center: an agency or interagency dispatch center, 911 call centers, emergency control or command dispatch centers, or any naming convention given to the facility and staff that handles emergency calls from the public and communication with emergency management/response personnel

Complexity: the characteristic of healthcare that requires vigilance

Compliance safety and health officer: an OSHA representative whose primary job is to conduct workplace inspections

Comprehensive Preparedness Guide 101: FEMA document used to develop emergency plans for all-hazard emergency operations planning for state, territorial, local, and tribal governments

Concentration: the ratio of the amount of a specific substance in a given volume or mass of solution to the mass or volume of solvent

Concept Plan (CONPLAN): a plan that describes the concept of operations for integrating and synchronizing federal capabilities to accomplish critical tasks; describes how federal capabilities will be integrated into and support regional, state, and local plans to meet the objectives described in the Strategic Plan

Conductivity: the property of a circuit that permits the flow of an electrical current

Confirmation bias: people, when uncertain, will tend to agree with authority

Consensus standards: a variety of standards developed according to a consensus of agreement among several organizations, stakeholders, or individuals

Consequence management: measures taken to protect public health and safety, restore essential government services, and provide emergency relief to governments, businesses, and individuals affected by the consequences of a chemical, biological, nuclear, and/or high-yield explosive situation

Contaminated: the presence or the reasonably anticipated presence of blood or other potentially infectious materials on an item or surface

Contaminated laundry: laundry that has been soiled with blood or other potentially infectious materials or may contain sharps

Contaminated sharps: any contaminated object that can penetrate the skin including, but not limited to, needles, scalpels, broken glass, broken capillary tubes, and exposed ends of dental wires

Contamination (radioactive): the deposit of unwanted radioactive material on the surfaces of structures, areas, objects, or people where it may be external or internal

Contingency plan: an emergency plan developed in expectation of a disaster; contingency plans are often based on risk assessments, the availability of human and material resources, community preparedness, and local and international response capabilities

Control bed rail: a bed rail that incorporates bed function controls for patient/staff activation

Control device: any control hardware providing a means for human intervention in the control of a robot or robot system, such as an emergency-stop button, a start button, or a selector switch

Controlling: measuring performance of work by monitoring outcomes

Controls program: the inherent set of control instructions that defines the capabilities, actions, and responses of the robot system not intended to be modified by the user or operator

Coordinate: to advance systematically an analysis and exchange of information among principals who have or may have a need to know certain information to carry out specific incident management responsibilities

Coordinated straight line motion: control wherein the axes of the robot arrive at their respective end points simultaneously, giving a smooth appearance to the motion

Corrective actions: implementation of procedures that are based on lessons learned from actual incidents or during realistic exercises

Corrosive: a substance that causes visible destruction or permanent change in human skin tissue at the site of contact

Cosmic radiation: radiation produced in space when heavy bombardment of the earth occurs

Cost-benefit analysis: evaluation of a situation that focuses on comparing expenditures with potential benefits, but not necessarily a dollar for dollar comparison

Counterterrorism Security Group: an interagency body convened on a regular basis to develop terrorism prevention policy and to coordinate threat response and law enforcement investigations associated with terrorism; group evaluates various policy issues of interagency importance regarding counterterrorism and makes recommendations to senior levels of the policymaking structure for decisions

Credentialing: authentication and verification of the certification and identity of designated incident managers and emergency responders

Crisis management: measures to identify, acquire, and plan to use the resources available to anticipate, prevent, and resolve a threat or act

Criteria standard: a standard against which performance can be measured

Critical infrastructure: systems, assets, and networks, whether physical or virtual, so vital to the United States that the incapacity or destruction of such systems and assets would have a debilitating impact on security, national economic security, national public health or safety, or any combination thereof

Critical mass: the minimum amount of fissile material that can achieve a self-sustaining nuclear chain reaction

Criticality: a relative measure of the consequences of a failure mode and its frequency of occurrence

Criticality (nuclear): a fission process in which the neutron production rate equals the neutron loss rate to absorption or leakage

Criticality analysis: a procedure by which each potential failure mode is ranked according to the combined influence of severity and probability of occurrence

Cubic feet per minute: measure of the volume of a substance flowing through air within a specified time period; used to measure air exchanged in ventilation systems

Culture of trust: a culture in which workers have a voice and choice in organizational matters

Cumulative dose: the total dose resulting from repeated or continuous exposures of the same portion of the body, or of the whole body, to ionizing radiation

Curie (ci): traditional measure of radioactivity based on the observed decay rate of 1 gram of radium

Cutaneous radiation syndrome: effects can be reddening and swelling of the exposed area creating blisters, ulcers on the skin, hair loss, and severe pain

D

Damage assessment: the process used to appraise or determine the number of injuries and deaths, damage to public and private property, and the status of key facilities and services such as hospitals and other healthcare facilities, fire and police stations, communications networks, water and sanitation systems, utilities, and transportation networks resulting from a man-made or natural disaster

Damper: control that varies airflow through an air inlet, outlet, or duct

Dead zone: zone that lies outside the sensing capability of sensors within a protected area; a dead zone may result from defective or improperly adjusted sensors or from interference, such as blocking objects or structures

Decay (radioactive): disintegration of the nucleus of an unstable atom by the release of radiation

Decay products: isotopes or elements formed and the particles and high-energy electromagnetic radiation emitted by the nuclei of radionuclide during radioactive decay

Decibel: a unit to express the relative intensity of a sound on a scale from 0 to 130 (average pain level); sound doubles every 10 decibels

Decomposition: breakdown of a chemical or substance into different parts or simpler compounds that can occur due to heat, chemical reaction, or decay

Decontamination: process of removing contaminants from the body or a surface

Defatting: removal of natural oils from the skin by the use of a fat dissolving solvent

Defense layer: building design or exterior perimeter barriers intended to delay attempted forced entry

Demand respirator: an atmosphere-supplying respirator that admits breathing air to the face piece only when a negative pressure is created inside the face piece by inhalation

Denaturant: a substance added to ethyl alcohol to prevent its being used for internal consumption

Denier: a term used in the textile industry to designate the weight per unit length of a filament

Density: the ratio of weight (mass) to volume of any substance; usually expressed as grams per cubic centimeter

Density mass: substance of mass per unit volume usually compared to water, which has a density of one

Department of Homeland Security (DHS): a new cabinet level department charged with strengthening the security and protecting the assets of the United States of America and its territories, with a primary mission of (1) preventing terrorist attacks within the United States, (2) reducing America's vulnerability to terrorism, and (3) minimizing the damage and recovery from attacks that do occur

Department operating center: an emergency operations center specific to a single department or agency

Deposition density: the activity of a radionuclide per unit area of ground

Dequervain's disease: the tendon sheath of both the long and the short abductor muscles of the thumb narrows

Dermatitis: inflammation of the skin caused by defatting of the dermis

Desiccant chemical: a substance that absorbs moisture

Desorption: the reverse process of absorption; agent will be "removed" from the surface (off gassing or out gassing)

Detection mechanism: the means or methods by which a failure can be discovered by an operator under normal system operation or can be discovered by the maintenance crew by some diagnostic action

Deterministic effects: effects that can be related directly to the radiation dose received with severity increasing as the dose increases

Detonation: a release of energy caused by the extremely rapid chemical reaction of a substance in which the reaction front advances into the unreacted substance at equal to or greater than sonic velocity; an explosive reaction that consists of the propagation of a shock wave through the explosive accompanied by a chemical reaction that furnishes energy to sustain the shock propagation in a stable manner, with gaseous formation and pressure expansion following shortly thereafter

Deuterium: a nonradioactive isotope of the hydrogen atom that contains a neutron in its nucleus in addition to the one proton normally seen in hydrogen

Device: any control hardware such as an emergency-stop button, selector switch, control pendant, relay, solenoid valve, or sensor

DFM: this abbreviation refers to a respirator filter cartridge suitable for use against dusts, fumes, or mist and is used in the new NIOSH regulation on respirator certification

Dielectric material: an electrical insulator in which an electric field can be sustained with a minimum dissipation of power

Directing: providing the necessary guidance to others during job accomplishment

Dirty bomb: a device designed to spread radioactive material by conventional explosives when the bomb detonates

Disaster: any natural catastrophe including any hurricane, tornado, storm, high water, wind-driven water, tidal wave, tsunami, earthquake, volcanic eruption, landslide, mudslide, snowstorm, and drought or, regardless of cause, any fire, flood, or explosion

Disaster recovery center (DRC): facility established in a centralized location within or near the disaster area at which disaster victims apply for disaster aid

Disinfectant: an agent with the ability to kill at least 95 percent of targeted microorganisms

Division: the partition of an incident into geographical areas of operation; divisions are established when the number of resources exceeds the manageable span of control of the operations chief; located within the Incident Command System organization between the branch and resources in the operations section

Doff: to take off or remove (e.g., PPE)

Domestic terrorism: the unlawful use, or threatened use, of force or violence by a group or individual based and operating entirely within the United States or Puerto Rico without foreign direction committed against persons or property to intimidate or coerce a government, the civilian population, or any segment thereof in furtherance of political or social objectives

Don: to put on, in order to wear (e.g., PPE)

Dose (radiation): radiation absorbed by a person's body

Dose coefficient: the factor used to convert radionuclide intake to dose

Dose equivalent: a quantity used in radiation protection to place all radiation on a common scale for calculating tissue damage

Dose quantity: radiation absorbed per unit of mass by the body or by any portion of the body

Dose rate: the radiation dose delivered per unit of time

Dose reconstruction: a scientific study that estimates doses to people from releases of radioactivity or other pollutants

Dose response: relationship between the amount of a toxic or hazardous substance and the extent of illness or injury produced in humans

Dosimeter: small portable instrument (such as a film badge, thermoluminescent dosimeter, or pocket dosimeter) for measuring and recording the total accumulated dose of ionizing radiation a person receives

Dosimetry: assessment (by measurement or calculation) of radiation dose

Drive powers: the energy source or sources for the robot actuators

Drop test: a test required by DOT regulations for determination of the quality of a container or finished product

Dry bulb temperature: temperature of air measured with a dry bulb thermometer in a psychomotor to measure relative humidity

Dry pipe: piping under pressure; when the head opens, air is released and water flows into the system

Dusts: solid particles generated by handling, crushing, grinding, rapid impact, detonation, and decrepitating of organic or inorganic materials such as rock, ore, metal, coal, wood, and grain

E

Ecoterrorism: sabotage intended to hinder activities that are considered damaging to the environment

Effective dose: a dosimetry quantity useful for comparing the overall health affects of irradiation of the whole body, effective dose is used to compare the overall health detriments of different radio nuclides in a given mix

Effective half-life: the time required for the amount of a radionuclide deposited in a living organism to be diminished by 50 percent as a result of the combined action of radioactive decay and biologic elimination

Effective standoff distance: a standoff distance at which the required level of protection can be shown to be achieved through analysis or can be achieved through building hardening or other mitigating construction or retrofit

Elastomer: a term used to describe any high polymer having the essential properties of vulcanized natural rubber

Electrochemistry: chemistry concerned with the relationship between electrical forces and chemical reactions

Electrode: a material used in an electrolytic cell to enable the current to enter or leave the solution

Electrolysis: decomposition of a chemical compound by means of an electric current

Electron: a particle of negative electricity, electrons surround the nucleus of an atom because of the attraction between their negative charge and the positive charge of the nucleus

Electron volt: unit of energy equivalent to the amount of energy gained by an electron when it passes from a point of low potential to a point one volt higher in potential

Electrosurgery: radio frequency energy to produce cutting and coagulation in body tissues

Electrosurgical unit: a device that produces radio frequency energy for electrosurgery procedures

Element: all isotopes of an atom that contain the same number of protons or in a nuclear device the fuel element (metal rod) containing fissile materials fuel element is a metal rod containing the fissile material

Embedded laser: a laser with an assigned class number higher than the inherent capability of the laser system in which it is incorporated

Emergency: any incident, whether natural or man-made, that requires responsive action to protect life or property

Emergency management: a subset of incident management, the coordination and integration of all activities necessary to build, sustain, and improve the capability to prepare for, protect against, respond to, recover from, or militate against threatened or actual natural disasters, acts of terrorism, or other man-made disasters

Emergency Management Assistance Compact: a Congressionally ratified organization that provides form and structure to interstate mutual aid during emergency event response

Emergency management committee: a preparedness entity established by an organization that has the responsibility for emergency management program oversight within the organization

Emergency management program: a program that implements the organization's mission, vision, management framework, and strategic goals and objectives related to emergencies and disasters

Emergency manager: a person who has the day-to-day responsibility for emergency management programs and activities; roles include coordinating mitigation, preparedness, response, and recovery

Emergency medical services (EMS): services, including personnel, facilities, and equipment required to ensure proper medical care for the sick and injured from the time of injury to the time of final disposition, including medical disposition within a hospital, temporary medical facility, or special care facility, release from site, or declared dead; in addition, emergency medical services specifically include those services immediately required to ensure proper medical care and specialized treatment for patients in a hospital and coordination of related hospital services

Emergency operations center (EOC): physical location at which the coordination of information and resources to support incident management (on-scene operations) activities normally takes place; an EOC may be a temporary facility or may be located in a more central or permanently established facility, perhaps at a higher level of organization within a jurisdiction, and may be organized by major functional disciplines (e.g., fire, law enforcement, and medical services), by jurisdiction (e.g., Federal, State, regional, tribal, city, county), or some combination thereof

Emergency operations plan: an ongoing plan maintained by various jurisdictional levels for responding to a wide variety of potential hazards, disasters, or emergency events

Emergency public information: information that is disseminated primarily in anticipation of an emergency or during an emergency that provides guidance or requires actions

Emergency response guide: a document that provides guidance on emergency response in a transportation incident involving a particular chemical

Emergency response team (ERT): an interagency team, consisting of the lead representative from each federal department or agency assigned primary responsibility for an Emergency Response Function and key members of the Federal Coordinating Officer's (FCO) staff, formed to assist the FCO in carrying out his/her coordination responsibilities; the ERT may be expanded by the FCO to include designated representatives of other federal departments and agencies as needed, and usually consists of regional-level staff

Emergency situation: any occurrence including equipment failure, rupture of containers, or failure of control equipment that results in an uncontrolled substantial release of a contaminant

Emergency stop: the operation of a circuit using hardware-based components that overrides all other robotic controls

Emergency support function: refers to a group of capabilities of federal departments and agencies to provide the support, resources, program implementation, and services that are most likely to be needed to save lives, protect property, restore essential services and critical infrastructure, and help victims return to normal following a national incident

Employee exposure: an exposure to a concentration of an airborne contaminant that would occur if the employee were not using respiratory protection

Emulsion: a permanent suspension or dispersion, usually of oil particles in water

Enabling device: a manually operated device that permits motion when continuously activated; releasing the device stops robot motion and motion of associated equipment that may present a hazard

Enclave: a secured area within another secured area

Enclosed laser device: any laser or laser system located within an enclosure that does not permit hazardous optical radiation emission from the enclosure

End-effector: an accessory device or tool specifically designed for attachment to the robot wrist or tool mounting plate to enable the robot to perform its intended task

End-of-service-life indicator: a system that warns the respirator user of the approach of the end of adequate respiratory protection

Endothermic: a term used to characterize a chemical reaction that requires absorption of heat from an external source

Energy sources: any electrical, mechanical, hydraulic, pneumatic, chemical, thermal, or other source

Engineering controls: the preferred method of controlling employee exposures in the workplace

Enriched uranium: any uranium in which the proportion of the isotope uranium-235 has been increased by removing uranium-238 mechanically

Entrapment: an event in which a patient is caught, trapped, or entangled in the spaces in or about the bed rail, mattress, or hospital bed frame

Envelope: space or maximum volume of space encompassing the maximum designed movements of all robot parts including the end-effector, work piece, and attachments

Enzyme complex: protein produced by living cells that starts up biochemical reactions

Epidemiology: study of the distribution and determinants of health-related states or events in specified populations, and the application of this study to the control of health problems

Ergonomics: a multidisciplinary activity that deals with interactions between workers and their total working environment plus stresses related to such environmental elements as atmosphere, heat, light, and sound, as well as tools and equipment in the workplace

Error tolerant system: a system with corrective actions to recover from errors

Escape-only respirator: a respirator intended to be used only for emergency exit

Etiologic agent: viable microorganism or its toxin which can cause human disease

Evacuation: organized, phased, and supervised withdrawal, dispersal, or removal of humans from dangerous or potentially dangerous areas

Evaluating: assessing the effectiveness for the purpose of improving

Evaporation rate: the rate at which a material is converted to a vapor at a given temperature and pressure when compared to the evaporation rate of a given substance

Excimer: a gas mixture used as the active medium in a family of lasers emitting ultraviolet light

Exhaust ventilation: the removal of air from any space, usually by mechanical means

Exothermic: a term used to characterize a chemical reaction that gives off heat as it proceeds

Experience rating: process of basing insurance or workers' compensation premiums on the insured record of losses

Explosimeter: a device that detects and measures the presence of gas or vapor in an explosive atmosphere

Explosion class 1: flammable gas/vapor

Explosion class 2: combustible dust

Explosion class 3: ignitable fibers

Explosion-proof can: an electrical apparatus designed so that the explosion of flammable gas or vapor inside an enclosure will not ignite flammable gas or vapor outside

Explosive: any chemical compound or chemical mixture that, under the influence of heat, pressure, friction, or shock, undergoes a sudden chemical change (decomposition) with the liberation of energy in the form of heat and light and accompanied by a large volume of gas

Explosive limit: the amount of vapor in the air which forms an explosive mixture

Exposure (radiation): a measure of ionization in air caused by x-rays or gamma rays only; the unit of exposure most often used is the Roentgen

Exposure level: the level or concentration of a physical or chemical hazard to which an individual is exposed

Exposure limit: the concentration of a substance under which it is believed that nearly all workers may be repeatedly exposed day after day without adverse effects

Exposure pathway: a route by which a radionuclide or other toxic material can enter the body

Exposure rate: a measure of the ionization produced in air by x-rays or gamma rays per unit of time

F

Face velocity: the average air velocity into the exhaust system measured at the opening into the hood or booth

Fahrenheit: the temperature scale commonly used in the United States, where the freezing point of water is 32 degrees Fahrenheit and the boiling point is 212 degrees Fahrenheit at sea level

Fail-safe interlock: an interlock where the failure of a single mechanical or electrical component of the interlock will cause the system to go into, or remain in, a safe mode

Failure cause: the physical or chemical process, design defects, part misapplication, quality defects, or other processes that are the basic reason for failure or which initiate the physical process by which deterioration proceeds to failure

Failure definition: a general statement of what constitutes a failure of the item in terms of performance parameters and allowable limits for each specified output

Failure effect: consequence(s) a failure mode has on the operation, function, or status of an item; failure effects are usually classified according to how the entire system is impacted

Failure mode: the way by which failure is observed; describes the way the failure occurs and its impact on equipment operation

Failure mode and effects analysis (FMEA): process by which each potential failure mode in a system is analyzed to determine the results, or effects thereof, on the system and to classify each potential failure mode according to its severity

Fallout (nuclear): minute particles of radioactive debris that descend slowly from the atmosphere after a nuclear explosion

Federal Emergency Management Agency (FEMA): an independent agency reporting to the President and tasked with responding to, planning for, recovering from and mitigating against disaster

Federal-to-federal support: support that may occur when a federal department or agency responding to an incident under its own jurisdictional authorities requests Department of Homeland Security coordination to obtain additional federal assistance; federal departments and agencies execute interagency or intra-agency reimbursable agreements, in accordance with the Economy Act or other applicable authorities as part of federal-to-federal support

FEMA: see Federal Emergency Management Agency

Field Assessment Team (FAST): a small team of pre-identified technical experts who conduct an assessment of response needs (not a PDA) immediately following a disaster; experts are drawn from FEMA, other agencies and organizations—such as the U.S. Public Health Service, U.S. Army Corps of Engineers, U.S. Environmental Protection Agency, and the American Red Cross—and the affected state(s); FAST operations are joint federal/state efforts

Film badges: a package of photographic film worn like a badge by persons working with or around radioactive material to measure exposure to ionizing radiation; the absorbed dose can be calculated from the degree of film darkening caused by the irradiation

Filter: an air-purifying component used in respirators to remove solid or liquid aerosols from the inspired air

Filtering face piece (dust mask): a negative pressure particulate respirator with a filter as an integral part of the face piece or with the entire face piece composed of the filtering medium

Finance/administration section: the ICS functional area that addresses the financial, administrative, and legal/regulatory issues for the Incident Management System

Fireman pole: a pole secured (floor and ceiling mooring) next to the bed that acts as a support for the patient to get into and out of the bed

First aid (OSHA): any one-time treatment and subsequent observation of minor scratches, cuts, burns, and splinters that normally does not require medical care

First receiver: employees at a hospital engaged in decontamination and treatment of victims who have been contaminated by a hazardous substance(s) during an emergency incident

First report: a state-mandated workers' compensation form used to report work-related injuries and illnesses

First responder: personnel who have responsibility to initially respond to emergencies such as firefighters, police officers, highway patrol officers, lifeguards, forestry personnel, ambulance attendants, and other public service personnel; the first personnel trained to arrive on the scene of a hazardous or emergency situation

First responder awareness level: individuals who might reasonably be anticipated to witness or discover a hazardous substance release and who have been trained to initiate an emergency response sequence by notifying the proper authorities of the release; they would take no further action beyond notifying the authorities

First responder operations level: individuals who respond to releases or potential releases of hazardous substances as part of the initial response to the site for the purpose of protecting nearby persons, property, or the environment from the effects of the release; OSHA mandates these individuals must receive at least eight hours of training or have sufficient experience to objectively demonstrate competency in specific critical areas

Fission: splitting of a nucleus into at least two other nuclei that releases a large amount of energy

Fit factor: a quantitative estimate of the fit of a particular respirator to a specific Individual; typically estimates the ratio of the concentration of a substance in ambient air to its concentration inside the respirator when worn

Fit test: the use of a protocol to qualitatively or quantitatively evaluate the fit of a respirator on an individual; fit testing can be qualitative or quantitative

Flame arrestors: mesh or perforated metal insert within a flammable storage can that protects its contents from external flame or ignition

Flame extension: the distance a flame will travel from an aerosol container when exposed to an ignition source

Flame retardant: substances applied to or incorporated in a combustible material to reduce or eliminate its tendency to ignite when exposed a low-energy flame

Flammable liquid: a liquid with a flash point below 100 degrees F (37.8 degrees C)

Flash back: a phenomenon characterized by vapor ignition and flame traveling back to the vapor source

Flash point: the temperature at which an organic liquid evolves a high enough concentration vapor at or near its surface to form an ignitable mixture with air

Flocculation: the process to make solids in water increase in size by biological or chemical means so they may be separated from water

Fluorocarbon: any of a broad group of organic compounds analogous to hydrocarbons in which all or most of the hydrogen atoms of a hydrocarbon have been replaced by fluorine

Flux: any material or substance that will reduce the melting or softening temperature of another material when added to it

Force protection: security program developed to protect service members, civilian employees, family members, facilities and equipment, in all locations and situations, through the planned and integrated application of combating terrorism, physical security, operations security, personal protective services supported by intelligence, counterintelligence, and other security programs

Fractionated exposure: an exposure to radiation that occurs in several small acute exposures, rather than continuously as in a chronic exposure

Frequency: electrical term indicating the number of wave cycles in a second, measured in units called hertz

Fumes: particulate matter consisting of the solid particles generated by condensation from the gaseous state, generally after violation from melted substances, and often accompanied by a chemical reaction, such as oxidation

Function: one of the five major activities in the Incident Command System: command, operations, planning, logistics, and finance/administration (a sixth function, intelligence/investigations, may be established, if required, to meet incident management needs); the term is also used when describing the activity involved (e.g., the planning function)

Functional approach: the functional approach is normally used when hardware items cannot be uniquely identified or when system complexity requires analysis from the top down

Functional area: a major grouping of the similar tasks that agencies perform in carrying out incident management activities

Functional block diagrams: diagrams that illustrate the operation and interrelationships between functional entities of a system as defined in engineering data and schematics

Fungi: organisms that lack chlorophyll and must receive food from decaying matter

Fusion: a reaction in which at least one heavier, more stable nucleus is produced from two lighter, less stable nuclei

G

Galvanizing: application of a protective layer of zinc to a metal, chiefly steel, to prevent or inhibit corrosion

Gamma rays: high-energy electromagnetic radiation emitted by certain radionuclides when their nuclei transition from a higher to a lower energy state, very similar to x-rays

Gas: a state of matter in which a material has very low density and viscosity; gases expand and contract greatly in response to changes in temperature and pressure

Gas discharge laser: a laser containing a gaseous lasing medium in a glass tube in which a constant flow of gas replenishes the molecules depleted by the electricity or chemicals used for excitation

Gas neutralizer: a product used in riot control operations to neutralize the effect of tear gases; usually packaged as an aerosol spray and issued to police personnel

Gas/vapor sterilization waste treatment: a technique that uses gases or vaporized chemicals such as ethylene oxide and formaldehyde as sterilizing agents

Gauge: thickness of the steel used to manufacture a drum; the lower the gauge, the thicker the material

Geiger counter: a radiation detection and measuring instrument consisting of a gas-filled tube containing electrodes, between which an electrical voltage but no current flows; Geiger counters are the most commonly used portable radiation detection instruments

General staff: a group of incident management personnel organized according to function and reporting to the incident commander; the general staff normally consists of the operations section chief, planning section chief, logistics section chief, and finance/administration section chief (an intelligence/investigations chief may be established, if required, to meet incident management needs)

Generator: EPA term for any person, organization, or agency whose act or process produces medical waste or causes waste to become subject to regulation

Genetic effects: hereditary effects (mutations) that can be passed on through reproduction because of changes in sperm or ova

Glacial: a term applied to a number of acids, which, in a highly pure state, have a freezing point slightly below room temperature

Golden hour: a principle that states unstable victims must be stabilized within one hour following injury to reduce the risk of death

Gram: a standard unit of mass (weight) equivalent to 1/453.49 pound

Gravimetric: a term used by analytical chemists to denote methods of quantitative analysis that depend upon the weight of the components in the sample

Gray (gy): a unit of measurement for the amount of energy absorbed in a material

Ground zero: the central point of a nuclear detonation (or other large blast); refers to the point on the ground below or above a nuclear detonation if the device is triggered in the air or underground

H

Half-life: the time any substance takes to decay by half of its original amount; see also biological half-life, decay constant, effective half-life, radioactive half-life

Hand antisepsis: refers to either antiseptic hand wash or antiseptic hand rub

Hand hygiene: a general term that applies to either handwashing, antiseptic hand wash, antiseptic hand rub, or surgical hand antisepsis

Handgrips: devices attached to either side of the bed to provide the patient/resident the ability to reposition themselves while in bed as well as an aid to enter and leave the bed

Hard target: a building, piece of critical infrastructure (i.e. dam, power plant, utility company, etc.) or other commercial or noncommercial entity that has rigid security measures in place including barriers, cameras, guards, etc.

Hazard: a potential or actual force, physical condition, or agent with the ability to cause human injury, illness, and/or death, and significant damage to property, the environment, critical infrastructure, agriculture and business operations, and other types of harm or loss

Hazard classes (DOT): the nine descriptive terms established by the United Nations committee of experts to categorize hazardous chemical, physical, and biological materials; categories are flammable liquids, explosives, gaseous oxidizers, radioactive materials, corrosives, flammable solids, poisons, infectious substances, and dangerous substances

Hazard control: a means of reducing the risk due to exposure to a hazard; such means may include ergonomic designing of workstations and equipment; arranging, safety-guarding and interlocking of equipment; barricading of pedestrian and vehicular traffic routes; controlling exposure to toxic materials; and wearing protective gear

Hazard control management: the practice of identifying, evaluating, and controlling hazards to prevent accidents

Hazard identification: a process to identify hazards and associated risk to persons, property, and structures and to improve protection from natural and human-caused hazards

Hazard Identification and Risk Assessment (HIRA): a process to identify hazards and associated risk to persons, property, and structures and to improve protection from natural and human-caused hazards; HIRA serves as a foundation for planning, resource management, capability development, public education, and training and exercises

Hazard operability study: a structured means of evaluating a complex process to find problems associated with operability or safety of the process

Hazard rating (NFPA): classification system that uses a four-color diamond to communicate health, flammability, reactivity, and specific hazard information for a chemical substance; a numbering system that rates hazards from zero (lowest) to four (highest)

Hazard vulnerability analysis (healthcare): the identification of potential emergencies and direct and indirect effects these emergencies may have on the healthcare organization's operations and the demand for its services

Hazard vulnerability analysis (HVA): a systematic approach to identifying all potential hazards that may affect an organization; assessing the probability of occurrence, and the consequences for each hazard, the organization creates a prioritized comparison of the hazard vulnerabilities

Hazardous material (HAZMAT): any substance or material that when involved in an accident and released in sufficient quantities, poses a risk to people's health, safety, and/or property; these substances and materials include explosives, radioactive materials, flammable liquids or solids, combustible liquids or solids, poisons, oxidizers, toxins, and corrosive materials; (DOT) a substance or material that has been determined by DOT to pose an unreasonable risk to health, safety, and property when transported in commerce, 49 CFR 171.8

Hazardous motion: any motion of machinery or equipment that could cause personal physical harm

Hazardous substance: any substance to which exposure may result in adverse effects on the health or safety of employees; includes substances defined under Section 101(14) of CERCLA; biological or disease-causing agents that may reasonably be anticipated to cause death, disease, or other health problems; any substance listed by the U.S. Department of Transportation as hazardous material under 49 CFR 172.101 and appendices; and substances classified as hazardous waste

Hazardous waste (EPA): any solid or combination of solid wastes that because of its physical, chemical, or infectious characteristics may pose a hazard when not managed properly

HAZCOM: the OSHA hazard communication standard (29 CFR 1910.1200)

HAZMAT: see Hazardous material

HAZWOPER: OSHA Standard on Hazardous Waste Operations and Emergency Response, 29 CFR 1910.120; paragraph (q) of this standard covers employers whose employees are engaged in emergency response to hazardous substance releases

Health hazard: a chemical for which there is statistically significant evidence that acute or chronic health effects may occur in exposed individuals

Health physics: a scientific field that focuses on protection of humans and the environment from radiation; health physics uses physics, biology, chemistry, statistics, and electronic instrumentation to help protect individuals from any damaging effects of radiation

Healthcare coalition: a group of individual healthcare organizations in a specified geographic area that agree to work together to enhance their response to emergencies or disasters; the healthcare coalition, being composed of relatively independent organizations that voluntarily coordinate their response, do not conduct command or control; operates consistent with multiagency coordination system (MAC system) principles to support and facilitate the response of its participating organizations

Hearing conservation: preventing or minimizing noise-induced deafness through the use of hearing protection devices, engineering methods, annual audiometric tests, and employee training

Helium neon laser: a laser in which the active medium is a mixture of helium and neon; its wavelength is usually in the visible range

High impact mat: a mat placed next to the bed that absorbs the shock if the patient falls from the bed

High level radioactive waste: the radioactive material resulting from spent nuclear fuel reprocessing; this can include liquid waste directly produced in reprocessing or any solid material derived from the liquid wastes having a sufficient concentration of fission products

High-efficiency particulate air filter: any filter with at least 99.97 percent efficiency in the filtration of air borne particles 0.3 microns in diameter or greater

High-reliability organizations: an organization or entity that operates in a culture with accepted risks but does so with everyone working to reduce frequency and severity of mishaps, accidents, errors, and incidents

Homeland security: a concerted national effort to prevent terrorist attacks within the United States, reduce America's vulnerability to terrorism, and minimize the damage and recover from attacks that do occur; homeland security includes federal, state, and local governments, the private sector, and individual citizens

Homeland Security Council (HSC): entity that advises the President on national strategy and policy during large-scale incidents; together with the National Security Council, ensures coordination for all homeland and national security-related activities among executive departments and agencies and promotes effective development and implementation of related policy

Hospital decontamination zone: a zone that includes any areas where the type and quantity of hazardous substance are unknown and where contaminated victims, contaminated equipment, or contaminated waste may be present; this zone is sometimes called the warm zone, contamination reduction zone, yellow zone, or limited access zone

Hospital incident command system: a generic crisis management plan expressly for comprehensive medical facilities that is modeled closely after the fire service incident command system

Hospital post-decontamination zone: an area considered uncontaminated, sometimes called the cold zone or clean area

Hot spot: any place where the level of radioactive contamination is considerably greater than the area around it

Hot zone: an area immediately surrounding an incident, which extends far enough to prevent adverse effects from the device/agent to personnel outside the zone; also referred to as the exclusion zone, real zone, or restricted zone; the area that the incident commander judges to be the most affected by the incident, including any area to which the contaminant has spread or is likely to spread, after giving consideration to the type of agent, the volume released, the means of dissemination, the prevailing meteorological conditions, and the potential effects of local topography; priorities within the hot zone may include conducting rescue and search, performing mitigation, and identifying weapons of mass destruction (WMD) or other physical obstacles to the entry point

HSPD 5: Homeland Security Presidential Directive that addresses the management of domestic incidents, including a national response plan to integrate all federal government domestic prevention, preparedness, response, and recovery plans into a single all discipline, all hazards plan; hospitals should be national incident management system compliant and develop an all hazards approach to emergency management

HSPD 6: Homeland Security Presidential Directive that addresses the integration and use of screening information to protect against terrorism and uses information as appropriate and to the full extent permitted by law to support (a) federal, state, local, territorial, tribal, foreign-government, and private-sector screening processes, and (b) diplomatic, military, intelligence, law enforcement, immigration, visa, and protective processes

HSPD 7: Homeland Security Presidential Directive that addresses critical infrastructure, identification, prioritization, and protection

HSPD 8: Homeland Security Presidential Directive that addresses national preparedness and calls for a national preparedness goal that establishes measurable priorities and targets and an approach to developing needed capabilities; it clearly states that hospital emergency medical facilities are considered emergency response providers as defined by the Department of Homeland Security Act

HSPD 9: Homeland Security Presidential Directive that addresses the defense of U.S. agriculture and food; the directive establishes a national policy to defend the agriculture and food system against terrorist attacks, major disasters, and other emergencies

HSPD 10: Homeland Security Presidential Directive that was issued in response to the fears of bioterrorism following the anthrax attacks of 2001, the threat of pandemic influenza, and the outbreak of severe acute respiratory syndrome; serves as the basis for the country's biodefense program; the directive calls upon hospitals to not only plan for more traditional hazards, such as explosive or incendiary threats, but also to be ready to respond to bioterrorism attacks

HSPD 12: Homeland Security Presidential Directive that addresses a common identification standard for federal employees and contractors; directive was issued in response to the fears of bioterrorism following the anthrax attacks of 2001 and to address the threat of pandemic influenza and severe acute respiratory syndrome

HSPD 20: Homeland Security Presidential Directive that addresses the national continuity policy that establishes a comprehensive national policy on the continuity of federal government structures and operations, and also creates the position of a single national continuity coordinator responsible for coordinating the development and implementation of federal continuity policies

HSPD 21: Homeland Security Presidential Directive that addresses the national strategy for public health and medical preparedness

Humidity (relative): the ratio of the amount of water vapor present in air at a given temperature to the maximum that can be held by air at that temperature

Hydrocarbon: any compound composed of carbon and hydrogen

Hydrophilic: a term that refers to substances that tend to absorb and retain water

Hydrophobic: a term that describes substances which repel water

Hygroscopic: a term used to describe solid or liquid materials that pick up and retain water vapor from the air

Hypersensitivity diseases: diseases characterized by allergic responses to animal antigens, often associated with indoor air quality conditions such as asthma and rhinitis

I

IDLH: see Immediately dangerous to life or health

Ignitable solid, liquid, or compressed gas: must have a flash point less than 140 degrees F

Ignition temperature: lowest temperature at which a substance can catch fire and continue to burn

Illumination: the amount of light a surface receives per unit area, expressed in lumens per square foot or foot candles

Immediately dangerous to life or health: an atmospheric concentration of any toxic, corrosive, or asphyxiates substance that poses an immediate threat to life or would interfere with an individual's ability to escape from a dangerous atmosphere

Immiscible: a term used to describe substances of the same phase that cannot be uniformly mixed or blended

Impact area: an area having designated boundaries within the limits of which all ordnance will detonate or impact

Incapacitating agent: an agent that produces physiological or mental effects, or both, that may persist for hours or days after exposure, rendering an individual incapable of performing his or her assigned duties

Incident: an actual or impending unplanned event with hazard impact, either caused by humans or natural phenomena, that requires action by emergency personnel to prevent or minimize loss of life or damage to property and/or natural resources

Incident Action Plan (IAP): an oral or written plan containing general objectives reflecting the overall strategy for managing an incident. It may include the identification of operational resources and assignments. It may also include attachments that provide direction and important information for management of the incident during one or more operational periods

Incident Command System (ICS): a standardized on-scene emergency management construct specifically designed to provide for the adoption of an integrated organizational structure that reflects the complexity and demands of single or multiple incidents, without being hindered by jurisdictional boundaries; used for all kinds of emergencies and applicable to small as well as large and complex incidents

Incident Commander (IC): the individual responsible for all incident activities, including the development of strategies and tactics and the ordering and the release of resources; has overall authority and responsibility for conducting incident operations and is responsible for the management of all incident operations at the incident site

Incident management: refers to how incidents are managed across all homeland security activities, including prevention, protection, and response and recovery

Incident management team: the Incident Commander and appropriate command and general staff personnel assigned to an incident

Incident objectives: statements of guidance and direction necessary for selecting appropriate strategies and the tactical direction of resources; incident objectives are based on realistic expectations of what can be accomplished when allocated resources have been effectively deployed

Incompatible: the term used to indicate that one material cannot be mixed with another without the possibility of a dangerous reaction

Indicator: a measurement used to evaluate program effectiveness within an organization

Indoor air quality: the study, evaluation, and control of indoor air quality related to temperature, humidity, and building contaminants

Industrial equipment: physical apparatus used to perform industrial tasks, such as welders, conveyors, machine tools, fork trucks, turn tables, positioning tables, or robots

Industrial robot: a reprogrammable, multifunctional manipulator designed to move material, parts, tools, or specialized devices through variable programmed motions for the performance of a variety of tasks

Industrial robot system: a system that includes industrial robots, the end-effectors, and the devices and sensors required for the robots to be taught or programmed, or for the robots to perform the intended automatic operations, as well as the communication interfaces required for interlocking, sequencing, or monitoring the robots

Inert: having little or no chemical affinity or activity

Infectious waste: waste-containing pathogens that can cause an infectious disease in humans

Infrared radiation: invisible electromagnetic radiation with wavelengths that lie within the range of 0.70 to 1000 micrometers

Ingestion: taking a substance into the body through the mouth

Inhalation: breathing of an airborne substance into the body; may be in the form of a gas, vapor, fume, mist, or dust

Inhibitor: a substance that is added to another substance to prevent or slow down an unwanted reaction or change

Innocuous: harmless

Inorganic: this term refers to a major and the oldest branch of chemistry; concerned with substances that do not contain carbon

Intelligence: the process by which analysis is applied to information and data to inform policy making, decision making, operations and tactical decisions; serves many purposes among which are the identification and elimination of threat sources, the investigations and resolution of threats, the identification and treatment of a security risk, and the elimination of a threat source

Interior structural firefighting: the physical activity of fire suppression, rescue, or both, inside of buildings or enclosed structures that are involved in a fire situation beyond the incipient stage

Interior zone: a protective zone established inside a perimeter zone. Also called a secondary zone

Interlock: an arrangement whereby the operation of one control or mechanism brings about or prevents the operation of another

Internal exposure: exposure to radioactive material taken into the body

Interoperability: the ability of emergency management/response personnel to interact and work well together, in the context of technology; also refers to having an emergency communications system that is the same or is linked to the same system that a jurisdiction uses for nonemergency procedures, and that effectively interfaces with national standards as they are developed

Intra-beam viewing: the viewing condition whereby the eye is exposed to all or part of a direct laser beam or a specular reflection

Iodine: a nonmetallic solid element; there are both radioactive and nonradioactive isotopes of iodine

Ion: an atom, group, or molecule that has either lost one or more electrons or gained one or more electrons

Ionization: the process of adding one or more electrons to, or removing one or more electrons from, atoms or molecules, thereby creating ions; high temperatures, electrical discharges, or nuclear radiation can cause ionization

Ionizing radiation: any radiation capable of displacing electrons from atoms, thereby producing ions; high doses of ionizing radiation may produce severe skin or tissue damage

Irradiation: exposure to radiation

Irradiation sterilization: the use of ionizing radiation for the treatment of infectious waste

Isolation zone: an area adjacent to a physical barrier, clear of all objects which could conceal or shield an individual

Isomer: one of two or more compounds having the same molecular weight and formula, but often having quite different properties and somewhat different structure

Isotonic: having the same osmotic pressure as the fluid phase of a cell or tissue

Isotope: a nuclide of an element having the same number of protons but a different number of neutrons; any of two or more forms of an element in which the weights differ by one or more mass units due to a variation in the number of neutrons in the nuclei

J

Jersey barrier: a protective concrete barrier initially and still used as a highway divider; now functions as an expedient method for traffic speed control at entrance gates and to keep vehicles away from buildings

Job hazard analysis: the breaking down of methods, tasks, or procedures into components to determine hazards

Joint Field Office (JFO): the primary federal incident management field structure; a temporary federal facility that provides a central location for the coordination of federal, state, tribal, and local governments and private sector and nongovernmental organizations with primary responsibility for response and recovery; the JFO structure is organized, staffed, and managed in a manner consistent with National Incident Management System principles and is led by the Unified Coordination Group; uses an Incident Command System structure, but does not manage on-scene operations, and instead focuses on providing support to on-scene efforts and conducting broader support operations that may extend beyond the incident site

Joint Information Center (JIC): an interagency entity established to coordinate and disseminate information for the public and media concerning an incident; may be established locally, regionally, or nationally depending on the size and magnitude of the incident

Joint Information System (JIS): mechanism that integrates incident information and public affairs into a cohesive organization designed to provide consistent, coordinated, accurate, accessible, timely, and complete information during crisis or incident operations; mission of the JIS is to provide a structure and system for developing and delivering coordinated interagency messages; developing, recommending, and executing public information plans and strategies on behalf of the Incident Commander; advising the Incident Commander concerning public affairs issues that could affect a response effort; and controlling rumors and inaccurate information that could undermine public confidence in the emergency response effort

Joint motion: a method for coordinating the movement of the joints such that all joints arrive at the desired location simultaneously

Joint Operations Center (JOC): an interagency command post established by the Federal Bureau of Investigation to manage terrorist threats or incidents and investigative and intelligence activities; to coordinate the necessary local, state, and federal assets required to support the investigation; and to prepare for, respond to, and resolve the threat or incident

Joint Task Force (JTF): based on the complexity and type of incident, and the anticipated level of DOD resource involvement, DOD may elect to designate a JTF to command federal (Title 10) military activities in support of the incident objectives; if established, consistent with operational requirements, its command and control element will be co-located with the senior on-scene leadership at the JFO to ensure coordination and unity of effort

Joule: unit of energy used to describe a single pulsed output of a laser; equal to one watt-second or 0.239 calories

Jurisdiction: a political subdivision with the responsibility for ensuring public safety, health, and welfare within its legal authorities and geographic boundaries

K

Ketone: a class of unsaturated and reactive compounds whose formula is characterized by a carbonyl group to which two organic groups are attached

Kevlar: a synthetic yellow-brown fiber of very high tensile strength, woven into bulletproof vests, and molded into solid sheets of lightweight armor (from aircraft to helmets); Kevlar is the brand name from DuPont

Kilogram: about 2.2 pounds

Kiloton: the energy of an explosion that is equivalent to an explosion of 1000 tons of TNT

Kinetic energy: the energy that a particle or an object possesses due to its motion or vibration

L

Lab pack: generally refers to any small container of hazardous waste in an over-packed drum; not restricted to laboratory wastes

Lacquer: a type of organic coating in which rapid drying is effected by evaporation of solvents

Laser: a term for light amplification by stimulated emission of radiation, a laser is a cavity with mirrors at the ends, filled with material such as crystal, glass, liquid, gas or dye; produces an intense beam of light with the unique properties of coherency, collimation, and monochromaticity

Laser medium: material used to emit the laser light and for which the laser is named

Laser safety officer: person with authority to monitor and enforce measures to control laser hazards and effect the knowledgeable evaluation and control of laser hazards

Laser system: an assembly of electrical, mechanical, and optical components of a laser; under federal standards it also includes the power supply

Latent period: time between exposure to a toxic material and the appearance of a resultant health effect

Layered security: a physical security approach that requires a criminal to penetrate or overcome a series of security layers before reaching the target; the layers might be perimeter barriers; building or area protection with locks, CCTV and guards; and point and trap protection using safes, vaults, and sensors

Leading: a person who creates an atmosphere and purpose that encourages people to succeed and achieve

Leak test: a test performed to detect leakage of a radiation source

Lens: a curved piece of optically transparent material that, depending on its shape, is used to either converge or diverge light

LEPC: Local Emergency Planning Committee

Level of analysis: the level of analysis applies to the system hardware or functional level at which failures are postulated

Liaison officer: a member of the command staff responsible for coordinating with representatives from cooperating and assisting agencies or organizations

Light-emitting diode: a semiconductor diode that converts electric energy efficiently into spontaneous and noncoherent electromagnetic radiation at visible and near-infrared wavelengths

Limited area: a restricted area within close proximity of a security interest; uncontrolled movement may permit access to the item, while escorts and other internal restrictions may prevent access to the item

Limiting aperture: maximum circular area over which radiance and radiant exposure can be averaged when determining safety hazards

Limiting device: a device that restricts the maximum envelope (space) by stopping or causing to stop all robot motion; independent of the control program and the application programs

Line organization: an organization with a chain of command hierarchy

Liquefied petroleum: a gas usually comprised of propane and some butane created as a byproduct of petroleum refining

Liquid crystal display: a constantly operating display that consists of segments of a liquid crystal whose reflectivity varies according to the voltage applied to the unit

Local exhaust ventilation: a ventilation system that captures/removes contaminants at the point produced before they escape into the work area

Long-term recovery: a process of recovery that may continue for a number of months or years, depending on the severity and extent of the damage sustained

Loose-fitting face piece: a respiratory inlet covering that is designed to form a partial seal with the face

Loss ratio: a fraction calculated by dividing losses of an organization and the amount of insurance premiums paid

Lost workdays: the number of workdays an employee is away from work beyond the day of injury or onset of illness

Lower explosive limit: the lowest concentration of a substance that will produce a fire or flash when an ignition source is present, expressed as a percent of vapor or gas in the air by volume

Low-level waste: radioactively contaminated industrial or research waste such as paper, rags, plastic bags, medical waste, and water-treatment residues

Lumbar: the section of the lower vertebral column immediately above the sacrum, located in the small of the back and consists of five large lumbar vertebrae

M

Major disaster: under the Robert T. Stafford Disaster Relief and Emergency Assistance Act, any natural catastrophe (including any hurricane, tornado, storm, high water, wind-driven water, tidal wave, tsunami, earthquake, volcanic eruption, landslide, mudslide, snowstorm, or drought) or, regardless of cause, any fire, flood, or explosion in any part of the United States that, in the determination of the President, causes damage of sufficient severity and magnitude to warrant major disaster assistance under the Stafford Act to supplement the efforts and available resources of states, local governments, and disaster relief organizations in alleviating the damage, loss, hardship, or suffering caused thereby

Management by exception: a manager makes a decision by reviewing key information and not all available information on a subject

Management by objective: a management theory where a manager and subordinates agree on a predetermined course of action or objective

Mass: the amount of material substance present in a body, irrespective of gravity

Mass casualty incident: an incident that generates a sufficiently large number of casualties whereby the available healthcare resources, or their management systems, are severely challenged or unable to meet the healthcare needs of the affected population

Mass effect incident: an incident that primarily affects the ability of an organization to continue its normal operations

Mass spectroscopy: process that identifies compounds by breaking them up into all combinations of ions and measuring mass-to-charge ratios at detector

Material Safety Data Sheet: a document that contains descriptive information on hazardous chemicals under the OSHA hazard communication standard; data sheets also provide precautionary information, safe handling procedures, and emergency first-aid procedures

Maximum permissible exposure: the level of laser radiation to which a person may be exposed without hazardous effect or adverse biological changes in the eye or skin

Measure: term used in the quality field for the collection of quantifiable data and information about performance, production, and goal accomplishment

Measures of effectiveness: defined criteria for determining whether satisfactory progress is being accomplished toward achieving the incident objectives

Medical surge: the ability to provide adequate medical evaluation and care in events that severely challenge or exceed the normal medical infrastructure of an affected community

Medical waste: any solid waste generated in the diagnosis, treatment, or immunization of humans or animals

Megaton (mt): the energy of an explosion that is equivalent to an explosion of one million tons of TNT; one megaton is equal to a quintillion (1018) calories; see also kiloton

Memorandum of agreement: a conditional agreement between two or more parties; one party's action depends on the other party's action

Memorandum of understanding: a formal agreement documenting the commitment of two or more parties to an agreed undertaking

Method of dissemination: the way a chemical agent or compound is finally released into the atmosphere

Microbe: minute organism, including bacteria, protozoa, and fungi, which is capable of causing disease

Micron: a unit of length in the metric system equivalent to one-millionth of a meter

Milligrams per cubic meter: unit used to measure air concentration of dust, gas, mist, and fume

Mindfulness: a culture that encourages alertness/vigilance to prevent error, mitigate harm, prevent injury, and limit damage to property or the environment

Mitigation: activities designed to reduce or eliminate risks to persons or property or to lessen the actual or potential effects or consequences of a hazard

Mobile robots: freely moving automatic programmable industrial robots

Mobilization: process and procedures used by all organizations for activating, assembling, and transporting all resources that have been requested to respond to or support a disaster incident

Molecular weight: the total obtained by adding together the weights of all the atoms present in a molecule

Molecule: a combination of two or more atoms that are chemically bonded; the smallest unit of a compound that can exist by itself and retain all of its chemical properties

Multi-agency coordination (MAC): a group of administrators/executives, or their appointed representatives, who are authorized to commit agency resources and funds, and can provide coordinated decision-making and resource allocation among cooperating agencies

Multi-agency coordination system(s) (MACS): multi-agency coordination systems provide the architecture to support coordination for incident prioritization, critical resource allocation, communications systems integration, and information coordination; elements of multi-agency coordination systems include facilities, equipment, personnel, procedures, and communications

Multi-jurisdictional incident: an incident requiring action from multiple agencies that each have jurisdiction to manage certain aspects of the incident; in the Incident Command System, these incidents will be managed under Unified Command

Munroe effect: the focusing of the force produced by an explosion resulting in an increased pressure wave

Mutagen: a substance or agent capable of changing the genetic material of a living cell

Muting: the deactivation of a presence-sensing safeguarding device during a portion of the robot cycle

Mutual aid agreement: written instrument between agencies and/or jurisdictions in which they agree to assist one another upon request, by furnishing personnel, equipment, supplies, and/or expertise in a specified manner

N

Naphtha: any of several liquid mixtures of hydrocarbons of specific boiling and distillation ranges derived from either petroleum or coal tar

Narcosis: stupor or unconsciousness caused by exposure to a chemical

National Disaster Medical System (NDMS): a federally coordinated system that augments the nation's medical response capability; purpose of the NDMS is to establish a single, integrated national medical response capability for assisting state and local authorities in dealing with the medical impacts of major peacetime disasters

National Incident Management System (NIMS): system that provides a proactive approach guiding government agencies at all levels, the private sector, and nongovernmental organizations to work seamlessly to prepare for, prevent, respond to, recover from, and mitigate the effects of incidents, regardless of cause, size, location, or complexity, in order to reduce the loss of life or property and harm to the environment

National Joint Terrorism Task Force (NJTTF): entity responsible for enhancing communications, coordination, and cooperation among federal, state, tribal, and local agencies representing the intelligence, law enforcement, defense, diplomatic, public safety, and homeland security communities by providing a point of fusion for terrorism intelligence and by supporting Joint Terrorism Task Forces throughout the United States

National Military Command Center (NMCC): facility that serves as the nation's focal point for continuous monitoring and coordination of worldwide military operations; directly supports combatant commanders, the Chairman of the Joint Chiefs of Staff, the Secretary of Defense, and the President in the command of U.S. Armed Forces in peacetime contingencies and war

National Operations Center (NOC): serves as the primary national hub for situational awareness and operations coordination across the federal government for incident management; NOC provides the Secretary of Homeland Security and other principals with information necessary to make critical national-level incident management decisions

National Preparedness Guidelines: guidance that establishes a vision for national preparedness and provides a systematic approach for prioritizing preparedness efforts across the nation; these Guidelines focus on policy, planning, and investments at all levels of government and the private sector, replace the Interim National Preparedness Goal, and integrate recent lessons learned

National Response Coordination Center (NRCC): as a component of the National Operations Center, serves as the Department of Homeland Security/Federal Emergency Management Agency primary operations center responsible for national incident response and recovery as well as national resource coordination; monitors potential or developing incidents and supports the efforts of regional and field components

National Response Framework (NRF): guides how the nation conducts all-hazards response; documents the key response principles, roles, and structures that organize national response; and describes how communities, states, the federal government, and private sector and nongovernmental partners apply these principles for a coordinated, effective national response; also describes special circumstances where the federal government exercises a larger role

National Security Council (NSC): advises the President on national strategy and policy during large-scale incidents; together with the Homeland Security Council, ensures coordination for all homeland and national security–related activities among executive departments and agencies and promotes effective development and implementation of related policy

Natural gas: a combustible gas composed largely of methane and other hydrocarbons obtained from natural earth fissures

Necrosis: death of plant or animal cells

Needleless systems: devices that do not use needles for the collection of bodily fluids or withdrawal of body fluids after initial venous or arterial access is established

Negative pressure: a condition caused when less air is supplied to a space than is exhausted from the space; air pressure in the space is less than that in surrounding areas

Negligence: failure to do what reasonable and prudent persons would do under similar or existing circumstances

Nerve agents: agents that effect the transmission of nerve impulses by reacting with the enzyme cholinesterase, permitting an accumulation of acetylcholine and continuous muscle stimulation; muscles tire due to over-stimulation and begin to contract

Neutralization: the reaction between equivalent amounts of an acid and a base to form a salt

Neutron: a small atomic particle possessing no electrical charge typically found within an atom's nucleus; neutral in their charge

Night vision: technology used in a variety of devices, using (passive) image intensifiers (intensification of residual light) and/or thermal (infrared) imagers to improve observation, target acquisition, or aiming in low light conditions; they can be coupled with (active) laser aiming lights (laser illuminators or designators, target markers, spot projectors), and take the form of handheld or helmet-mounted binocular and monocular goggles, pocket scopes, rifle-mounted weapon sights, or armored vehicle periscopes

Nitrogen oxide: compound produced by combustion

Nomenclature: names of chemical substances and the system used for assigning them

Nonexclusive zone: an area around an asset that has controlled entry but shared or less restrictive access than an exclusive zone

Nonionizing radiation: radiation that has lower energy levels and longer wavelengths than ionizing radiation; examples include radio waves, microwaves, visible light, and infrared

Nonlethal weapon: weapons used by friendly forces designed to incapacitate the target or otherwise neutralize hostile forces rather than to kill or seriously injure; examples include gas, such as tear gas, and stun grenades

Nonpersistant agent: an agent that remains in the target area(s) for a relatively short period of time; the hazard, predominantly vapor, will exist for minutes or, in exceptional cases, hours after dissemination of the agent (as a general rule, a nonpersistent agent's duration will be less than 12 hours)

Nonstochastic effects: effects that can be related directly to the radiation dose received

Nucleus: central part of an atom that contains protons and neutrons

Nuclide: a general term applicable to all atomic forms of an element
Numerically controlled machine tools: tools operated by a series of coded instructions comprised of numbers, letters of the alphabet, and other symbols

O

Occupational illness: illness caused by environmental exposure during employment
Occurrence: incident classified as major or minor that results from apparent or foreseen causal factors
Odor threshold: minimum concentration of a substance at which most people can detect and identify its characteristic odor
Operating envelope: the portion of the restricted envelope (space) that is actually used by the robot while performing its programmed motions
Operations level: see First responder operations level
Operations plan: a plan developed by and for each federal department or agency describing detailed resource, personnel, and asset allocations necessary to support the concept of operations detailed in the concept plan
Optical cavity (resonator): space between the laser mirrors where lasing action occurs
Optical density: logarithmic expression of the attenuation afforded by a filter
Optical fiber: a filament of quartz or other optical material capable of transmitting light along its length by multiple internal reflections and emitting it at the end
Order of magnitude: a term used in science to indicate a range of values representing numbers, dimensions, or distances which start at any given value and end at 10 times that value
Ordnance: weapons, ammunition, or other consumable armament
Organic: any compound containing the element carbon; describes substances derived from living organisms
Organizing: arranging work or tasks to be performed in the most efficient manner
Outcomes: results reached due to performance or nonperformance of a task, job, or process
Output power: energy per second measured in watts emitted from the laser in the form of coherent light
Overt culture: the formal, expected, published, visible, or anticipated culture of an organization
Overt threat: a terrorist act that is done out in the open without regard to possible discovery
Oxidant: an oxygen-containing substance that reacts chemically to produce a new substance
Oxidation: reaction in which electrons are transferred from one atom to another, either in the uncombined state or within a molecule
Oxidizers: materials that may cause the ignition of a combustible material without the aid of an external ignition source
Oxygen-deficient atmosphere: an atmosphere with oxygen content below 19.5 percent by volume
Ozone: reactive oxidant that contains three atoms of oxygen

P

Parenteral: piercing mucous membranes or the skin barrier through such events as needle ticks, human bites, cuts, and abrasions
Particulates: fine solid or liquid particles found in air and other emissions
Parts per million: a unit for measuring the concentration of a gas or vapor in contaminated air; used to indicate the concentration of a particular substance in a liquid or solid
Pasteurization: heat treatment of liquid or semi-liquid food products for the purpose of killing or inactivating disease-causing bacteria
Pathways: the routes by which people are exposed to radiation or other contaminants; the three basic pathways are inhalation, ingestion, and direct external exposure

Patient assessment: an assessment that provides ongoing information necessary to develop a care plan, to provide the appropriate care and services for each patient

Patient safety science: helps create systems that do no harm

PEL: permissible exposure limit; OSHA limit for employee exposure to chemicals (29 CFR 1910.1000) based on a time-weighted average of hours for a 40-hour work week

Pendant: any portable control device, including teach pendants, that permits an operator to control the robot from within the restricted envelope space of a robot

Pendant control: a means used by either the patient or the operator to control the drives that activate various bed functions; attached to the bed by a cord

Penetrating radiation: any radiation that can penetrate the skin and reach internal organs and tissues

Perimeter: the edge or boundary of property or location

Periodic law: states that the arrangement of electrons in the atoms of any given chemical element, and the properties determined by this arrangement, are closely related to the atomic number of that element

Periodic table: a systematic classification of the chemical elements based on the periodic law

Permeation rate: invisible process by which a hazardous chemical moves through a protective material

Persistent activity: activity defined as the prolonged or extended antimicrobial activity that prevents or inhibits the proliferation or survival of microorganisms after application of the product

Persistent agent: an agent that remains in the target area for longer periods of time; hazards from both vapor and liquid may exist for hours, days, or in exceptional cases, weeks or months after dissemination of the agent (as a general rule, a persistent agent's duration will be greater than 12 hours)

Personal protection: equipment designed to protect individuals against injury from firearms, nuclear or conventional explosives, and chemical and/or biological agents

Personal protective equipment (PPE): examples include protective suits, gloves, foot covering, respiratory protection, hoods, safety glasses, goggles, and face shields

pH: a scale indicating the acidity or alkalinity of aqueous solutions

Physical hazard of a chemical: a chemical validated as being or having one of the following characteristics: combustible liquid, compressed gas, explosive, flammable, organic peroxide, oxidizing qualities, pyrophoric, unstable, or water reactive

Planning: actions taken to predetermine the best course of action

Planning section: functional area responsible for the collection, evaluation, and dissemination of operational information related to the incident and for the preparation and documentation of the incident action plan and its support

Plume: smoke from the use of electrosurgery, lasers, and aerosols

Poison: solid or liquid substance that is known to be toxic to humans

Polychlorinated biphenyls: a pathogenic and teratogenicity industrial compound used as a heat transfer agent; they accumulate in human or animal tissue

Polymerization: a chemical reaction in which one or more small molecules combine to form larger molecules

Polyvinyl chloride: a member of the family of vinyl resins

Positive pressure: a respirator in which the pressure inside the respiratory inlet covering exceeds the ambient air pressure outside the respirator

Powered air-purifying respirator (PAPR): a PAPR uses a battery-powered blower to force air through a filter or purifying cartridge before blowing the cleaned air into the respirator face piece

Pre-action: the main water control valve is opened by an actuating device

Pre-emptive attack: an attack initiated on the basis of incontrovertible evidence that an enemy attack is imminent

Pre-filter: a filter used in conjunction with a cartridge on an air-purifying respirator

Preparedness: actions that involve a combination of planning, resources, training, exercising, and organizing to build, sustain, and improve operational capabilities; the process of identifying the personnel, training, and equipment needed for a wide range of potential incidents, and developing jurisdiction-specific plans for delivering capabilities when needed for an incident

Presence-sensing safeguarding device: a device designed, constructed, and installed to create a sensing field or area to detect an intrusion into the field or area by personnel, robots, or other objects

Pressure-demand respirator: a positive pressure atmosphere-supplying respirator that admits breathing air to the face piece when the positive pressure is reduced inside the face piece by inhalation

Prevention: actions to avoid an incident or to intervene to stop an incident from occurring; involves actions to protect lives and property, and applying intelligence and other information to a range of activities that may include such countermeasures as deterrence operations; heightened inspections; improved surveillance and security operations; investigations to determine the full nature and source of the threat; public health and agricultural surveillance and testing processes; immunizations, isolation, or quarantine; and, as appropriate, specific law enforcement operations aimed at deterring, preempting, interdicting, or disrupting illegal activity and apprehending potential perpetrators and bringing them to justice

Priority I: patients with correctable life-threatening illnesses or injuries such as respiratory arrest or obstruction, open chest or abdomen wounds, femur fractures, or critical or complicated burns

Priority II: patients with serious but non-life-threatening illnesses or injuries such as moderate blood loss, open or multiple fractures (open increases priority), or eye injuries

Priority III: patients with minor injuries such as soft tissue injuries, simple fractures, or minor to moderate burns

Priority zero (or IV): patients who are dead or fatally injured; fatal injuries include exposed brain matter, decapitation, and incineration

Process: method of interrelating steps, events, and mechanisms to accomplish an action or goal

Process flow diagram: sequence of events diagram

Protocol: a set of established guidelines for actions that may be designated by individuals, teams, functions, or capabilities under various specified conditions

Proton: basic unit of mass that is a constituent of the nucleus of all elements, the number present being the atomic number of a given element

Public information: processes, procedures, and systems for communicating timely, accurate, accessible information on an incident's cause, size, and current situation; resources committed; and other matters of general interest to the public, responders, and additional stakeholders (both directly affected and indirectly affected)

Public Information Officer (PIO): a member of the command staff responsible for interfacing with the public and media and/or with other agencies with incident-related information requirements

Pulse duration: the "on" time of a pulsed laser; may be measured in terms of milliseconds, microseconds, or nanoseconds

Pyrolysis: a chemical change brought about by heat alone

Pyrophoric: a chemical that will ignite spontaneously in air at a temperature of 130 degrees F or below

Q

Qualitative analysis: examination of a sample of a material to determine the kinds of substances present and to identify each constituent

Qualitative fit test: a pass/fail fit test to assess the adequacy of respiratory fit that relies on the individual's response to the test agent

Quantitative fit test: an assessment of the adequacy of respirator fit by numerically measuring the amount of leakage into the respirator

Quarantine: restriction of the activities of well persons or animals who have been exposed to a case of communicable disease during its period of communicability (i.e., contacts) to prevent disease transmission during the incubation period if infection should occur; absolute or complete quarantine is the limitation of freedom of movement of those exposed to a communicable disease for a period of time not longer than the longest usual incubation period of that disease; modified quarantine is a selective, partial limitation of freedom of movement of contacts, commonly on the basis of known or presumed differences in susceptibility and related to the danger of disease transmission

Quaternary ammonium compounds: chemical substances used to disinfect or sanitize by rupturing the cell walls of microorganisms

R

Rad (radiation absorbed dose): a basic unit of absorbed radiation dose

Radiation warning symbol: a symbol prescribed by OSHA; magenta on a yellow background, displayed where certain quantities of radioactive materials are present or where certain doses of radiation could be received

Radio assay: a test to determine the amounts of radioactive materials through the detection of ionizing radiation

Radioactivity: the spontaneous decay or disintegration of an unstable atomic nucleus, usually accompanied by the emission of ionizing radiation

Radiography: medical use of radiant energy (such as x-rays and gamma rays) to image body systems

Radioisotope: isotopes of an element that have an unstable nucleus, commonly used in science, industry, and medicine

Reactivity: susceptibility of a substance to undergo chemical reaction and change, which could result in an explosion or fire

Reagent: any chemical compound used in laboratory analyses to detect and identify specific constituents of the material being examined

Reciprocal accountability: between front-line staff and leadership

Recommend exposure limit: a NIOSH chemical exposure limit recommendation

Recovery: the phase of comprehensive emergency management that encompasses activities and programs implemented during and after response that are designed to return the entity to its usual state or to a "new normal"

Red team: a technique for assessing vulnerability that involves viewing a potential target from the perspective of an attacker to identify its hidden vulnerabilities, and to anticipate possible modes of attack

Reflection: return of radiant energy (incident light) by a surface, with no change in wavelength

Refraction: change of direction of propagation of any wave, such as an electromagnetic wave, when it passes from one medium to another in which the wave velocity is different

Regional Response Coordination Centers: located in each FEMA region, these multi-agency coordination centers are staffed by Emergency Support Functions in anticipation of a serious incident in the region or immediately following an incident; operate under the direction

of the FEMA Regional Administrator, coordinate federal regional response efforts, and
     maintain connectivity with state emergency operations centers, state fusion centers, fed-
     eral executive boards, and other federal and state operations and coordination centers that
     have potential to contribute to development of situational awareness

Relative humidity: the ratio of the quantity of water vapor present in air to the quantity that would
     saturate the air at any specific temperature

Relative risk: the ratio between the risks for disease in an irradiated population to the risk in an
     unexposed population

Release zone: an area in and immediately surrounding a hazardous substance release; assumed to
     pose an immediate health risk to all persons, including first responders

Reliability block diagrams: diagrams that define the series dependence, or independence, of all
     functions of a system or functional group for each life-cycle event

Relief valve: a valve designed to release excess pressure within a system without damaging the
     system

Rem: roentgen equivalent man; the unit of dose of any ionizing radiation that produces the same
     biological effect on human tissue as one roentgen of x-rays

Resiliency: the ability of an individual or organization to quickly recover from change or misfortune

Resin: naturally occurring water-insoluble mixtures of carboxylic acids, essential oils, and other
     substances formed in numerous varieties of trees and shrubs

Resonator: mirrors (or reflectors) making up the laser cavity including the laser rod or tube; the
     mirrors reflect light back and forth to build up amplification

Resource Conservation and Recovery Act: legislation used by the EPA to regulate waste materi-
     als, including hazardous wastes, from generation through final disposal

Resource management: a system for identifying available resources at all jurisdictional levels to
     enable timely and unimpeded access to resources needed to prepare for, respond to, or
     recover from an incident

Resources: personnel and major items of equipment, supplies, and facilities available or potentially
     available for assignment to incident operations and for which status is maintained; under
     the National Incident Management System, resources are described by kind and type and
     may be used in operational support or supervisory capacities at an incident or at an emer-
     gency operations center

Respiratory inlet covering: the portion of a respirator that forms the protective barrier between the
     user's respiratory tract and an air-purifying device or breathing air source

Response: activities that address the direct effects of an incident; includes immediate actions to
     save lives, protect property, and meet basic human needs

Response force: the people who respond to an act of aggression; depending on the nature of the
     threat, the response force could consist of guards, special reaction teams, military or civil-
     ian police, an explosives ordnance disposal team, or a fire department.

Restricted area: any area with access controls that is subject to these special restrictions or controls
     for security reasons

Reversible: a chemical reaction that can proceed first to the right and then to the left when the
     conditions change

Reynaud's syndrome: a condition where the blood vessels in the hand constrict from cold tempera-
     ture, vibration, emotion, or unknown causes

Right to know: phrase that relates to an employee's right to know about the nature and hazards of
     agents used in the workplace, and/or to the right of communities

Risk: the probability of injury, illness, disease, loss, or death under specific circumstances

Risk assessment: an evaluation of the risk to human health or the environment by hazards; risk
     assessments can look at either existing hazards or potential hazards

Risk migration: risk mitigated in one part of a system can move to another part

Roentgen: a unit of exposure to x-rays or gamma rays

Roentgen equivalent man: a unit of equivalent dose that relates the absorbed dose in human tissue to the effective biological damage of the radiation

S

Safe haven: secure areas within the interior of the facility; should be designed such that it requires more time to penetrate by terrorist than it takes for the response force to reach the protected area to rescue the occupants

Safeguard: a barrier guard, device, or safety procedure designed for the protection of personnel

Safety: human actions to control, reduce, or prevent accidental loss

Safety belt: a belt worn to prevent falls when working in high places; a belt used to secure passengers in vehicles or airplanes

Safety can: an approved container of not more than five-gallon capacity with a spring-closing lid and a spout cover designed to safely relieve internal pressure when exposed to fire

Safety hat: a hard hat worn to protect a worker from head injuries, flying particles, and electric shock

Safety procedure: an instruction designed for the protection of personnel

Salt: one of the products resulting from a reaction between an acid and a base

Sanitize: to destroy common microorganisms on a surface to a safe level

Scanning laser: a laser having a time-varying direction, origin, or pattern of propagation with respect to a stationary frame of reference

Section: the organizational level having responsibility for a major functional area of incident management such as operations, planning, logistics, finance/administration, and intelligence

Self-contained breathing apparatus (SCBA): a respirator that provides fresh air to the face piece from a compressed air tank

Semiconductor laser: type of laser which produces its output from semiconductor materials

Sense making: a process that transforms raw experience into intelligible views by making sense of new or changing circumstances

Sensitivity: the ability of an analytical method to detect small concentrations of radioactive material

Sensitizer: a substance that may cause no reaction in a person during initial exposure, but will cause an allergic response upon further exposure

Sensor: a device that responds to physical stimuli such as heat, light, sound, pressure, magnetism, or motion

Serious injury: an injury classification that includes disabling work injuries and injuries such as eye injuries, fractures, hospitalization for observation, loss of consciousness, and any other injury that requires medical treatment by a physician

Service: to adjust, repair, maintain, and make fit for use

Service life: the period of time that a respirator, filter or sorbent, or other respiratory equipment provides adequate protection to the wearer

Service robots: machines that extend human capabilities

Severity: consequences of a failure as a result of a particular failure mode

Severity classification: a classification assigned to provide a qualitative measure of the worst potential consequences resulting from design error or item failure

Sharp end: point of vulnerability in care delivery where failure is visible, where expertise is applied, and adverse events are experienced

Sharps: objects that can penetrate the skin, such as needles, scalpels, and lancets

Shielding: material between a radiation source and a potentially exposed person that reduces exposure

Shock wave: a transient pressure pulse that propagates at supersonic velocity

Short-term exposure limit: an OSHA measurement of the maximum concentration for a continuous 15-minute exposure period

Short-term recovery: a process of recovery that is immediate and overlaps emergency response actions; includes such actions as providing essential public health and safety services, restoring interrupted utility and other essential services, reestablishing transportation routes, and providing food and shelter for those displaced by a disaster (although called "short term," some of these activities may last for weeks)

Shrapnel: high-speed metal fragments from a shell or bomb explosion; can be quite lethal to personnel, and can also cause considerable damage to aircraft; fragments from exploding munitions that can acquire velocities comparable to those of rifle bullets and cause great impact effects; objects that are attached to the outside or included inside a device to increase the blast damage and/or injure/kill personnel (the device/container walls themselves can also function in this manner)

Sick building syndrome: a situation where building occupants experience acute health or discomfort that appears to be linked to time spent in the building, but no specific illness or cause can be determined

Single failure point: a failure of an item that would result in failure of the system and is not compensated by redundancy or alternative operational procedure

Situation report: document that contains confirmed or verified information and explicit details (who, what, where, and how) relating to an incident

Situational awareness: the ability to identify, process, and comprehend the critical elements of information about an incident

Sludge: a solid material that collects as the result of air or water treatment processes

Solubility: the percentage of a material (by weight) that will dissolve in water at a specified temperature

Solvent: a substance that dissolves or disperses another substance

Somatic effects: the effects of radiation that are limited to the exposed person

Source: means either laser or laser-illuminated reflecting surface

Span of control: the number of individuals a supervisor is responsible for, usually expressed as the ratio of supervisors to individuals (under the National Incident Management System, an appropriate span of control is between 1:3 and 1:7, with optimal being 1:5)

Special needs population: populations whose members may have additional needs before, during, and after an incident in functional areas, including but not limited to maintaining independence, communication, transportation, supervision, and medical care; individuals in need of additional response assistance may include those who have disabilities; who live in institutionalized settings; who are elderly; who are children; who are from diverse cultures; who have limited English proficiency or are non-English speaking; or who are transportation disadvantaged

Specific gravity: the weight of a material compared to the weight of an equal volume of water

Spectrum: a range of frequencies within which radiation has some specified characteristic, such as audio-frequency spectrum, ultraviolet spectrum, and radio spectrum

Stafford Act: the Robert T. Stafford Disaster Relief and Emergency Assistance Act, P.L. 93-288, as amended; this Act describes the programs and processes by which the federal government provides disaster and emergency assistance to state and local governments, tribal nations, eligible private nonprofit organizations, and individuals affected by a declared major disaster or emergency; covers all hazards, including natural disasters and terrorist events

Staging area: any location in which personnel, supplies, and equipment can be temporarily housed or parked while awaiting operational assignment

Standard industrial classification: a classification developed by the office of management and budget used to assign each establishment an industry code that is determined by the product manufactured or service provided

Standard operating procedure: complete reference document or an operations manual that provides the purpose, authorities, duration, and details for the preferred method of performing a single function or a number of interrelated functions in a uniform manner

Standard procedure: a written instruction that establishes what action is required, who is to act, and when the action is to take place

Standoff zone: the area between the protected structure and the perimeter barrier protecting the asset against potential threats

Staphylococcus: any of various spherical parasitic bacteria which occur in grape-like clusters and cause infections

Static pressure: the potential pressure exerted in all directions by a fluid at rest

Status report: information specifically related to the status of resources (e.g., the availability or assignment of resources)

Steam sterilization: treatment method for infectious waste using saturated steam within a pressurized vessel such as an autoclave

Sterilize: the use of a physical or chemical procedure to destroy all microbial life including highly resistant spores

Stochastic effect: effect that occurs on a random basis independent of the size of dose of radiation

Stoichiometry: study of the mathematics of the material and energy balances (equilibrium) of chemical reactions

Strategic: elements of incident management are characterized by continuous long-term, high-level planning by senior level organizations

Strategic Guidance Statement and Strategic Plan: documents that together define the broad national strategic objectives; delineate authorities, roles, and responsibilities; determine required capabilities; and develop performance and effectiveness measures essential to prevent, protect against, respond to, and recover from domestic incidents

Strategic Information and Operations Center (SIOC): the focal point and operational control center for all federal intelligence, law enforcement, and investigative law enforcement activities related to domestic terrorist incidents or credible threats, including leading attribution investigations; serves as an information clearinghouse to help collect, process, vet, and disseminate information relevant to law enforcement and criminal investigation efforts in a timely manner

Strategy: the general plan or direction selected to accomplish objectives

Streptococcus: any of various rounded, disease-causing bacteria that occur in pairs or chains

Strontium (sr): a silvery, soft metal that rapidly turns yellow in air; sr-90 is one of the radioactive fission materials created within a nuclear reactor during its operation, and emits beta particles during radioactive decay

Substandard: a condition that deviates from what is acceptable, normal, or correct and is a potential hazard

Supplied-air respirator (SAR): a respirator that provides breathing air through an airline hose from an uncontaminated compressed air source to the face piece; the face piece can be a hood, helmet, or tight fitting face piece

Surface burst: a nuclear weapon explosion that is close enough to the ground for the radius of the fireball to vaporize surface material; fallout from a surface burst contains very high levels of radioactivity

Surge capability: the ability to manage patients requiring unusual or very specialized medical evaluation and care

Surge capacity: the ability to evaluate and care for a markedly increased volume of patients—one that challenges or exceeds normal operating capacity

Survey: comprehensive study or assessment of a facility, workplace, or activity for insurance or loss control purposes

System: a clearly described functional structure, including defined processes, that coordinates otherwise diverse parts to achieve a common goal

Systematic: striving toward goal accomplishment in a planned manner using predetermined steps or procedures

T

Tactical: ICS elements characterized by the execution of specific actions or plans in response to an actual incident

Tactics: deployment and directing of resources on an incident to accomplish the objectives designated by strategy

Teach mode: the control state that allows the generation and storage of positional data points effected by moving the robot arm through a path of intended motions

Teamwork: based on trust, communication, and innovation

Tear gas: a chemical agent typically in liquid form and released as an aerosol liquid or gas; upon contact with the target persons, it produces disorientation, nausea, a copious flow of tears and irritation of the eyes, and other disabling effects of temporary duration

Tendinitis: a condition where the muscle–tendon junction becomes inflamed

Tenosynovitis: a condition that results in the inflammation of the tendons and their sheaths

Teratogen: a substance or agent that can cause malformations in the fetus when a pregnant female is exposed

Terrorism: any premeditated, unlawful act dangerous to human life or public welfare that is intended to intimidate or coerce civilian populations or governments

Thermoluminescent dosimeter: a badge that contains a thermoluminescent, a chip worn by persons working with or around radioactive materials

Tiger team: a team of experts who assess the security measures by conducting unannounced penetration attempts such as trying to circumvent access controls or bypassing other security protection

TLV: an ACGIH-published threshold limit value of an airborne concentration of a hazardous/toxic substance to which workers may be repeatedly exposed day after day without adverse effect

Toxic substance: any substance that can cause acute or chronic injury or illness to the human body

Toxicity: potential of a substance to have a harmful effect and a description of the effect and the conditions or concentration under which the effect takes place

Transparency: openness about error and learning from error

Triage: the process of screening and classifying sick, wounded, or injured persons to determine priority needs in order to ensure the efficient use of resources

Trigger finger: a condition caused by any finger being frequently flexed against some type of resistance

Two-person rule: a security strategy that requires two people to be present in order to gain access to a secured area to prevent unobserved access by any individual

U

Ultraviolet radiation: electromagnetic radiation with wavelengths between soft x-rays and visible violet light

Undetectable failure: a postulated failure mode in the FMEA for which there is no failure detection method by which the operator is made aware of the failure

Unified command: agencies working together through their designated incident commanders or managers at a single location to establish a common set of objectives and strategies and a single incident action plan

Uniform fire code: regulations consistent with nationally recognized good practice for safeguarding life and property from the hazards of fire and explosion that arise from the storage, handling, and use of hazardous substances, materials, and devices

Unity of command: principle of management stating that each individual involved in incident operations will be assigned to only one supervisor

Universal precautions: an OSHA term for the method of infection control in which all human blood and certain other materials are treated as infectious for bloodborne pathogens

Unstable: a chemical that, when in the pure state, will vigorously polymerize, decompose, condense, or become self-reactive under conditions of shock, pressure, or temperature

Upper explosive limit: highest concentration of a substance that will burn or explode when an ignition source is present, expressed in percent of vapor or gas in the air by volume

Uranium: a naturally occurring radioactive element its a hard, silvery-white, shiny metallic ore that contains a minute amount of uranium-234

User seal check: an action conducted by the respirator user to determine if a respirator is properly seated to the face

V

Vapor: the gaseous form of a substance that is normally in the solid or liquid state at room temperature and pressure

Vapor density: the weight of a vapor or gas compared to the weight of an equal volume of dry air

Vapor pressure: the pressure exerted by a saturated vapor above its own liquid in a closed container

Vector: organism that carries disease, such as insects or rodents

Vesicant agent: an agent that acts on the eyes and lungs and blisters the skin

Viscosity: the property of a liquid that causes it to resist flow or movement in response to external force applied to it

Visible radiation (light): electromagnetic radiation that can be detected by the human eye

Volatile: the tendency or ability of a liquid to vaporize

Volatile organic compounds: compounds that evaporate from many housekeeping, maintenance, and building products made from organic chemicals

Vomiting agent: compounds that cause irritation of the upper respiratory tract and involuntary vomiting

W

Water-reactive: chemical that reacts with water to release a gas that either is flammable or presents a health hazard

Wavelength: the distance in the line of advance of a wave from any point to a like point on the next wave; usually measured in angstroms, microns, micrometers, or nanometers

Weapons of mass destruction (WMD): weapons that are capable of a high order of destruction and/or of being used in such a manner as to destroy large numbers of people; can be high explosives or nuclear, biological, chemical, and radiological weapons, but exclude the means of transporting or propelling the weapon where such means is a separable and divisible part of the weapon

Whole body count: the measure and analysis of the radiation being emitted from a person's entire body, detected by a counter external to the body

Whole body exposure: an exposure of the body to radiation, in which the entire body, rather than an isolated part, is irradiated by an external source

Wood alcohol: methyl alcohol

Work practice controls: any controls that reduce the likelihood of exposure by altering the manner in which a task is performed

X

Xenobiotic: a man-made substance, such as plastic, found in the environment

X-ray: electromagnetic radiation caused by deflection of electrons from their original paths, or inner orbital electrons that change their orbital levels around the atomic nucleus

Z

Zone: a section of an alarmed, protected, or patrolled area; often means a space having one or more sensors

Appendix 13
AHRQ Patient Safety Tools and Resources

The Agency for Healthcare Research and Quality (AHRQ) offers the following tools for healthcare organizations, providers, policymakers, and patients to improve patient safety in health care settings.

The **Hospital Survey on Patient Safety Culture** examines patient safety culture from a hospital staff perspective and allows hospitals to assess their safety culture and track changes over time. Hospitals that administer the patient safety culture survey can voluntarily submit their data to the Comparative Database, a resource for hospitals wishing to compare their survey results to similar types of hospitals (AHRQ Publication No. 04-0041).

Hospital Survey on Patient Safety Culture: Comparative Database Reports give benchmark data collected voluntarily from more than 1000 U.S. hospitals. Survey results from these hospitals are averaged over the entire sample by topical composite or individual survey item. Two appendixes report the average responses, which are broken down by hospital or respondent characteristics.

The **Medical Office Survey on Patient Safety Culture** measures issues relevant to patient safety in the ambulatory medical office setting. Pilot tested in approximately 100 medical offices, the survey lets providers and staff members assess their safety culture, identify areas where improvement is needed, track changes in patient safety, and evaluate the effect of interventions. Researchers can also use the survey to assess patient safety culture improvement initiatives (AHRQ Publication No. 08(09)-0059).

The **Medical Office Survey on Patient Safety Culture: 2012 Comparative Database Report** presents data from 23,679 staff within 934 U.S. medical offices that completed the Medical Office Survey on Patient Safety Culture so offices can compare their patient safety culture to other medical offices. The full report contains detailed comparative data for various medical office characteristics (number of providers, specialty, ownership, and region) and staff positions (AHRQ Publication No. 12-0052).

The **Nursing Home Survey on Patient Safety Culture** uses provider and staff perspectives to assess their nursing home's safety culture, identify areas where improvement is needed, track changes in patient safety, and evaluate the impact of interventions. The survey also lets researchers assess safety culture improvement initiatives in nursing homes (AHRQ Publication No. 08(09)-0060).

The **Nursing Home Survey on Patient Safety Culture: 2011 User Comparative Database Report** is based on data from 226 nursing homes in the United States and provides initial results that nursing homes can use to compare their patient safety culture to other U.S. nursing homes. The report consists of a narrative description of the findings and four appendixes presenting data by nursing home characteristics and respondent characteristics (AHRQ Publication No. 11-0030).

The **Pharmacy Survey on Patient Safety Culture** focuses on patient safety culture. AHRQ sponsored the survey, which was designed specifically for community pharmacy staff and asks for their opinions about the culture of patient safety in their pharmacy (AHRQ Publication No. 12(13)-0085).

Patient Safety Organizations (PSOs) were created by the Patient Safety and Quality Improvement Act to improve the quality and safety of healthcare by encouraging clinicians and healthcare organizations to voluntarily report patient safety events without fear of legal discovery. PSOs offer a secure environment to identify and reduce the risks associated with patient care. As independent, external experts, PSOs collect, analyze, and aggregate patient safety data locally, regionally, and nationally to develop insights into the underlying causes of patient safety events (Web: http://www.pso.ahrq.gov).

Patient safety, quality and risk managers, clinicians, and others use **Common Formats** to collect patient safety event information in a standard way, using common language, definitions, technical requirements for electronic implementation, and reporting specifications. Common Formats optimize the opportunity for the public and private sectors to learn more about trends in patient safety with the purpose of improving healthcare quality. AHRQ has developed Common Formats for hospitals and nursing homes (including skilled nursing facilities) to collect data for all types of adverse events, near misses, and unsafe conditions (Web: http://www.pso.ahrq.gov).

Measures of healthcare quality that make use of readily available hospital administrative data, the **Quality Indicators**™ can be used to highlight potential quality concerns, identify areas that need further study and investigation, and track changes over time. AHRQ distributes the Quality Indicators through free software programs that can help hospitals identify quality of care events that might need further study. The current AHRQ Quality Indicators modules represent various aspects of quality:

- **Patient Safety Indicators** reflect quality of care inside hospitals, as well as geographic areas, to focus on potentially avoidable complications and iatrogenic events.
- **Prevention Quality Indicators** identify hospital admissions in geographic areas that evidence suggests may have been avoided through access to high-quality outpatient care.
- **Inpatient Quality Indicators** reflect quality of care inside hospitals, as well as across geographic areas, including inpatient mortality for medical conditions and surgical procedures.
- **Pediatric Quality Indicators** use indicators from the other three modules with adaptations for use among children and neonates to reflect quality of care inside hospitals, as well as geographic areas, and identify potentially avoidable hospitalizations.

A **Toolkit for Hospitals: Improving Performance on the AHRQ Quality Indicators**™ helps hospitals understand AHRQ's Quality Indicators that use hospital administrative data to assess the quality of care provided, identify areas of concern in need of further investigation, and monitor progress over time. The toolkit is a general guide to using improvement methods and focuses on the 17 Patient Safety Indicators and the 28 Inpatient Quality Indicators to improve quality and patient safety.

The **Hospital Consumer Assessment of Healthcare Providers and Systems** is a survey instrument for measuring patients' perspectives on hospital care. The 27-question survey contains patient perspectives on care and patient rating items that encompass key topics, including communication with doctors and nurses, responsiveness of hospital staff, pain management, communication about medicines, discharge information, and cleanliness and quietness of the hospital environment. The survey also includes screener questions and demographic items that are used for adjusting the mix of patients across hospitals and for analytical purposes. (Web: http://www.hcahpsonline.org).

The **Comprehensive Unit-Based Safety Program (CUSP) Toolkit** includes training tools to make care safer by improving the foundation of how physicians, nurses, and other clinical team members work together. It builds the capacity to address safety issues by combining clinical best practices and the science of safety. Created for clinicians by clinicians, the CUSP toolkit is modular and modifiable to meet individual unit needs and was proven effective through a national project that reduced central line-associated blood stream infections by 41 percent. Each module includes teaching tools and resources to support change at the unit level, presented through facilitator notes that take you step-by-step through the module, presentation slides, tools, and videos.

The **Re-Engineered Discharge (RED) Toolkit** is designed to assist hospitals, including those that serve diverse populations, in implementing RED. A variety of forces are pushing hospitals to improve their discharge processes to reduce preventable readmissions. Researchers at the

Boston University Medical Center developed and tested a RED process, which was effective at reducing readmissions and post-hospital emergency department visits (AHRQ Publication No. 12(13)-0084).

Preventing Falls in Hospitals: A Toolkit for Improving Quality of Care focuses on overcoming the challenges associated with developing, implementing, and sustaining a fall prevention program. The toolkit features an implementation guide for the team that is putting the new prevention strategies into practice and also has links to tools and resources.

The **Falls Management Program: A Quality Improvement Initiative for Nursing Facilities** presents an interdisciplinary quality improvement initiative designed to assist nursing facilities in providing individualized, person-centered care and improving their fall care processes and outcomes through educational and quality improvement tools.

Making Health Care Safer II: An Updated Critical Analysis of the Evidence for Patient Safety Practices is the result of a panel of patient safety experts who assessed the evidence behind 41 patient safety strategies and identified 10 strategies that health systems should adopt now. The strategies can help prevent harmful events such as med errors, bed sores, and healthcare-associated infections. Making Health Care Safer II updates Evidence-based Practice Center report (#43), which was published in 2001 and provided the first systematic assessment of patient safety practices.

The **Emergency Severity Index (ESI): A Triage Tool for Emergency Department Care, Version 4** is a five-level emergency department triage algorithm that provides clinically relevant stratification of patients into five groups from one (most urgent) to five (least urgent) on the basis of acuity and resource needs. The ESI helps hospital emergency departments rapidly identify patients in need of immediate attention, better identify patients who could safely and more efficiently be seen in a fast-track or urgent care center rather than the main emergency department, and more accurately determine thresholds for diversion of ambulance patients from the emergency department. The 2012 edition of the Implementation Manual includes a pediatrics section and many other updates (AHRQ Publication No. 12-0014).

Improving Patient Flow and Reducing Emergency Department Crowding: A Guide for Hospitals presents step-by-step instructions for planning and implementing patient flow improvement strategies to alleviate crowded emergency departments. It addresses creating a patient flow team, measuring performance, identifying strategies, preparing to launch, facilitating change, and sharing results (AHRQ Publication No. 11(12)-0094).

Medications at Transitions and Clinical Handoffs (MATCH) Toolkit for Medication Reconciliation, based on the MATCH website, incorporates the experiences and lessons learned by healthcare facilities that have implemented MATCH strategies to improve their medication reconciliation processes for patients as they move through healthcare settings (AHRQ Publication No. 11(12)-0059).

The **Guide to Patient and Family Engagement in Hospital Quality and Safety** will help hospitals work as partners with patients and families to improve quality and safety. It contains four strategies to help hospitals partner with patients and families, and it has an implementation handbook and tools for patients, families, and clinicians for each strategy. The four strategies are helping hospitals recruit and work with patient and family advisors, communicating with patients and families throughout their hospital stay to improve quality, implementing nursing bedside change of shift reports, and engaging patients and families in discharge planning (AHRQ 13-0033).

The **Toolkit for Reduction of Clostridium difficile Through Antimicrobial Stewardship** assists hospital staff and leadership in developing an effective antimicrobial stewardship program (ASP) with the potential to reduce *Clostridium difficile* infection (*C. difficile*), a serious public health problem that has recently increased in both incidence and severity. An ASP is a systematic approach to developing coordinated interventions to reduce overuse and inappropriate selection of antibiotics, and to achieve optimal outcomes for patients in cost-efficient ways. ASPs targeted to *C. difficile* reduction show promise because increased rates of *C. difficile* are associated with inappropriate antibiotic use.

The **Preventing Pressure Ulcers in Hospitals Toolkit** assists hospital staff in implementing effective pressure ulcer prevention practices through an interdisciplinary approach to care. The toolkit draws on literature on best practices in pressure ulcer prevention and includes both validated and newly developed tools.

Preventing Hospital-Acquired Venous Thromboembolism: A Guide for Effective Quality Improvement is based on quality improvement initiatives undertaken at the University of California, San Diego Medical Center and Emory University Hospitals in Atlanta. This guide assists quality improvement practitioners in leading an effort to improve prevention of one of the most serious problems facing hospitalized patients: hospital-acquired venous thromboembolism (AHRQ Publication No. 08-0075).

Developing a Community-Based Patient Safety Advisory Council provides approaches for hospitals and other healthcare organizations to use to develop a community-based advisory council that can drive change for patient safety through education, collaboration, and consumer engagement (AHRQ Publication No. 08-0048).

Mistake-Proofing the Design of Health Care Processes is illustrated with numerous examples and explains how to apply the industrial engineering concept of mistake-proofing to processes in hospitals, clinics, and physicians' offices. (AHRQ Publication No. 07-0020)

Team Strategies and Tools to Enhance Performance and Patient Safety (TeamSTEPPS®) is a set of tools to help train clinicians in teamwork and communication skills to reduce risks to patient safety (AHRQ Publication No. 06-0020-0).

TeamSTEPPS Rapid Response Systems Guide includes PowerPoint presentations, teaching modules, and video vignettes for training hospital staff who work with Rapid Response Systems, in which hospitals use groups of clinicians to bring critical care expertise to patients requiring immediate treatment (AHRQ Publication No. 08(09)-0074-CD).

TeamSTEPPS Enhancing Safety for Patients with Limited English Proficiency Module helps healthcare organizations develop and deploy a customized plan to train staff in teamwork skills and lead a medical teamwork improvement initiative for working with patients who have difficulty communicating in English. Comprehensive curricula and instructional guides include short case studies and videos illustrating teamwork opportunities and successes (AHRQ Publication No. 12(13)-0068-DVD).

Research suggests that adverse events affect patients with limited English proficiency more frequently, are often caused by communication problems, and are more likely to result in serious harm compared to those that affect English-speaking patients. **Improving Patient Safety Systems for Patients With Limited English Proficiency: A Guide for Hospitals** focuses on how hospitals can better identify, report, monitor, and prevent medical errors in patients with limited English proficiency (AHRQ Publication No. 12-0041).

TeamSTEPPS Long-Term Care Version adapts the core concepts of the TeamSTEPPS program to reflect the environment of nursing homes and other long-term care settings such as assisted living and continuing care retirement communities. The examples, discussions, and exercises are tailored to address and improve teamwork in the long-term care environment (AHRQ Publication No. 12(13)-0004-DVD).

Improving Patient Safety in Long-Term Care Facilities is intended for use in training frontline personnel in nursing home and other long-term care facilities. The educational materials are presented in three modules: Module One addresses detecting changes in a resident's condition, Module Two addresses communicating changes in a resident's condition, and Module Three addresses falls prevention and management. The Instructor Guide comprises all three modules, including suggested slides and pre- and posttests to gauge the student's knowledge level before and after training. Separate student workbooks are available for each module (AHRQ Publication No. 12-0001-1).

Transforming Hospitals: Designing for Safety and Quality presents information about three model hospitals that incorporated evidence-based design elements into their construction and renovation projects. This DVD shows hospital leaders how evidence-based design can improve the quality and safety of hospital services. It is an especially useful tool for hospitals that are planning capital construction projects or renovations (AHRQ Publication No. 07-0076-DVD).

Patient Safety and Quality: An Evidence-Based Handbook for Nurses is a three-volume handbook in which nurses will find peer-reviewed discussions and reviews of issues and literature regarding patient safety and quality healthcare. Each of the 51 chapters and three leadership vignettes presents an examination of the state of the science behind quality and safety concepts and challenges nurses to use evidence to change practices and engage in developing the evidence base to address critical knowledge gaps (AHRQ Publication No. 08-0043-CD).

Advances in Patient Safety: New Directions and Alternative Approaches is a four-volume set of 115 articles that describe patient safety findings, investigative approaches, process analyses, lessons learned, and practical tools to prevent patients from being harmed. It includes articles by AHRQ-funded patient safety researchers on topics such as reporting systems, risk assessment, safety culture, medical simulation, health information technology, and medication safety (AHRQ Publication No. 08-0034).

Advances in Patient Safety: From Research to Implementation is a four-volume set of 140 articles that describe accomplishments between 1999 and 2004 by federally funded programs in understanding medical errors and implementing programs to improve patient safety. Included are articles with a research and methodological focus, articles that address implementation issues, and tools to improve patient safety (AHRQ Publication No. 05-0021-CD).

20 Tips to Help Prevent Medical Errors tells patients what they can do to get safer care and addresses medicines, hospital stays, surgery, medical tests, and more (AHRQ Publication No. 11-0089).

Index